Unto Others

Unto Others

The Evolution and Psychology of Unselfish Behavior

Elliott Sober

David Sloan Wilson

Harvard University Press
Cambridge, Massachusetts
London, England

Printed in the United States of America

Second printing, 1999

First Harvard University Press paperback edition, 1999

Library of Congress Cataloging-in-Publication Data

Sober, Elliott.
Unto others : the evolution and psychology of unselfish behavior / Elliott Sober, David Sloan Wilson.
p. cm.
Includes bibliographical references and index.
ISBN 0-674-93046-0 (cloth)
ISBN 0-674-93047-9 (pbk.)
1. Altruism. 2. Egoism. 3. Group selection (Evolution). 4. Evolution (Biology). I. Wilson, David Sloan. II. Title.
BF637.H4S65 1998
171′.8—dc21 97-42185

To altruists everywhere, especially those who are unsure as to what their motives really are

Acknowledgments

Maybe it should come as no surprise that we have received so much help in writing a book about altruism. We want to thank the following battalion of people. They come from many different disciplines and helped in many different ways: Richard Alexander, André Ariew, Martin Barrett, Daniel Batson, John Beatty, Len Berkowitz, Howard Bloom, Chris Bohm, Tom Bontly, Deric Bownds, Robert Boyd, Robert Brandon, Felix Breden, Don Campbell, Noël Carroll, Anne Clark, Robert Colwell, Richie Davidson, Jeff Dean, Lee Dugatkin, Nancy Eisenberg, Berent Enç, Murvet Enç, Ted Everett, Robert Frank, Steve Frank, Peter Godfrey-Smith (for the title), Charles Goodnight, Leslie Graves, Jim Griesemer, David Gubernick, William Hamilton, Henry Harpending, Dan Hausman, Jack Hirschleifer, Harmon Holcomb III, Todd Hughes, Robert Jeanne, John Kelly, Bruce Knauft, Hugh LaFollette, Andrew Levine, John Maynard Smith, Rick Michod, Don Moskowitz, Gregory Mougin, Steven Orzack, Stuart Peck, Jane Piliavin, Greg Pollock, Will Provine, William Puka, David Queller, H. Kern Reeve, Peter Richerson, Steve Rissing, Tom Seeley, Larry Shapiro, Alan Sidelle, Barbara Smuts, Dennis Stampe, Chris Stephens, Kim Sterelny, William Talbott, Peter Taylor, Frans de Waal, Mike Wade, Denis Walsh, Doris Williams, George Williams, Edward Wilson, Vero Wynne-Edwards.

We also must extend our thanks to entities that are not individuals; perhaps this also is to be expected in a book about multilevel selection theory. We are grateful for support from Binghamton University, the National Science Foundation, and the University of Wisconsin at Madison. Finally, we want to acknowledge the support and encouragement that have come from two groups that we count as nearest and most dear—our families.

Contents

Unto Others

Introduction: Bentham's Corpse

The mummified corpse of Jeremy Bentham occupies a cabinet the size of a telephone booth in University College, London. The head went cheesy some time ago and was replaced by a wax substitute; the original is in a box that sits discreetly between Bentham's feet. Bentham's will decreed that his body was to be preserved and carried into meetings of the University's Board of Trustees (Runes 1959, p. 250). Bentham thought that his stern countenance would inspire future generations to live up to the standards that he and John Stuart Mill advocated in their theory of morality and politics. Bentham and Mill together created the philosophy of *utilitarianism*—the view that people should promote the greatest happiness for the greatest number of individuals. Bentham hoped that a mummy at a meeting would encourage those voting to do the right thing.

It is a matter of conjecture just how much influence Bentham's corpse has exercised over University College in the years since his death, but it is a certainty that the ideas Bentham defended have profoundly affected the wider culture. Bentham thought that all human activity should aim to maximize pleasure and minimize pain. Underlying this claim about what people *ought* to do was Bentham's picture of how the human mind works *in fact*. According to Bentham (1789), "nature has placed mankind under the governance of two sovereign masters, pain and pleasure." Bentham, here, is endorsing *psychological hedonism*—the theory that avoiding pain and attaining pleasure are the only ultimate motives that people have; everything else that we want, we want solely as a means to achieving those twin ends.

Hedonism is a specific version of a more general theory, *psychological egoism,* which claims that every individual's ultimate goal is to benefit him- or herself. Egoism maintains that when we care about what happens to others, we do so only as a means to increasing our own welfare. The view denies that people ever have altruistic ultimate motives. Egoism does not say whether we should rejoice or despair at this feature of the human mind. It claims only to describe how things are in fact.[1] In all our social interactions, we are driven by a single question—"What's in it for me?"

It would be difficult to exaggerate the pervasive influence that hedonism and egoism have had and continue to have on people's thinking. For many, egoism seems obvious, a matter of common sense. People are often unsurprised when others act with ruthless selfishness but find it quite remarkable when others sacrifice themselves for the sake of someone else. If someone says that human beings are by nature selfish, people frequently regard this pronouncement as proceeding from a clear-eyed realism; however, if someone says that human beings are by nature benevolent, people often smile indulgently, thinking that the assertion reflects a propensity to view the world through rose-colored glasses.

Why does psychological egoism have such a grip on our self-conception? Does our everyday experience provide conclusive evidence that it is true? Has the science of psychology demonstrated that egoism is correct? Has philosophy? All these questions must be answered in the negative, or so we will argue. The influence that psychological egoism exerts far outreaches the evidence that has been mustered on its behalf.

Egoism is easy enough to refute when it is given a simplistic formulation. For example, if the egoist claims that the only ultimate goal that people have is to maximize their access to consumer goods, it is not hard to describe behaviors that show that this is false. But if the egoist says that human beings strive for internal, psychological benefits, the proposal is harder to prove wrong. When people sacrifice their own interests to help someone else, the egoist maintains that they do so in order to feel good about themselves and to avoid feeling guilty. Egoism is a mansion with many rooms. There seems to be room enough in the theory to explain helping behavior and the existence of desires concerning the welfare of others; both are explained as instruments for promoting self-interest.[2] As a result, the concept of altruism remains an endangered species (Campbell 1994).

Psychological egoism is hard to disprove, but it also is hard to prove. Even if a purely selfish explanation can be imagined for every act of helping, this doesn't mean that egoism is correct. After all, human behavior also is consistent with the contrary hypothesis—that some of our ultimate goals are altruistic. Psychologists have been working on this problem for decades and philosophers for centuries. The result, we believe, is an impasse—the problem of psychological egoism and altruism remains unsolved. A new approach is needed. The novel perspective that we will explore is provided by the theory of evolution.

Bentham died in 1832, more than two decades before the publication in 1859 of Darwin's *Origin of Species.* John Stuart Mill lived well past this watershed event, though he never accepted the theory of evolution by natural selection.[3] Darwin's theory gave rise to a fundamental puzzle about the behavior of organisms. The basic idea of natural selection is that characteristics evolve because they help the individuals who possess them to survive and reproduce. A herd of zebra, for example, will gradually increase in running speed because faster zebras do a better job evading predators. Faster zebras are *fitter*—they are more able to survive and tend to have more offspring than slower ones. If offspring resemble their parents, the frequency of fast zebras—the proportion of good runners in the herd—will increase. Notice that a zebra that runs fast benefits itself—not other zebras, not lions, not the whole ecosystem. In this example, natural selection favors those who help themselves. It therefore appears that helping *other* individuals to survive and reproduce at the expense of one's *own* survival and reproduction is the very thing that natural selection will eliminate. In short, natural selection appears to be a process that promotes selfishness and stamps out altruism.

Darwin was aware that organisms in nature sometime behave in ways that appear altruistic. For example, a honeybee sacrifices its life for the colony when it uses its barbed stinger to attack intruders to the nest. And many of the most praiseworthy human qualities—honesty, charity, trust, and heroism—appear to benefit others at expense to self. Darwin explained these characteristics by saying that natural selection sometimes acts on *groups,* just as it acts at other times on *individuals.* An altruist may have fewer offspring than a nonaltruist within its own group, but groups of altruists will have more offspring than groups of nonaltruists. In a famous passage from *The Descent of Man,* Darwin used the principle of group selection to explain the evolution of human morality:

> It must not be forgotten that although a high standard of morality gives but a slight or no advantage to each individual man and his children over the other men of the same tribe, yet that an increase in the number of well-endowed men and advancement in the standard of morality will certainly give an immense advantage to one tribe over another. There can be no doubt that a tribe including many members who, from possessing in a high degree the spirit of patriotism, fidelity, obedience, courage, and sympathy, were always ready to aid one another, and to sacrifice themselves for the common good, would be victorious over most other tribes; and this would be natural selection. At all times throughout the world tribes have supplanted other tribes; and as morality is one important element in their success, the standard of morality and the number of well-endowed men will thus everywhere tend to rise and increase. (Darwin 1871, p. 166)

Although Darwin never discussed how important a role group selection played in the history of life, his practice was to appeal to this process only rarely. Darwin's successors were less abstemious, invoking the process widely and often uncritically. According to Allee (1951), dominance hierarchies exist to minimize within-group conflict, so that the entire group can be more productive. According to Wynne-Edwards (1962), individual organisms restrain themselves from consuming food and from reproducing, so that the population can avoid crashing to extinction. And according to Dobzhansky (1937), whole species maintain genetic diversity to cope with new environmental challenges; like savvy investors, they diversify their portfolios because the future is uncertain. Many biologists happily invoked these and other group-level explanations while at the same time explaining the evolution of camouflage, disease resistance, and other traits as individual-level adaptations. Biologists often simply chose the level of explanation that they found more intuitive, appealing to individual adaptation on Mondays, Wednesdays, and Fridays, and to group adaptation on Tuesdays, Thursdays, and Saturdays. Theorizing was unconstrained, and adaptationist explanation was similarly unrestrained.

All this changed in the 1960s, when group selection was attacked by a number of biologists. The most thorough and devastating critique was G. C. Williams's 1966 book, *Adaptation and Natural Selection.* Williams touched a nerve, and his vigorous rejection of adaptations that exist for the good of the group spread quickly through the community of evolutionary biologists. For the next decade, group selection theory was widely regarded as not just false but as off-limits, as far as serious evolutionary thought was concerned. At best, group adaptation was regarded as a theoretical possibility but as so enormously unlikely that alternative explanations should be preferred whenever possible. The following passage (Ghiselin 1974, p. 247) illustrates the fervor with which altruistic, group-level explanations were rejected in favor of accounts that appeal to selfishness:

> The economy of nature is competitive from beginning to end . . . The impulses that lead one animal to sacrifice himself for another turn out to have their ultimate rationale in gaining advantage over a third . . . Where it is in his own interest, every organism may reasonably be expected to aid his fellows . . . Yet given a full chance to act in his own interest, nothing but expediency will restrain him from brutalizing, from maiming, from murdering—his brother, his mate, his parent, or his child. Scratch an "altruist," and watch a "hypocrite" bleed.

The interpretation of human behavior was similarly transformed. In *The Biology of Moral Systems,* Alexander (1987, p. 3) shows the extent to which biologists abandoned the idea that genuinely self-sacrificial behaviors are part of our evolutionary legacy:

> I suspect that nearly all humans believe it is a normal part of the functioning of every human individual now and then to assist someone else in the realization of that person's own interests to the actual

> net expense of those of the altruist. What this greatest intellectual revolution of the century [i.e., the individualistic perspective in evolutionary biology] tells us is that, despite our intuitions, there is not a shred of evidence to support this view of beneficence, and a great deal of convincing theory suggests that any such view will eventually be judged false.

The concept of selfishness that became established in evolutionary biology, like the concept of egoism in psychology, is a mansion with many rooms. It claims to explain such apparently altruistic traits as the bee's barbed stinger and human morality. These and other characteristics are said to be only apparently altruistic because individuals who help others receive benefits in return or promote their "genetic self-interest" by helping copies of their own genes that are found in the bodies of others. Genuinely altruistic traits that evolve by group selection became an endangered species in evolutionary biology during the 1960s and 70s, just as genuine psychological altruism has long been an endangered species in the social sciences.

The concepts of psychological egoism and altruism concern the motives that people have for acting as they do. The act of helping others does not count as (psychologically) altruistic unless the actor thinks of the welfare of others as an ultimate goal. In contrast, the evolutionary concepts concern the effects of behavior on survival and reproduction. Individuals who increase the fitness of others at the expense of their own fitness are (evolutionary) altruists, regardless of how, or even whether, they think or feel about the action. Many researchers are careful to draw this distinction between the psychological and evolutionary concepts. Nonetheless, the concepts of selfishness in biology and the social sciences are often thought to be compatible and to reinforce each other. If evolutionary altruism is absent in nature, why should psychological altruism be present in human nature?

In this book, we will thoroughly explore the concepts of altruism and selfishness in evolutionary biology, psychology, and philosophy. In contrast with the views just outlined, our argument builds a strong case for both evolutionary and psychological altruism. However, the relationship between these two concepts is not simple. The case for evolutionary altruism requires showing that group selection has been an important force in evolution. The case for psychological altruism requires showing that an ultimate concern for the welfare of others is among the psychological mechanisms that evolved to motivate adaptive behavior. Both arguments are evolutionary, but they are sufficiently different that we have divided the book in two.

We mentioned before that group selection was once regarded as both thoroughly confused and thoroughly refuted. Nevertheless, it would be a mistake for the reader to regard us as two heretics crying out in the wilderness. During the 1970s, a robust theory of group selection emerged that could withstand the earlier criticisms. Readers who think they are familiar with the subject may be surprised to learn that even G. C. Williams, the icon of the individual selection movement, has accepted the evidence for group selection as the best explanation of important biological adaptations such as female-biased sex ratios and reduced virulence in disease organisms. In short, rather than defending a heretical new theory, we will be reporting and extending a transition in evolutionary thought that is already in full swing.

If group selection has become respectable again, the reader may well wonder why the news is not generally known. One reason is that ten or twenty years is not a long time for certain kinds of scientific change, especially when the subject is as emotionally loaded as altruism and selfishness. The rejection of group selection during the 1960s was based on an evaluation of the theories and evidence available at the time. Unfortunately, the verdict has been transmitted more faithfully through the years than the reasons behind it. Many evolutionary biologists learned just one thing about group selection during their graduate training—"Don't do it!" They avoid the hypothesis partly because it seems scandalous, and partly because they sometimes feel unqualified to evaluate the arguments. As a result, the modern theory of group selection has developed in partial isolation, even within the field of evolutionary biology. Articles that treat group selection as uncontroversial appear in the most respected journals alongside other articles that continue to treat it as a bogeyman. One of the purposes of our book is to present the arguments for and against group selection in enough detail so that readers—biologists and nonbiologists alike—can judge for themselves.

Although the modern theory of group selection is already well developed and empirically supported, the psychological question about altruistic ultimate motives remains open. Some psychologists think that experimental evidence now exists to decide the question, but many of their colleagues disagree. Indeed, some have suggested that psychological experiments are incapable of distinguishing between altruistic and selfish ultimate motives. It is our ambition, in this book, to outline an evolutionary theory of psychological motives that can solve this problem. Since this is a relatively new enterprise, our case for psychological altruism is more provisional than our case for evolutionary altruism.

The idea that human behavior is governed entirely by self-interest and that altruistic ultimate motives don't exist has never been supported by either a coherent theory or a crisp and decisive set of observations. The entire debate has been characterized by an intellectual pecking order in which an egoistic explanation for a given behavior, no matter how contrived, is favored over an altruistic explanation, even in the absence of empirical evidence that discriminates between the two approaches. It is interesting that a similar pecking order existed during the 1960s for the subject of evolutionary altruism, which made the case against group selection appear much stronger than it actually was. Intellectual pecking orders are sometimes justified—for example, when one of the approaches appears very weak on theoretical grounds—but the group selection debate moved beyond this stage in the 1970s and now is conducted on an even playing field. Alternative theories have equal status and generate different predictions that can be tested empirically. The debate over psychological altruism will never make real progress until it undergoes the same transition. We think that our analysis of psychological altruism will help move the debate onto the same type of even playing field. If psychological mechanisms are partially designed by natural selection to motivate adaptive behaviors, there is good reason to expect these psychological mechanisms not to funnel all behavior through the narrow tube of egoistic ultimate motives.

At the risk of sounding defensive, we feel we should address a criticism that is often leveled at advocates of altruism in psychology and of group selection in biology. It is frequently said that people endorse such hypotheses because they *want* the world to be a friendly and hospitable place. The defenders of egoism and individualism who advance this criticism thereby pay themselves a compliment; they pat themselves on the back for staring reality square in the face. Egoists and individualists are objective, they suggest, whereas proponents of altruism and group selection are trapped by a comforting illusion.

This criticism is made so often that it is tempting to reply in kind, with conjectures about the psychological benefits that defenders of egoism and individualism extract from believing their pet theories. However, speculations about the motives that prompt someone to defend a theory are irrelevant. They are *ad hominem*. The point is to discover which theories are *true*. What is needed is a focused attention on the evidence for *theories*, not on the psychological quirks of *theorists*.

In any event, it is worth saying here that our goal in this book is not to paint a rosy picture of universal benevolence. Group selection does provide a setting in which helping behavior directed at members of one's own group can evolve; however, it equally provides a context in which hurting individuals in other groups can be selectively advantageous. Group selection favors within-group niceness *and* between-group nastiness. Group selection theory does not abandon the idea of competition that forms the core of the theory of natural selection; rather, it provides an additional setting in which competition can occur. Not only do individuals compete with other individuals in the same group; in addition, groups compete with other groups.[4]

Similar remarks apply to the story we will tell about psychological altruism. We will not suggest that everyone has a thoroughgoing and saintly dedication to helping others—that people always treat the well-being of others as an end in itself and never think of their own welfare. Rather, our objective will be to show that concern for others is *one* of the ultimate motives that people *sometimes* have. Even if we are right, our view leaves plenty of room for the hypothesis that individuals spend a good deal of time looking out for number one.

This book draws on four disciplines—evolutionary biology, social psychology, anthropology, and philosophy. In discussing material from each of these fields, we have tried to begin at the beginning. Our goal is not to address the handful of people who already are conversant with all four areas, but to reach people who know something about only one, or even about none of them at all. Beginning at the beginning also has the virtue, we feel, of forcing one to rethink fundamentals. This has benefited our own thinking about altruism; we think it will benefit our readers as well.

The significance of a book is the result of an interaction between its contents and the diverse conceptual frameworks of its readers. We anticipate that our readers will come from at least three very different conceptual backgrounds; these can be labeled *individual-level functionalism, group-level functionalism,* and *anti-functionalism.* To avoid needless controversy, we want to describe how our argument will relate to these three points of view.

We have already described individual-level functionalism; it is the view that individuals are the primary functional units. Group behavior is "just" the product of interactions among individuals, and groups are not functionally organized in their own right. As G. C. Williams (1966) put the point, a fleet herd of deer is just a herd of fleet deer—the group runs fast not because this benefits the group but because it benefits each individual. The individualistic tradition in evolution and methodological individualism in the human sciences are examples of individual-level functionalism. Against this background, the primary message of our book is that groups, too, can be functional units and that individuals sometimes behave more like organs than like organisms.

Although individual-level functionalism is the dominant tradition that we are opposing, group-level functionalism represents a long-standing point of view; it embodies the opposite belief that groups are the primary functional units. Herds of deer run fast, and have other characteristics, because those traits benefit the herd. Outside of science and cross-culturally, the idea that individuals exist to benefit their society may be more common than the idea that society is merely a collection of selfish individuals. Group-level functionalism also was common among the founding fathers of sociology, anthropology, and social psychology, who often treated culture and society as organic wholes that obey their own higher-order laws. Though less common today, it still exists as a minority view and is even the dominant tradition in some subdisciplines of biology and the human sciences. Our book is a mixture of good and bad news for group-level functionalism. The good news is that we can offer the first robust theory of group-level functionalism. The bad news is that it is not nearly as grandiose as many group-level functionalists would like. One can never simply assume that higher-level units such as cultures, societies, or biological ecosystems must be well-functioning organic wholes. Higher-level functional organization always requires special conditions and is vulnerable to subversion from within. It is important for readers inclined toward group-level functionalism to appreciate the bad news as much as the good news if they are to proceed realistically in developing their ideas.

Even though natural selection is the evolutionary process that explains functional organization, it is important to recognize that natural selection is not the only force in evolution and that organisms are not perfectly adapted to their environments. It follows that functionalism will never provide a complete explanation of any entity—whether that entity is an individual or a group. Enormous disagreement currently exists among evolutionary biologists concerning the importance of adaptation and natural selection in the history of life (see, e.g., Gould and Lewontin 1979). The controversy becomes even more impassioned when evolutionary ideas are applied to human beings. Antifunctionalists think that natural selection has been overemphasized as an evolutionary force and that many traits have no more function than the moon or the color of the sky. Potential readers of this book who share this view may think that we are advocating a vulgar biological imperialism when we attempt to apply Darwin's theory of natural selection to human behavior. We are especially eager to retain these readers; we think that our framework is more harmonious with their views than they may initially suspect. We therefore need to explain how it is possible to indulge freely in adaptationist thinking without being committed to the idea that natural selection is the only important force in evolution or in human behavior.

Adaptationism is sometimes understood as a claim about nature—that organisms are well adapted (or even perfectly adapted) to their environments. At other times, however, adaptationism is understood as a method for investigating nature. This is the idea that a useful procedure for studying an organism is to ask, "What would the organism be like if it were well adapted to its environment?" Posing this question does not commit one to the position that the organism actually *is* well adapted. Perhaps the population inhabits a novel environment and has not had time to adapt. Perhaps the most adaptive behaviors never arose by mutation. Perhaps maladaptive behaviors spread by the process of random genetic drift. Perhaps the entire species has been taken over by a process of cultural evolution that is insensitive to biological fitness. If any of these possibilities obtains, then the exercise in adaptationist thinking will fail to describe the actual organism. Even so, this kind of failure can be highly instructive because it allows deviations from the optimal phenotype to be discovered and interpreted (Sober 1993b; Orzack and Sober 1994). Thus, even after we acknowledge that there is more to evolution than natural selection (and more to human nature than evolution), it still will be useful to consider what the organism would be like if it were well adapted to its environment.

We will use this adaptationist methodology in connection with both evolutionary and psychological altruism. The evolutionary problem requires that we supplement the question about organisms with a comparable question about groups. What would a *group* be like if it were well adapted to its environment? The answer to this question about groups often conflicts with the answer to the question about individuals. In addition, there are a number of interesting cases in which the *evidence* conflicts with both. Some organisms display traits that appear to involve *compromises* between purely individual adaptation, purely group adaptation, and other factors. The failures of an adaptationist model can be as instructive as its successes.

With respect to the issue of psychological altruism, our argument will be that natural selection is unlikely to have given us purely egoistic motives. Even if we are right, this does not *prove* that psychological egoism must be false, since the question may still be asked as to whether nonselective processes have blocked or reversed the effects of natural selection. Still, we will suggest that our analysis of the problem provides evidence for the reality of psychological altruism. The goal of our adaptationist methodology in this instance is not to explain a trait already known to be present but to predict whether altruistic ultimate motives exist, where this question has not been answered by direct observation. We are not in the business of inventing a just-so story to explain something that everyone can plainly see.

The problem of altruism, in both its evolutionary and its psychological settings, is fascinating on several levels. It relates to the life experiences we all have—our feelings for others and the feelings that others have for us. The existence of altruism is something we care about, and not just for theoretical reasons. Although this question of motivation matters to us personally, most of us find, upon reflection, that it is not easy to answer. The ultimate motives of others are often obscure, even if their short-term goals are reasonably clear. And when we look within our own hearts, many of us realize that our ultimate motives are far from obvious, even to ourselves. Perhaps we start caring about the altruism question for personal reasons, but a sustained curiosity forces us to turn to theory if we want an answer.

Once we turn to biology and psychology, the problem becomes even more fascinating. The reason is not that these sciences quickly solved the problem, so that our only task now is to describe how researchers neatly and concisely applied the scientific method with triumphant results. Quite the contrary—the scientific work is interesting precisely because the question turned out to be subtle, confusing, and difficult. It is wonderful when science progresses by a straight march to the truth. But science is even more interesting when its progress is less direct. To understand the process of science, we need to consider not just the destination reached at the end of the journey but the false starts and detours encountered along the way.

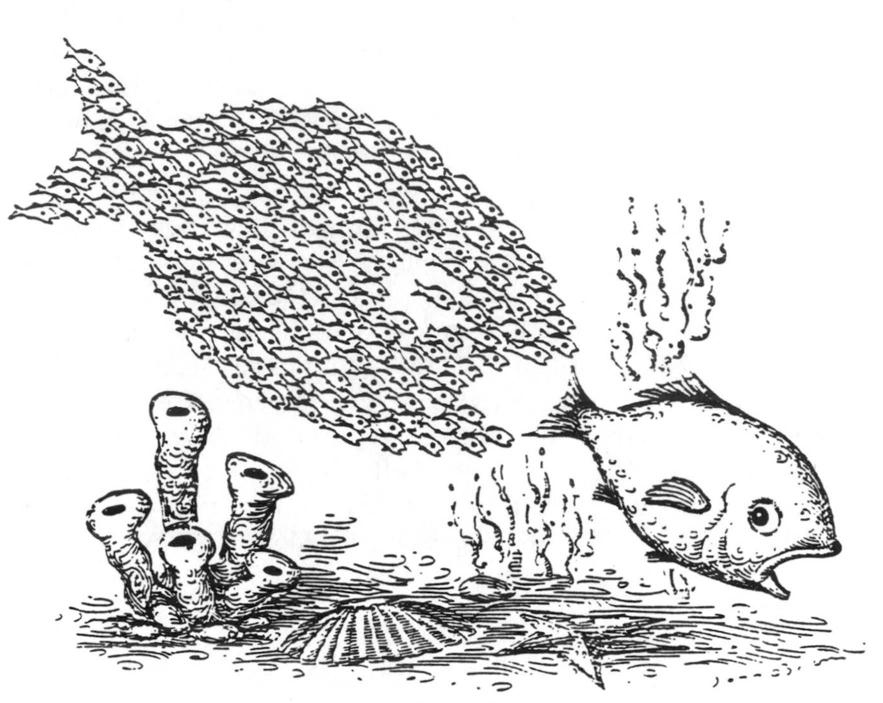

➤ I ➤

Evolutionary Altruism

1

Altruism as a Biological Concept

The concept of altruism in everyday language seems to require an element of both action and motive. People who never help others are seldom considered altruists. On the other hand, when people do help others, we want to know why before we call them altruistic. Even saints could be regarded as selfish if they perceived their lives of sacrifice as tickets to heaven.

Despite the importance of motives in conventional definitions, evolutionary biologists define altruism entirely in terms of survival and reproduction. A behavior is altruistic when it increases the fitness of others and decreases the fitness of the actor. The challenge for the evolutionary biologist is to show how such self-sacrificial behaviors can evolve, regardless of how or even whether the individual thinks or feels as it performs the behavior.

It may seem strange that the biological concept of altruism can remain intuitive when the vital element of motives has been removed. Actually, the strangeness cuts both ways. Philosophical and psychological discussions of altruism often concentrate so heavily on motives that the actual act of helping is ignored. We intend to pay equal attention to actions and motives in this book, but we must begin by separating them. Our first task is to show how behaviors that benefit others at the expense of self can evolve. Our second task is to understand the psychological mechanisms that evolved to motivate these adaptive behaviors.

To see why evolutionists cannot resist talking about altruism, consider the trematode parasite *Dicrocoelium dendriticum,* which spends the adult stage of its life cycle in the liver of cows and sheep (Wickler 1976). The eggs exit with the feces of the mammalian host and are eaten by land snails, which serve as hosts for an asexual stage of the parasite life cycle. Two generations are spent within the snail before the parasite forms yet another stage, the cercaria, which exits the snail enveloped in a mucus mass that is ingested by ants. About fifty cercariae enter the ant along with its meal. Once inside, the parasites bore through the stomach wall and one of them migrates to the brain of the ant (the subesophagal ganglion), where it forms a thin-walled cyst known as the brain worm. The other cercariae form thick-walled cysts. The brain worm changes the behavior of the ant, causing it to spend large amounts of time on the tips of grass blades. Here the ant is more likely to be eaten by livestock, in whose bodies parasites may continue their life cycle. This is one of many fascinating examples of parasites that manipulate the behavior of their hosts for their own benefit. For our purposes, however, the example is interesting because the brain worm, which is responsible for putting the ant in the path of a grazing animal, loses its ability to infect the mammalian host. It sacrifices its life and thereby helps to complete the life cycle of the *other* parasites in its group. It is hard to resist calling this kind of behavior altruistic,[1] even if the parasite doesn't think or feel anything about its fate.

How to Study Evolutionary Change

A Model of Altruistic Behavior

The question of whether and how altruism can evolve has received an enormous amount of attention from evolutionary biologists. E. O. Wilson (1975, p. 3) even called it "the central theoretical problem of sociobiology." Evolutionists are fascinated by altruistic behaviors, not only because they might be important in nature but also because they appear so difficult to explain from the Darwinian perspective. After all, natural selection evolves traits that cause individuals to have *more* offspring than their competitors, not fewer. There is a selective advantage in being selfish, just as there is a selective advantage in having strong teeth and keen eyesight.

Box 1.1 presents the standard model of altruistic behavior that has been developed by evolutionary biologists. A bit of algebra is useful for precision but the basic idea can easily be described in words. Consider a population that is composed of altruists (A) and nonaltruists (S). All individuals have a certain number of offspring (a measure of fitness) in the absence of altruism. In addition, each altruist behaves in a way that decreases its own number of offspring and increases the number of offspring of a single recipient in the population. Altruists can benefit from other altruists in their group, but they also experience the cost of their own self-sacrificial behavior. Selfish types do not experience any cost and can benefit from all altruists in the group. Thus, altruists suffer a double disadvantage; not only do they incur a direct cost from performing an altruistic behavior, but they can receive donations only from *other* altruists, whereas a selfish individual can receive donations from *all* altruists. It should be obvious that selfish types always have more offspring than altruists and will be favored by natural selection. This model therefore captures the essence of what we already said in words: Altruism is the very opposite of the survival of the fittest.

Box 1.1. A mathematical model of altruistic behavior

The fitness of an individual includes both its ability to survive and its ability to reproduce. In this model, an altruistic behavior influences reproduction only, so number of offspring serves as a measure of fitness. Consider a population containing n individuals. There are two genetically encoded traits, altruism (A) and selfishness (S), which occur in frequencies p and $(1 - p)$, respectively. The group therefore contains np altruists and $n(1 - p)$ nonaltruists. All individuals have the same average number of offspring (X) in the absence of the altruistic behavior. Each altruist behaves in a way that causes itself to have c fewer offspring and a single other member of the group to have b more offspring. The fitnesses of altruists (W_A)

and nonaltruists (W_S) can then be specified by the following equations:

$$W_A = X - c + [b(np - 1)/(n - 1)] \quad (1.1)$$

$$W_S = X + [bnp/(n - 1)] \quad (1.2)$$

Each altruist experiences the cost of its altruistic behavior $(-c)$ but also serves as a possible recipient of benefits from the $(np - 1)$ other altruists in the group. Since the other altruists are dispensing their benefits among $(n - 1)$ individuals (to all n members except themselves), the total expected benefit that each altruist experiences is $b(np - 1)/(n - 1)$. Nonaltruists do not experience the cost of altruism and also serve as possible recipients from all np altruists. Thus, not only do altruists suffer a direct cost $(-c)$ but they also serve as possible recipients from fewer altruists than do selfish individuals ($np - 1$ vs. np). It is obvious that W_A is always less than W_S, so altruists will always be selected against within this population.

Suppose that the parameters in the model have the following values:

Population size (n)	100
Frequency of altruists (p)	0.5
Baseline fitness (X)	10
Benefit to recipient (b)	5
Cost to altruist (c)	1

The altruistic type increases the fitness of a single recipient in its group by $b = 5$ units at a cost to itself of $c = 1$ unit. Each altruist can receive a benefit b from the 49 other altruists in the group, while selfish types can be recipients from all 50 altruists in the group. These numbers enable us to calculate the fitnesses of altruists and nonaltruists:

$$\text{Fitness of altruist: } W_A = 10 - 1 + 5(49)/99 = 11.47$$

$$\text{Fitness of nonaltruist: } W_S = 10 + 5(50)/99 = 12.53$$

Everyone's fitness is increased by the presence of altruists in the group, but the selfish S types benefit more than the altrustic A types. From these figures we can calculate the population size n' and the frequency of altruists p' among the offspring.

$$\text{Total number of offspring: } n' = n[pW_A + (1 - p)W_S] = 1200$$

$$\text{Frequency of altruists among offspring: } p' = npW_A/n' = 0.478$$

The population cannot grow to infinity, so we assume that mortality operates on all types equally, returning the population to a size of $n = 100$ without changing the new frequency of altruism ($p' = .478$). In this fashion, the altruists decline in frequency every generation and ultimately go extinct.

In the numerical example described in Box 1.1 (from Wilson 1989), we suppose that the initial population size is 100 individuals, evenly divided between altruists and nonaltruists. All individuals have 10 offspring in the absence of altruism. Each altruist bestows an additional five offspring on a single recipient at a cost of one offspring to itself. From the equations displayed in Box 1.1, we can calculate that the average A type produces 11.47 offspring and the average S type produces 12.53 offspring. For simplicity, assume that each individual reproduces by asexual reproduction, with offspring exactly resembling their parents. A total of 1,200 offspring have been produced and the proportion of altruists among the progeny is 0.478, a decline from the parental value of 0.5. Since populations cannot grow to infinity, we also assume that mortality occurs equally among the A and S types and that the population of offspring is thereby reduced to a size of 100. At this point we expect approximately 48 A types and 52 S types to survive. If this procedure is repeated many times—as it will if natural selection operates over many generations—the A type continues to decline in frequency and ultimately becomes extinct.

Before continuing, we should point out that some assumptions of this model are unrealistic. Most species are sexual, so why do we assume asexual reproduction? The answer is that the model is simpler if we do, and it is important to keep models simple. If we include sexual reproduction and get the same basic results, then we can leave sexual reproduction out for illustrative purposes. Another assumption is a one-to-one relationship between genes and behavior. Evolutionary biologists are quick to admit that this kind of genetic determinism does not exist for most species. Behaviors are caused by complex internal mechanisms that interact with the environment. These mechanisms may be genetically influenced, but recognizing this is a far cry from assuming that individuals are genetically programmed to be altruistic or selfish. Nevertheless, evolutionists frequently assume genetic determinism for the same reason that they assume asexual reproduction; it is a simplifying assumption that they hope does not alter the basic conclusions of the model. We don't have a gene for pulling our hand away from fire, but our complex psychological mechanisms make us behave as if we do in just those circumstances in which such a gene would evolve.

Simplifying assumptions are both the soul and the Achilles heel of mathematical models. It is absolutely essential to keep a model as simple as possible to explore a given subject, such as the evolution of altruism. It often turns out, however, that a simplifying assumption leaves something out of the model that does make a difference and needs to be included to make sense of the subject. Knowing what to include and what to leave out is a subtle skill that makes theoretical biology as much an art as a science. In later chapters we will show that the assumption of genetic determinism is less innocent than it first appears (some readers may already regard it as guilty!), insofar as it obscures important possibilities for the evolution of altruism. Nevertheless, the basic problem can be exhibited most simply by assuming genetic determinism. We therefore will play along with this simplifying assumption for a while before adding more complex and reasonable connections between genes and behavior.

Returning to our model, suppose we change the equations so that A types provide one additional offspring for themselves and two offspring for everyone else in the population.[2] Despite the fact that A types in this scenario increase their own fitness, they still go extinct in the model because they increase the fitness of others even more. This conclusion is well illustrated by the fable about the little red hen, who does all the work of making bread while her companions do nothing.

If we change the end of the story so that her companions succeed in their freeloading ways, then everyone will gather around the table and eat the bread together. The little red hen will get some bread, perhaps even enough to repay her efforts, but her net benefit will always be less than that of her companions, who share the benefits without paying the costs. If we put this kind of interaction into our evolutionary model, the little red hen will have fewer offspring and her kind will go extinct as surely as if she decreased her absolute fitness on behalf of others. In general, evolutionary success depends on *relative* fitness (Williams 1966). It doesn't matter how many offspring you have; it only matters that you have more than anyone else.

How Altruism Can Evolve

It might seem from our model that altruism can never evolve and that evolution is a process that inherently promotes selfishness. On the contrary, it is easy to show that altruism can evolve when more than one group is present. Figure 1.1 shows the simplest example of a population that is divided into two groups. The fitnesses of the A and S types are determined by the equations in Box 1.1 and equal numbers of altruists and nonaltruists exist in the total population, as in our first numerical example. In this case, however, one group has 20 percent altruists while the other group has 80 percent altruists. Looking at each group separately, we reach the same conclusion as before—selfish types will have more offspring than altruistic types. Adding the progeny of both groups together, however, we get the opposite answer; altruistic types have more offspring than selfish types.[3] This outcome is strange enough that we urge the reader to inspect the numbers (see Box 1.2). There is no magic or mysticism here; the altruists increase globally, despite decreasing in frequency within each group, because the two groups contribute different numbers of individuals to the global population.

The success of altruism in Figure 1.1 is an example of a statistical phenomenon known as Simpson's paradox (Simpson 1951; discussed in Sober 1984, 1993b). A nonbiological example of Simpson's paradox might help explain its counterintuitive nature. During the 1970s, the University of California at Berkeley was suspected of discriminating against women in its graduate admission policies (Cartwright

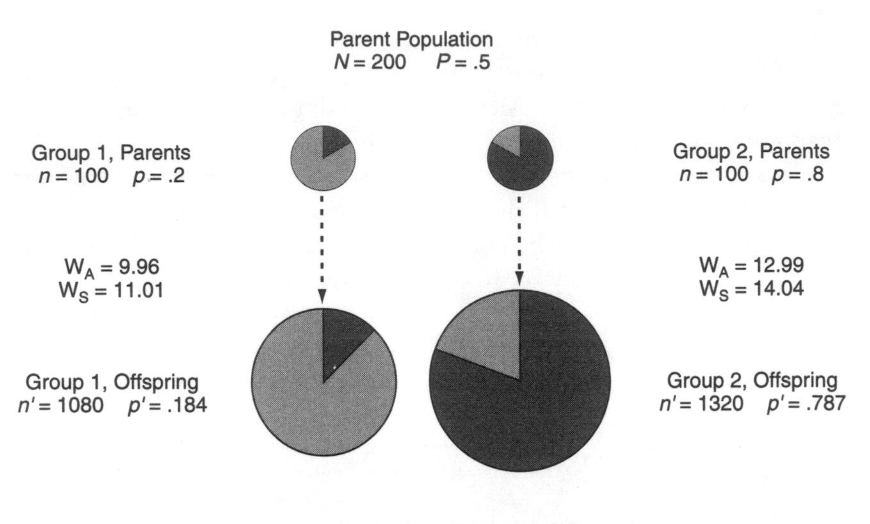

Figure 1.1. The evolution of altruistic and selfish types in an asexual population that is divided into two groups of size $n = 100$ (smaller circles). The altruistic type (black) makes up 20 percent of group 1 and 80 percent of group 2. In the offspring generation, the altruists decline in frequency within each group (the size of the black "slice" grows smaller), but the group with more altruists grows larger (to a population of 1,320) than the group with fewer altruists (1,080). As a result, altruists increase in frequency in the *global* population of 2,400 offspring. Adding the progeny from the two groups is biologically justified only if the groups periodically merge and re-form or otherwise compete in the formation of new groups.

1978). The percentage of women applicants who were admitted was less than the percentage of men, and the difference was large enough that it could not be attributed to chance. The University conducted a department-by-department inquiry and found in each department that women were admitted no less often than men. Evidently, women did worse than men overall, but not in any department.

This paradoxical finding fell into place when it was noticed that women tended to apply to departments with low acceptance rates. To see how this can happen, imagine that 90 women and 10 men apply to a department with a 30 percent acceptance rate. This department does not discriminate and therefore accepts 27 women and 3 men.

Box 1.2. The parameters of a two-group model

	Group 1	*Group 2*
n	100	100
p	0.2	0.8
W_A	$10 - 1 + 5(19)/99 = 9.96$	$10 - 1 + 5(79)/99 = 12.99$
W_S	$10 + 5(20)/99 = 11.01$	$10 + 5(80)/99 = 14.04$
n'	1080	1320
p'	0.184	0.787

	Global population
N	$100 + 100 = 200$
P	$[0.2(100) + 0.8(100)]/200 = 0.5$
N'	$1080 + 1320 = 2400$
P'	$[0.184(1080) + 0.787(1320)]/2400 = 0.516$

Another department, with a 60 percent acceptance rate, receives applications from 10 women and 90 men. This department doesn't discriminate either and therefore accepts 6 women and 54 men. Considering both departments together, 100 men and 100 women applied, but only 33 women were accepted, compared with 57 men. A bias exists in the two departments combined, despite the fact that it does not exist in any single department, because the departments contribute unequally to the total number of applicants who are accepted. In just the same way, altruists can increase in frequency in the two groups combined, despite the fact that they decrease in frequency within each group, because the groups contribute unequally to the total number of offspring.

Returning to the evolution of altruism, we need to emphasize that adding the progeny from the two groups is just statistical sleight of hand unless it can be justified biologically. If the two groups are permanently isolated from each other, natural selection will eliminate the altruists within each group, as we have already shown. The global increase in the frequency of altruists illustrated in Figure 1.1 will be a transient phenomenon of little interest. Suppose, however, that the progeny of both groups disperse and then physically come together before forming new groups of their own. In this case our adding procedure is appropriate and the increased frequency of altruists shown in Figure 1.1 will become the average frequency for the next generation. If the process is repeated over many generations, altruists will gradually replace the selfish types, just as the selfish types replaced the altruists in the one-group example. Of course, we still must explain how, generation after generation, altruists tend to find themselves living with altruists, and selfish individuals tend to associate with other selfish individuals. We will provide an explanation and biological examples in what follows. For the moment, however, we will consolidate our gains with the following general statement: *Altruism can evolve to the extent that altruists and nonaltruists become concentrated in different groups.*

Group Selection

In Figure 1.1, the global frequency of altruists increases even though the local frequency declines within each group. What is required to produce this interesting (and for many people counterintuitive) result? First, there must be more than one group; there must be a *population of groups*. Second, the groups must *vary* in their proportion of altruistic types. Third, there must be a direct relationship between the proportion of altruists in the group and the group's output; groups with altruists must be *more fit* (produce more individual offspring) than groups without altruists. Fourth, although the groups are isolated from each other by definition (the S types in group 1 do not benefit from the A types in group 2), there must also be a sense in which they are *not* isolated (the progeny of both groups must mix or otherwise compete in the formation of new groups). These are the necessary conditions for altruism to evolve in the multigroup model. To be sufficient, the differential fitness of groups (the force favoring the altruists) must be strong enough to counter the differential fitness of individuals within groups (the force favoring the selfish types).

These conditions are similar to the ones laid down in standard formulations of Darwin's theory of natural selection, which requires *a population of individuals* that *vary* in heritable characteristics, with some variants *more fit* than others. The analogy extends to the fourth condition, since individuals are isolated units but nevertheless compete in the creation of new individuals. Thus, natural selection can operate at more than one level of the biological hierarchy, as Darwin clearly appreciated in his discussion of human morality that we quoted in the Introduction. Individual selection favors traits that maximize relative fitness within single groups. Group selection favors traits that maximize the relative fitness of groups. Altruism is maladaptive with respect to individual selection but adaptive with respect to group selection. Altruism can evolve if the process of group selection is sufficiently strong.

Before continuing, let's use the concept of group selection to explain the evolution of altruism in the brain worm. Imagine that the parasite population consists of two types, one that is prone to become a brain worm (A) and one that is not (S). Also imagine that the brain-worm type originates as a single mutation.[4] The mutant egg is ingested along with other eggs by a snail. For purposes of our example, we shall assume that the average snail eats five eggs, so our mutant egg now finds itself in a miniature population (in the snail) with four nonmutant eggs. The brain-worm trait is not expressed while the parasites are within the snail, so the miniature population grows by asexual reproduction into a much larger population in which the mutant type continues to exist at a frequency of $1/5 = 0.2$. This population leaves the snail embedded in a mucus capsule and is ingested by an ant. For simplicity, we assume that all ants become infected by ingesting 50 parasites from a single mucus capsule.

At this point in the life cycle, we can envision the parasite population as living in a large number of isolated groups (the ants). Group size is 50. The vast majority of groups consist entirely of S-types, but one group contains 40 S and 10 A types ($p = 0.2$). Now the behavioral difference between A and S manifests itself. One of the A types burrows into the subesophagal ganglion and becomes the brain worm. The others form thick-walled cysts, as do all the S types. It is in the nature of this particular trait that only one individual in the group can perform the altruistic act. Since the brain worm is destined to die, the population size has been reduced to 49, with 9 A and 40 S types. The new frequency of A types is $p' = 0.183$, a decline from the original value of 0.2. The A type obviously has the lower relative fitness within the group, and for this reason it intuitively seems altruistic. It is interesting, however, that the *degree* of sacrifice is much less than it first appeared, because the extreme sacrifice of the brain worm is diluted by the survival of the A types that did *not* become the brain worm. In any case, we can conclude with confidence that A, the altruistic trait, is selected against within groups. To think about this situation in more familiar terms, imagine that you are in a group of fifty people in which a dangerous activity must be performed by a single person. Ten of you decide to draw straws to see who will perform the dangerous act. The other forty say, "We think it's *great* that you ten are going to draw straws! Go for it!" It is obvious that the ten individuals who take a chance are altruistic and that the forty are freeloaders.

Next, we must calculate fitness at the group level. Some ants will be eaten by livestock even when they do not camp out on the tips of grass blades, so the brain worm is not necessary for the parasite to complete its life cycle. We therefore assume that the brain worm increases group fitness by increasing the chance of being eaten by livestock from a value of E_1 to a value of $E_2 > E_1$. All parasites (except the brain worm) that are eaten by livestock have the same number of offspring. This particular trait influences fitness through survival rather than fecundity.

Now that we know the relative fitnesses within and between groups, we can put them together to see which level of selection prevails. For A types, the good news is that they always have the higher probability of being ingested by livestock (E_2). The bad news is that they must enter a lottery in which one will die. The fitness of the average A type (equal, here, to the probability of not becoming a brain worm and being ingested by livestock) in the global population is therefore $W_A = 0.9(E_2)$. For S types, the good news is that they sometimes find themselves in groups containing A types and therefore benefit from the altruistic behavior without paying a cost. The bad news is that they sometimes find themselves in groups without altruists. If we let q be the fraction of S types that live in mixed groups, then the fitness of the average S type in the global population is $W_S = q(E_2) - (1 - q)(E_1)$. When A originates as a single mutant, the proportion of S types that live in groups with A is vanishingly small (q is close to zero), so S experiences all the bad news and virtually none of the good news.[5] Even though the A types decline in frequency within single groups, they will increase in frequency in the global population when $(0.9)E_2 > E_1$, or $E_2/E_1 > 1.11$. Thus, even a modest increase in the chance of being ingested by livestock (a group-level benefit) can outweigh the suicidal behavior of the brain worm (an individual-level cost). A behavior that superficially seems impossible to explain from the evolutionary perspective is in fact easy to explain, once we realize that natural selection operates at the group level as well as at the individual level.

This example shows how certain assumptions of our two-group model, which may have seemed unlikely when presented in abstract form, can be biologically plausible. The model requires altruistic and selfish types to become concentrated into different groups. This is accomplished biologically by reproduction within the snails, which concentrates the progeny of the mutant altruist into a single group. The model requires groups to be isolated as far as the benefits of altruism are concerned but nevertheless to compete in the formation of new groups. Both of these conditions emerge naturally from the population structure of parasites. The brain worm benefits only the other individuals within its own group, but groups with brain worms outcompete groups without brain worms in their passage to the next stage of the life cycle. The mathematical adding of groups in Figure 1.1 has an exact biological counterpart in the mixing of parasites from many groups in the livers of cows and sheep.

We have shown that the brain worm can evolve from a low frequency in the population, but we have not yet shown that it can drive the selfish type extinct. Suppose that the population consists entirely of A types and we introduce a single mutant S type. During the ant stage of the life cycle, the vast majority of groups consist entirely of A types, but one group contains 40 A and 10 S (p = 0.8). The A types are still selected against within groups, although the chance of becoming the brain worm is now 1/40 rather than 1/10. There is still genetic variation among groups, but this variation does not result in fitness differences at the group level because all groups have at least 1 A type. As a result, there can be no group selection and S will increase in frequency in the global population. Both A and S can "invade the population" at a low frequency. This means that S will increase in frequency when the proportion of A types is close to 100 percent and that A will increase in frequency when the proportion of S types is close to 100 percent. As a result, both traits are retained in the population—or, in other words, the population evolves to a *stable polymorphism.* The equilibrium frequency is reached when the advantages of freeloading within groups are exactly counterbalanced by the disadvantages of being in groups that lack altruists.

Is it true that *Dicrocoelium dentriticum* actually consists of a mix of altruists and nonaltruists? Can it be shown that some ants containing the parasite do not camp out on grass blades because all of the parasites within are freeloaders? Alas, evolutionary biology is not like physics, with an army of empirical scientists eager to test the predictions that theorists formulate. Wickler (1976) called attention to the brain worm as an example of altruism, Wilson (1977a) predicted the likelihood of a polymorphism, and no relevant studies have appeared since then. No one has estimated the basic parameters of the model, such as the number of eggs ingested by snails and the number of mucus masses ingested by ants. The brain worm remains a fascinating *prima facie* example of altruism from the field of natural history, but the conceptually relevant details have only been guessed.

Altruism Is Only One Kind of Group-Level Adaptation

We have shown that altruism can evolve by group selection. It is important to realize, however, that group selection favors *any* behavior that increases the relative fitness of groups. Altruism fits this definition, but it also has the additional feature of decreasing the relative fitness of individuals within groups. In other words, evolutionary altruism is a two-dimensional concept that includes both benefits to others *and* costs to self. A trait that increases group fitness without decreasing relative fitness within groups would also evolve by group selection, and yet it would not be altruistic by our definition. We might even imagine some traits that are favored both within and between groups. In sum, altruism constitutes a subset of group-advantageous traits.

Not only is altruism a subset of traits favored by group selection; it is a distinctly inferior subset. Returning to the brain worm, imagine a new mutant parasite that manages to manipulate the behavior of the ant without killing itself, thereby benefiting the group at no personal cost. The new type would obviously replace the more altruistic A type.

Self-sacrifice on behalf of the group *can* evolve by group selection, but it is never advantageous *per se.* This is an important point, because it shows that the extreme altruism that cries out for explanation might be rare in nature for two very different reasons—(a) because group selection is seldom strong enough to evolve such behaviors, or (b) because there is usually a way to benefit the group without such extreme self-sacrifice.

The difference between altruism and group-level adaptation is illustrated by the fanciful cartoon that begins Part I of this book. It is clearly adaptive for the school of small fish to organize itself into the shape of a large fish in order to chase away would-be predators. This behavior is unlikely to have evolved by chance or to be the coincidental by-product of behaviors that maximize the relative fitness of individuals within the group. It presumably evolved by a selection process in which schools varied in their propensity to take on different shapes, with those resembling large fish contributing more offspring to future generations. But it is not obvious that altruism is required for this process to occur. That would depend on how the costs and benefits of assuming the fish shape are distributed among members of the group. Perhaps some fish voluntarily assume the most dangerous positions in the school, in which case altruism would be involved. However, it is also possible that all positions are equally dangerous, or the fish have a way of randomizing their position, or some fish are forced by others to assume the most dangerous positions. Regardless of the degree of altruism involved, we can appreciate that the school is well-adapted at the group level.

The Averaging Fallacy

It might seem from our account that the evolution of altruism is easily explained and that "the central theoretical problem of sociobiology" has been solved. On the contrary, however, anyone familiar with the evolution literature knows that altruism remains a controversial subject. Group selection, the mechanism that we have proposed to explain the evolution of altruism, is often rejected as an important evolutionary force. Behaviors that seem to benefit others at the expense of self are often said to be only "apparently" altruistic, with "genuine" altruism in nature remaining elusive.

To understand the nature of these debates, let us return to our two-group model in Figure 1.1. It shows that altruism declines in frequency in each group but nevertheless evolves because the altruistic group is more fit than the nonaltruistic group. Another way to show that altruism evolves is by simply averaging the fitnesses of individuals across groups. The 20 A types in group 1 have 9.96 offspring each and the 80 A types in group 2 have 12.99 offspring each, for an average of 12.38 offspring. The 80 S types in group 1 have 11.01 offspring each and the 20 S types in group 2 have 14.04 offspring each, for an average of 11.62 offspring. The average A type is more fit than the average S type in the global population and therefore the A trait evolves.

This method of calculating fitness does not change any facts about the model, and its simplicity has a certain appeal. In fact, it is easy to conclude that A types evolve by "individual selection" and are selfish after all because the average A type is more fit than the average S type. In short, a single trait can appear to be altruistic *or* selfish, depending on whether fitnesses are compared within groups or are averaged across groups and then compared.

Why not use fitness averaged across groups to define individual selection? One problem is that the definition encompasses everything that evolves, regardless of the type of selection process involved. In the one-group model (Box 1.1), the average S type is more fit and therefore evolves by individual selection. In the two-group model (Figure 1.1), the average A type is more fit and therefore evolves by individual selection. If we altered the two-group model so that there is less variation between groups, the average S type would again become the fitter trait and would evolve by individual selection. The averaging approach makes "individual selection" a synonym for "natural selection." The existence of more than one group and fitness differences between the groups have been folded into the definition of individual selection, defining group selection out of existence. Group selection is no longer a process that can occur in theory, so its existence in nature is settled *a priori.* Group selection simply has no place in this semantic framework.

Another reason to reject the averaging approach is that it fails to identify the separate causal processes that contribute to the evolutionary outcome. When altruism evolves, there typically are two processes at work. Between-group selection favors the evolution of altruism; within-group selection favors the evolution of selfishness. These two processes oppose each other. If altruism manages to evolve, this indicates that the group-selection process has been strong enough to overwhelm the force pushing in the opposite direction. When this two-level process occurs in a population, an appropriate causal analysis should describe what is going on. The summary statement that the trait that evolved had the higher average fitness does not include any of these details. It is neutral on what process or processes are responsible for that result. When one trait is more fit than another, this may be due to pure individual selection, to pure group selection, or to a mixture of the two. The description of the effect fails to specify what the causes were.

This is a problem that any science is prey to; it is not peculiar to the theory of natural selection. Suppose Sam pushes a billiard ball east, while Aaron pushes it west. If Aaron pushes harder, the ball moves west. There is a single resultant, but there are two component causes. To represent this Newtonian problem in terms of vectors, you draw an arrow pointing east and a longer arrow pointing west; these represent the component forces. Then you add the two vectors together, yielding an arrow that points west; this represents the resultant force. If you cared only about predicting the ball's trajectory, knowing the resultant would suffice. But if the point is to understand the processes at work, the resultant is not enough. Simpson's paradox shows how confusing it can be to focus only on net outcomes without keeping track of the component causal factors. This confusion is carried into evolutionary biology when the separate effects of selection within and between groups are expressed in terms of a single quantity.[6]

Given these problems, we feel confident calling the averaging approach the "averaging fallacy" when it is used to define altruism and selfishness in an evolutionary model (Sober 1984). Does anyone actually commit the averaging fallacy, defining group selection out of existence? In one sense, the answer to this question is "no." Every major thinker on the topic has treated group selection as a process that can occur in principle and has focused on the question of how strongly it operates in nature (e.g., Williams 1966, 1992; Dawkins 1976, 1982, 1989; Maynard Smith 1964, 1976; Alexander and Borgia 1978; Alexander 1987). No one has ever seriously maintained that group selection is just plain impossible.

In another sense, however, the answer is "yes"—profoundly, abundantly, confusingly "yes." Even though the averaging fallacy is not endorsed in its general form, it frequently occurs in specific cases. In fact, we will make the bold claim that the controversy over group selection and altruism in biology can be largely resolved simply by avoiding the averaging fallacy. The evolution of altruism is easily explained, genuine altruism exists in nature, and "the central theoretical problem of sociobiology" has been solved.

Our claim may sound outrageous, but it becomes more reasonable when one realizes that the averaging approach is seductively simple and that groups can vary tremendously in their spatial and temporal scale. Some groups are physically isolated populations that persist for many generations. Other groups are physically isolated but persist for only a fraction of a life cycle, as in our brain-worm example. Other groups are socially defined, such as a hunting pack. We will consider the nature of groups in more detail in Chapter 2. For now, suffice it to say that a biologist who does not commit the averaging fallacy with respect to some kinds of groups may well commit the fallacy with respect to other kinds of groups.

Now we can begin to see why the controversy over altruism and group selection involves more than the straightforward examination of empirical facts against a single theoretical background. It also involves different ways of seeing the evolutionary process—call them paradigms if you like—that are incompatible with each other. It is easy enough to understand both paradigms and to translate from one to the other. Nevertheless, if "individual selection" as defined by one paradigm is used to argue against "group selection" as defined by the other paradigm, only confusion and futile controversy can result.

We will review two examples in detail to document our claim that altruism exists in nature and can be clearly recognized as long as we avoid the averaging fallacy. Unlike the brain worm, these examples have been studied in enough detail to provide a compelling picture of natural selection as a multilevel process. However, the import of the examples as proof of group selection and altruism has been obscured by the averaging fallacy and a failure to distinguish between paradigms. Even today, decades after the evidence should have been accepted by the normal standards of science, many evolutionary biologists regard group selection as an unproven theory. To fully illustrate our point, therefore, we must consider not only the relevant biological facts but also the history of the scientific discourse in which the facts have been described.

Example 1: The Evolution of Sex Ratios

A Brief History of Group Selection Theory

Prior to the 1960s, group selection and altruism were poorly articulated concepts. Darwin's original clarity of thought and sparing use of group selection to explain the evolution of certain traits had been lost. The broad intellectual traditions of individual-level functionalism and group-level functionalism that we discussed in the Introduction flourished in biology, as in other disciplines. Some biologists were inclined to see societies and ecosystems as resembling a single organism, while others were inclined to see them as merely the by-product of individual interactions. Both camps felt that Darwin's theory justified their views, but their arguments seldom led to clear-cut hypotheses that could be tested empirically.

The clearest views on group selection were developed in the 1930s and 40s by the three founding fathers of population genetics—Ronald Fisher, J. B. S. Haldane, and Sewall Wright, who were largely responsible for building the entire mathematical foundation of evolutionary theory. Each considered the problem of group selection, but only briefly, providing sketches of models rather than fully developed treatments. In a sense, there *was* no theory of group selection prior to the 1950s, only the seeds of a theory. In addition, the three fathers of population genetics disagreed among themselves about the importance of group selection, much as their colleagues who were less gifted theoretically did. Wright thought that group selection was probably an important force in nature, whereas Fisher and Haldane accorded it little significance.

Enter George Williams, a tall man with a dry sense of humor and a craggy face that reminds one of Abe Lincoln or the statues on Easter Island. Williams went to the University of Chicago as a postdoctoral student in the 1950s. Chicago was a bastion of group-level functionalism, and Williams attended a lecture by Alfred Emerson, a termite biologist who interpreted all of nature on the model of a termite colony. As he later recounted the event to one of us (DSW), "If this was evolutionary biology, I wanted to do something else—like car insurance." Williams began work on a book that was meant to clarify the uses and misuses of adaptationism in evolutionary biology. When *Adaptation and Natural Selection* was published in 1966, it became a modern classic.

As Williams was writing, another book was published that provided the perfect foil for his arguments. V. C. Wynne-Edwards was a Scottish biologist who studied red grouse *(Lagopus lagopus scoticus),* a ground-dwelling bird that lives on the moors. Every year, a fraction of the red grouse population claims the best territories and breeds while the rest are pushed into marginal habitat and often die. Wynne-Edwards interpreted this social system as an adaptation that evolved to prevent the grouse population from overexploiting its food supply. In addition, he thought that most species in nature face the same problem. In *Animal Dispersion in Relation to Social Behavior,* Wynne-Edwards (1962) interpreted a vast array of social behaviors as adaptations for regulating population size. For example, birds sing in the morning and zooplankton migrate to surface waters at night to assess their density and regulate their reproduction accordingly. Wynne-Edwards's writing conveyed the electric quality of someone who believed he had discovered a major principle of evolution.

Wynne-Edwards was aware that social behaviors of this sort are potentially altruistic and can be exploited at the individual level. He also understood that a process of group selection was required to explain their evolution. Nevertheless, he spent only a few pages discussing group selection, appealing mostly to the work of Sewall Wright. Wynne-Edwards wrote as if Wright had solved the problem of group selection, whereas Wright had only scratched the surface. The degree to which Wynne-Edwards relied on group selection is illustrated by the following passage from his book (p. 20):

> Evolution at this level can be ascribed, therefore, to what is here termed group-selection—still an intraspecific process, and, for everything concerning population dynamics, much more important than selection at the individual level . . . Where the two conflict, as they

> do when the short-term advantage of the individual undermines the safety of the race, group-selection is bound to win, because the race will suffer and decline, and be supplanted by another in which antisocial advancement of the individual is more rigidly inhibited. In our own lives, of course, we recognize the conflict as a moral issue, and the counterpart of this must exist in all social animals.

Wynne-Edwards's work provoked a broadside of attacks from evolutionary biologists who knew that the problem of group selection had not been solved and in fact had barely been addressed. Williams was by no means the only critic, but his book-length discussion of adaptation as a concept that must be handled with care was devastatingly effective. His rejection of group selection was as complete as Wynne-Edwards's acceptance:

> It is universally conceded by those who have seriously concerned themselves with this problem that . . . group-related adaptations must be attributed to the natural selection of alternative groups of individuals and that the natural selection of alternative alleles within populations will be opposed to this development. I am in entire agreement with the reasoning behind this conclusion. Only by a theory of between-group selection could we achieve a scientific explanation of group-related adaptations. However, I would question one of the premises on which the reasoning is based. Chapters 5 to 8 will be primarily a defence of the thesis that group-related adaptations do not, in fact, exist. A *group* in this discussion should be understood to mean something other than a family and to be composed of individuals that need not be closely related. (Williams 1966, pp. 92–93)

Williams's exclusion of family groups from this sweeping claim will be discussed in Chapter 2. Although Williams and Wynne-Edwards had vastly different opinions about the importance of group selection, it is important to stress that *they agreed on what group selection is*—natural selection based on the differential fitness of groups.

Williams's case against group selection relied largely on discussions of broad categories of behavior, such as bird song, territoriality, and dominance. Williams argued that these behaviors should be interpreted by invoking the principle of parsimony. He contended that individual-level explanations are simpler and therefore preferable to group-level explanations whenever both are available. We will discuss the issue of parsimony in more detail in Chapter 3. For now, suffice it to say that arguments based on parsimony are sometimes legitimate but are often inconclusive devices for answering scientific questions. They certainly do not substitute for a critical test of hypotheses that make different and mutually exclusive predictions.

In addition to his often insightful conceptual analysis and his arguments based on parsimony, Williams also devised a critical test of group selection theory, one that relied on clear-cut predictions that could be confirmed or rejected with empirical data. His test involved the proportion of sons and daughters that are produced at birth.

Sex Ratios and Multilevel Selection

In sexually reproducing species, the maximum rate of population growth depends on the sex ratio. Imagine a population in which all females have ten offspring, half of which are daughters. A population initiated by a single fertilized female will consist of 5 females and 5 males during the first generation ($n = 10$), 25 females and 25 males during the next generation ($n = 50$), 125 females and 125 males during the third generation ($n = 250$), and so on. Now imagine a population in which 9 daughters are produced for every son. A population initiated by a single fertilized female will consist of 9 females and 1 male during the first generation ($n = 10$), 81 females and 9 males during the second generation ($n = 90$), 729 females and 81 males during the third generation ($n = 810$), and so on (we assume that the males are capable of fertilizing all the females). The female-biased population grows faster because population growth is limited by the number of eggs, not the number of sperm. The difference between the populations quickly becomes large because the growth rates are exponential.

Although a female-biased sex ratio benefits the group, it does not increase the relative fitness of individuals within the group. To see this, imagine a group that is initiated by two fertilized females, S and A, who produce sex ratios of 1:1 and 9:1, respectively. Female S has 5 daughters and 5 sons, while female A has 9 daughters and 1 son, for a total of 14 females and 6 males in the next generation. When these individuals mate and reproduce, each female will have ten offspring but the average male will have 140/6 = 23 offspring. Male fitness is higher than female fitness because the average male sires the offspring of more than two females. If we evaluate the fitnesses of the original S and A females by counting their grandchildren, we discover that S has 5(10) + 5(23) = 165 grandoffspring while A has only 9(10) + 1(23) = 113 grandoffspring. In this population males have more offspring than females because they are in the minority, so the female who produced more sons is fitter than the female who produced a preponderance of daughters. Producing an excess of daughters, therefore, is an act of altruism—it benefits the group while decreasing the mother's fitness. In general, any deviation from a 1:1 sex ratio at birth will cause the minority sex to have the higher relative fitness within groups, at which point genes that return the sex ratio to 1:1 will be favored. This basic prediction was first derived by Ronald Fisher in the 1930s.

Williams realized that group selection should evolve a sex ratio highly biased in favor of females if groups do best by maximizing their productivity, whereas individual selection should evolve an even sex ratio. However, rapid growth may not be advantageous for the group if it causes overexploitation of resources, as Wynne-Edwards stressed. Williams reasoned that if sex ratio is an adaptation to regulate population size, it should evolve to be facultatively male- or female-biased, depending on whether the population is above or below its optimum. In any case, the sex ratio should not be consistently even, as predicted by individual selection theory.

Here at last were some relatively clear-cut predictions that could be tested with empirical data. Williams (1966, p. 151) evaluated the information on sex ratio that he could find in the scientific literature and drew the following conclusion:

> Despite the difficulty of obtaining precise and reliable data, the general answer should be abundantly clear. In all well-studied animals of obligate sexuality, such as man, the fruit fly, and farm animals, a sex ratio close to one is apparent at most stages of development in most populations. Close conformity with the theory is certainly the rule, and there is no convincing evidence that sex ratios ever behave as a biotic adaptation [i.e., evolve by group selection].

Williams was so pleased with his empirical test that in the concluding section of his book (p. 272) he stated, "I would regard the problem of sex ratio as solved."[7]

After *Adaptation and Natural Selection,* group selection was a dead issue for most evolutionary biologists. Dead but not forgotten. On the contrary, the rejection of group selection was celebrated as a scientific advance, comparable to the rejection of Lamarckism, that allowed biologists to close the book on one set of possibilities and concentrate their attention elsewhere. The memory of group selection had to be kept alive as an example of how not to think. It became almost mandatory for the authors of journal articles to assure their readers that group selection was not being invoked. A generation of graduate students learned to avoid group selection almost as if it were prohibited by one of the ten commandments. We wish we were exaggerating, but many of our evolutionary colleagues will recall what Stephen Jay Gould (1982, p. xv) describes as "the hooting dismissal of Wynne-Edwards and group selection in any form during the late 60s and most of the 70s." As late as the 1980s, one of our colleagues who was interested in group selection received the following advice from a very distinguished evolutionary biologist: "There are three ideas that you do not invoke in biology—Lamarckism, the phlogiston theory, and group selection."

Evidence for Group Selection, Obscured by the Averaging Fallacy

One year after the publication of Williams's book, an article appeared in the journal *Science* by W. D. Hamilton (1967) entitled "Extraordinary Sex Ratios." Hamilton had already become famous for his theory of inclusive fitness, which we will describe in Chapter 2. In this equally ground-breaking paper, Hamilton provided many examples of female-biased sex ratios from the scientific literature and developed a theory to explain their evolution. He framed his theory in terms of a parasite species that is searching for hosts. Each host is colonized by a certain number of females, whose progeny mate randomly among themselves before dispersing in search of new hosts. All mating takes place within groups prior to dispersal by the females.

Even without describing the details of Hamilton's model, it should be obvious that it resembles our two-group model and the life cycle of the brain worm. The population is subdivided into a large number of groups (hosts) that are isolated from each other as far as the effects of sex ratio are concerned. Genes that cause female-biased sex ratios increase the size of the group prior to dispersal, but genes that cause even sex ratios get a larger than average slice of the pie within each group. The groups are not permanently isolated, because their progeny periodically disperse and engage in population-wide competition for new hosts. Hamilton's model is a group selection model, a mathematical version of Williams's hypothesis about sex ratio. Furthermore, Hamilton's perusal of the scientific literature revealed many examples of female-biased sex ratios that provide evidence for group selection in nature. These examples were especially common among small invertebrate species that occupy ephemeral habitats for periods of a few generations before dispersing more widely, as would be expected from Hamilton's theory. So many invertebrate species fit this description that the number of species with female-biased sex ratios is probably in the hundreds of thousands or even millions.

Unfortunately, even though Hamilton's theory was accepted on the strength of the evidence he provided, female-biased sex ratios were not interpreted as evidence for group selection. Hamilton presented his model in a way that averaged the fitness of alternative types across all groups in the global population, rather than focusing on relative fitness within and among groups. He called the type with the highest average fitness the "unbeatable strategy," an appropriate term, since it refers to the sex ratio that evolves. However, focusing on the end result obscures the fact that the unbeatable strategy (which might be 60 percent daughters, for example) is the outcome of component forces pushing in opposite directions—within-group selection favoring an even sex ratio and between-group selection favoring the maximum proportion of females that can be fertilized by a minority of males.

Virtually all evolutionary biologists who read Hamilton's paper interpreted the "unbeatable strategy" as an adaptation that evolves by individual selection. No one seemed to connect Hamilton's paper with Williams's use of sex ratio, only a year earlier, as a critical test of group selection theory. The study of female-biased sex ratios became a hot topic, the theory was elaborated, and dozens of additional examples were documented. At the very time that group selection theory was entering its dark age, the empirical evidence that *should* have counted as evidence *for* group selection was accepted as a triumph of individual selection theory!

Actually, there is one outstanding exception to this statement—W. D. Hamilton himself. Even though his analysis was based on fitness averaged across groups, he mentioned the opposing roles of group and individual selection in the forty-third footnote of his paper:

> The combination 0,0 [a group of two individuals, each producing maximally female biased sex ratios] gives the highest possible payoff to the group . . ., so this "solution" should be favored by a "group selectionist." From what has been said, the "solution" 1/2,1/2 [a group of two individuals, each producing even sex ratios] . . . should be favored by the extreme believer in a biological *bellum omnium contra omnes*. It is pleasing, therefore, to find that what turns out to be the true solution in this case, 1/4,1/4 [a group of two individuals who produce 1/4 sons], lies exactly midway between the others, both in position and in payoff. (p. 487; our comments within square brackets)

In other words, the "unbeatable strategy" is a moderately female-biased sex ratio that reflects an equilibrium between the even sex ratio favored by individual selection and the highly female-biased sex ratio favored by group selection. Hamilton continued to recognize the role of group selection in the evolution of female-biased sex ratios in subsequent papers (e.g., Hamilton 1975, 1979). Oddly, his view had little impact on other evolutionary biologists. The connection between Hamilton's sex ratio theory and Williams's original test was not noticed until Robert Colwell, then at the University of California at Berkeley, forcefully called attention to it in an article published in the journal *Nature* entitled "Group Selection Is Implicated in the Evolution of Female-Biased Sex Ratios" (Colwell 1981; see also Wilson and Colwell 1981). Some authorities on sex ratio accepted Colwell's interpretation immediately (e.g., Charnov 1982) but others couldn't believe that an adaptation as well-documented as female-biased sex ratios might count as evidence for a theory as heretical as group selection. Even now the dust has not entirely settled. Nevertheless, it is gratifying to know that Williams, who originally proposed the test, has come to accept female-biased sex ratios as empirical evidence for group selection (Williams 1992, p. 49): "I think it desirable, in thinking about organisms for which the haystack model is descriptive, to realize that selection in female-biased Mendelian populations favors males, and that it is only the selection among such groups that can favor the female bias." The haystack model that Williams refers to is Maynard Smith's (1964) model of group selection, which is similar to Hamilton's sex ratio model and will be described in Chapter 2.

To summarize, female-biased sex ratios provide one well-documented example of altruism that has evolved by group selection. However, the averaging fallacy has allowed female-biased sex ratios to be interpreted as an individual-level adaptation by many evolutionary biologists, despite the fact that both Williams and Hamilton successfully avoided the averaging fallacy.

Example 2: The Evolution of Virulence

Our second example of altruism involves the effects of parasites and diseases on their hosts.[8] The early literature on this subject was replete with the naive adaptationism that Williams criticized. Hosts and their diseases were assumed to evolve toward a harmonious relationship. Virulence was viewed as an evolutionary mistake because it killed the disease-causing population along with the host. The existence of virulent diseases was attributed to an early stage of coevolution, which would progress toward commensalism. These sentiments, which have long been rejected by evolutionary biologists, are still common in health-related disciplines that have had little contact with evolutionary biology. For example, Lewis Thomas (1972, p. 553) wrote that "disease usually represents the inconclusive negotiations for symbiosis . . . a biological misinterpretation of the borders."

It would be lovely if this were true. Unfortunately, even if the pathogenic population within a host could evolve as a single adaptive unit, it would not necessarily evolve a harmonious relationship with its host.[9] We have already seen that the brain worm has evolved to steer its ant host to its death to facilitate the parasite's passage to the next stage of its life cycle. Sheer reproduction provides another reason for diseases to evolve in ways that have a negative effect on their hosts. In many cases, the more copies a disease organism makes of itself, the sicker the host gets.[10] If disease reproduction shortens the lifetime of the disease population in the present host, this cost must be weighed against the benefits of producing progeny that will reach new hosts. As far as evolution in the disease population is concerned, we should expect a level of virulence that is optimal for the disease, rather than a level that is optimal (least pathogenic) for the host.

Ewald (1993) and others have used this cost-benefit argument to predict the optimal level of virulence for various kinds of diseases. Suppose that you are infected by a disease that is transmitted through the air. If the disease makes you too sick, you will stop mingling with other people and retire to your bed. From the standpoint of the disease, sending you to your room is almost as bad as killing you, because your behavior has vastly reduced its chances of reaching new hosts. It would be "smarter" for the disease organisms to reproduce less, so that you feel better, so that they will be less isolated from other potential hosts. On the other hand, imagine that you are infected with a disease that is transmitted through feces and causes diarrhea. Even if the disease makes you feel so sick that you retire to your bed, your feces will be dutifully carried from your room and sent into the environment, just as if you were well. It may well be "smart" for this disease to reproduce as much as possible, even to go as far as killing you, as long as in the process you release a torrent of progeny that can infect new hosts. As it turns out, intestinal diseases tend to be more virulent than airborne diseases (Ewald 1993).

This cost-benefit argument is only part of the story, however, because it assumes that the group of disease organisms inside a host acts like a single well-adapted unit, adjusting its virulence to maximize its transmission to new hosts. In reality, the group of disease organisms within a single host is a genetically variable population in its own right, within which natural selection can occur. Genetic variability may occur because the host was invaded by more than one strain of disease organism, or because of variability that arises *in situ* by mutation or recombination. Genotypes that have the highest relative fitness within a single host can easily cause the group as a whole to behave maladaptively. To assume that diseases evolve an optimal degree of virulence is to assume that they have evolved entirely by group selection. This is only slightly less naive than assuming that diseases evolve a harmonious relationship with their hosts.

To appreciate the conflicts between levels of selection in disease-causing organisms, let's continue the previous example by imagining an airborne disease that is optimally virulent. This organism—a virus, say—reproduces enough to make you broadcast millions of progeny with every sneeze, but not enough to make you retire to your bed. Now suppose that a mutant strain appears that reproduces twice as fast as the optimal strain. The mutant strain makes you feel so bad that you take to your bed and the billions of progeny that are broadcast with every sneeze perish before they reach a new host. Nevertheless, the hypervirulent strain still enjoys a tremendous advantage over the optimally virulent strain within you, a single host. With a generation time of an hour or less, the hypervirulent strain in your body will replace the optimal strain in a matter of weeks.

It might seem that the hypervirulent strain is "stupid" and should be eliminated by natural selection. The whole point of Williams's book, however, was to show that biological adaptations must be defined carefully in terms of relative fitness; they do not always appear "smart" from the human perspective. The hypervirulent strain, producing much more of its kind with each generation than the milder strain produces, gets the biggest slice of the pie (the environment provided by the host) and will ultimately replace the optimally virulent strain in all hosts containing both types. Only group selection can counter this trend by causing groups with less virulent strains to outcompete groups with more virulent strains.

Richard Lewontin (1970) of Harvard University was the first to realize that the evolution of virulence provides a test of group selection theory. For disease organisms whose reproduction negatively affects the host, group selection favors one trait (optimal virulence), individual selection favors another trait (virulence that is above the optimum from the standpoint of the group), and it should be possible to settle the issue of what evolves by studying the biological facts. An experiment that took place on a grand scale in Australia provided such a test. A type of virus called *Myxoma* was introduced by the government to control the rabbit population there. At first the virus was devastatingly effective, but later it became less virulent. According to Lewontin (pp. 14–15):

> When rabbits from the wild were tested against laboratory strains of virus, it was found that the rabbits had become resistant, as would be expected from simple individual selection. However, when virus recovered from the wild was tested against laboratory rabbits, it was discovered that the virus had become less virulent, which cannot be explained by individual selection.

The evolution of decreased virulence in the *Myxoma* virus was documented in unusual detail because it was part of a biological control program. However, Lewontin felt that it illustrated a more general process that operates in many disease organisms, "despite the complete lack of selective advantage of avirulence within demes [= hosts]."

Alternative Explanations—or, The Averaging Fallacy Returns

The *Myxoma* virus and other examples of reduced virulence (e.g., Herre 1993) would seem to provide compelling evidence for group selection, but they did not go unchallenged. It would be possible to argue against group selection by questioning the biological facts; for example, by asking whether reduced virulence might be favored by within-group selection after all. Most of the arguments against reduced virulence as evidence for group selection accepted the biological facts, however, and were made on the basis of the averaging fallacy.

Here is Alexander and Borgia's (1978, p. 453) interpretation of the *Myxoma* virus:

> Lewontin states that reduction of "virulence" in the virus "cannot be explained by individual selection." . . . Even if virulence is equated with rate of multiplication, interdemic selection in the usual sense is not indicated. Lewontin refers to each rabbit as "a deme from the standpoint of the virus." But the virus reproduces without sexuality, so that a rabbit infected by a single virus particle, or a few identical particles, will contain a clone of genetically identical viruses except for mutational changes. Lewontin's group selection model actually requires that, as a rule, less-and more-virulent viruses be

> mixed in the same rabbits . . . If the population of rabbits is composed largely of individuals infected with pure more-virulent and pure less-virulent strains (i.e. clones), the relevant selection on the virus might be more appropriately described as occurring at the individual level. Thus, if this virus has evolved for a long time in pure clones, a clone of identical viruses would be no more appropriately regarded as a group or deme than would be the cells of a metazoan organism, since each member of a clone should evolve to sacrifice as much for a clone member as for itself. Because vertebrates and other familiar organisms are almost devoid of clones of individuals, it is difficult for us even to think in terms of the extremes of altruism that are predicted by consistent cloning.

To understand the nature of this argument, recall that group selection requires genetic variation among groups. If all groups are exactly alike, there can be no group selection. The more groups vary in fitness, the stronger group selection becomes. The ultimate in variation among groups is the complete segregation of altruists and nonaltruists from each other. At this point, selfish types never profit from having altruists in their group and within-group selection disappears. Alexander and Borgia appear to agree with this reasoning, except for the last step. For them, group selection fails to occur when groups do not vary at all, and also when they vary to a maximal degree. Alexander and Borgia are not suggesting a selective advantage for avirulence within groups but are merely redefining group selection to exclude the case of maximum variation among groups.

The kernel of truth in this passage from Alexander and Borgia is that individuals in groups will evolve to be maximally altruistic toward each other when groups are genetically pure. The groups will become "superorganisms" without any internal conflicts of interest. Of course, this is exactly what we should expect if group selection is the only force operating on the species. We may want to relabel the unit an individual rather than a group because it has become so well organized (just as we can refer to an individual as a maximally altruistic group of cells), but we certainly cannot deny that selection has acted at the level of that unit. By calling the group an individual and saying that it evolves by individual selection, Alexander and Borgia are using the term "individual selection" in two completely different ways: (a) to refer to selection within groups when there are mixed groups and (b) to refer to selection among groups when there is complete segregation among groups.

Two more aspects of this passage warrant comment. First, Alexander and Borgia provide no evidence that the *Myxoma* virus or any other disease is genetically uniform within each host. The mere suggestion that avirulence could be explained without invoking group selection was enough to settle the issue for many evolutionary biologists; empirical evidence was somehow not required. Second, Alexander and Borgia continue to call avirulence altruistic even though they think it evolves by individual selection. In other publications, Alexander (1974, 1979, 1987, 1992) distinguishes between "genotypic altruism," which evolves by group selection, and "phenotypic altruism," which evolves by individual selection and is therefore "genotypically selfish." Alexander would say that avirulence is phenotypically altruistic and genotypically selfish because individuals are really just helping their genetically identical clone-mates. In general, Alexander readily admits that phenotypic altruism evolves and only challenges the existence of genotypic altruism. This is part of a more general tendency among evolutionary biologists to distinguish between "genuine altruism," which evolves by group selection, and "apparent altruism," which is "really selfish" because it evolves by individual selection. This distinction is unnecessary in the case of avirulence, which evolves by group selection even when the strains are maximally segregated into different groups.

Numerous other authors have questioned the role of group selection in the evolution of avirulence. In most cases, group selection is briefly mentioned and then dismissed as a ghost from the past. Then individual selection is defined to reflect both relative fitness within groups and the differential productivity of groups. Bremermann and Pickering (1983, p. 411) provide a nice example:

> While an individual's restraint benefits its within-host competitors through increased host longevity, the model does not invoke group selection. In the model, selection favors an individual's restraint when such behavior increases the individual's total number of propagules. Concurrent increases in the absolute and relative fitness of an individual's within-host competitors can be consequences of such individual selection.

It is worth noting that what Bremermann and Pickering call individual selection would be classified as group selection by Alexander and Borgia, since it involves competition between virulent and avirulent types within single hosts. The only thing these two analyses share is the claim that group selection is not involved.

In recent years, biologists who study the evolution of virulence have finally returned to Lewontin's original view (e.g., Bull 1994; Ewald 1993; Frank 1996a). According to Bull (1994, p. 1425), "both levels operate together in what may be regarded as a classic group selection hierarchy, the groups being the populations of parasites within hosts . . . To understand the complexities of this two-level evolutionary process, it is easiest to dissect the process into separate between-host and within-host components, studying each component before assembling them." Additional empirical studies have been added to the *Myxoma* example (e.g., Herre 1993). Group selection theory has important practical implications, since hospital procedures and public health practices can be designed to alter the population structure of disease organisms to favor the evolution of low-virulence strains (Ewald 1993). Even G. C. Williams has commented on the subject as part of his recent effort to develop a theory of Darwinian medicine (Williams and Nesse 1991; Nesse and Williams 1994). We will let group selection's most famous critic have the last word on the evolution of virulence:

> Bacterial pathogens may complete a million cycles of fission within the lifetime of one human host, and there may be more pathogens in one individual than the earth's human population. Even in one host, a pathogen can be expected to produce highly improbable mutations many times and to evolve significantly in response to even minute selection forces. . . . It has been realized for many years that some bacteria rapidly acquire high levels of antibiotic resistance . . . and that resistant strains can locally replace susceptible ones in a few weeks. A less commonly appreciated phenomenon, the evolution of virulence, will be emphasized here . . . As an extreme example of within-host selection of virulence, imagine two pathogen clones competing within a host. One uses optimal exploitation, which results in the maximum number of propagules dispersed during the lifetime of the host. The other uses maximal (lethal) exploitation, which converts host resources to propagules at the maximum possi-

> ble rate. The host will disperse more of the lethal type than its restrained competitor. The cost of the host's death is borne equally by the two competitors, whereas only the more virulent benefits from a greater rate of transmission . . . In highly virulent cases of cholera and shigellosis propagules may be dispersed at more than a hundred times the rate in less virulent cases. The host's final output of both strains, of course, may be less than the long-term output from the less virulent type when it is the sole exploiter. The evolutionary outcome will depend on relative strengths of within-host and between-host competition in pathogen evolution. This is a clear example of group vs. individual (clone) selection for altruism, for which many formal models have been proposed. (Williams and Nesse 1991, p. 8)

At the Heart of the Controversy

We have reviewed the examples of sex ratio and disease virulence in detail for two reasons. First, they show that group selection is more than "just a theory" and has been documented as well as any theory in evolutionary biology. Second, they provide microcosms of the more general controversy over group selection and altruism that has been raging among evolutionary biologists for thirty years. Consider the following epitaph for group selection theory, written by Richard Dawkins in 1982 (p. 115):

> As for group selection itself, my prejudice is that it has soaked up more theoretical ingenuity than its biological interest warrants. I am informed by the editor of a leading mathematics journal that he is continually plagued by ingenious papers purporting to have squared the circle. Something about the fact that this has been proved to be impossible is seen as an irresistible challenge by a certain type of intellectual dilettante. Perpetual motion machines have a similar fascination for some amateur inventors. The case of group selection is hardly analogous: it has never been proved to be impossible, and never could be. Nevertheless, I hope I may be forgiven for wondering whether part of group selection's romantic appeal stems from the authoritative hammering the theory has received ever since Wynne-Edwards (1962) did us the valuable service of bringing it out into the open.

This account is about as accurate as the school version of American history in which our goverment could do no wrong. Dawkins makes it sound as if group selection was rejected in the 1960s by a straightforward scientific procedure and that nothing has happened since then to alter the verdict. The real history of the controversy was quite different, and far more interesting.[11]

In *The Structure of Scientific Revolutions,* Thomas Kuhn (1970) used the term "normal science" to describe situations in which scientists share a general theoretical framework and are able to evaluate competing hypotheses by testing predictions based on them. Data are gathered, and scientists use the results to decide which hypotheses are true. We have shown that the group selection controversy includes an important element of normal science. Within- and between-group selection are alternative hypotheses that make different predictions about what evolves by natural selection. Williams and Lewontin devised tests for sex ratio and virulence, respectively, and the evidence weighed in favor of group selection, not as the *only* evolutionary force but as a *significant* evolutionary force.

Normal science did its job, but somehow it failed to have the right impact. Acceptance of female-biased sex ratios and reduced virulence as evidence for group selection has been a tortuous process that still, after decades, is not complete. Worse, these and other empirical tests have not forced a general reassessment of group selection theory. Many evolutionary biologists continue to play the "group selection is dead" song from the 1960s with the same fondness that they have for the Beatles. Little wonder, then, that scholars from other disciplines who are interested in evolution have heard almost nothing about these scientific developments.

To explain these aspects of the group selection controversy, we must look beyond the usual workings of normal science. Kuhn introduced the concept of competing paradigms to explain some of the additional factors that can complicate the scientific endeavor. When the fundamental concepts of opposing theories are so different that one cannot even be expressed in terms of the other, they are said to represent "mutually incommensurable paradigms." Because the theories are so conceptually different, normal scientific testing procedures do not resolve the dispute. The replacement of one paradigm by another occurs by different rules and constitutes a scientific revolution.

Kuhn forever changed the image of science as a straightforward march to the truth. Unfortunately, his concept of paradigms is full of ambiguities that other philosophers have been unable to resolve. It does not increase our understanding of group selection to call it a new paradigm or to say that it is an example of a scientific revolution. We have to look more closely at the specific factors that prevented normal science from running its course in this case, which may or may not be unique to the group selection controversy.

One reason that the conflict between group and individual selection was not resolved as a problem of normal science was the availability of a perspective that averages the fitnesses of individuals across groups. This perspective does not change any facts about the evolutionary process but merely looks at them in a different way. The same groups are present and the same evolutionary forces are operating. Relative fitness within and among groups can be examined at any time, but the focus is on the end product of what evolves. If we call the end product the "unbeatable strategy" and associate it with "individual selection," then we have defined group selection out of existence. Group selection theory and the averaging approach are perfectly compatible when it comes to predicting what evolves, and it is easy to translate from one to the other. When "individual selection" as defined by the averaging approach is used to argue against "group selection" as defined by group selection theory, however, only futile and never-ending controversy can result.

Two perspectives that view the same subject in different ways come close to what Kuhn meant by paradigms, but additional considerations are required to explain the enduring nature of the group selection controversy. After all, no one endorses the averaging fallacy in its general form. Even Dawkins states in his epitaph that group selection "has never been proved to be impossible, and never could be." Invoking the averaging fallacy in the case of sex ratio and virulence is a mistake that is easily avoided once one is aware of it; one simply checks for the presence of groups and sees if the trait evolves by virtue of a fitness advantage within groups or a fitness advantage between groups. Williams, Hamilton, Colwell, Lewontin, and others were able to avoid the averaging fallacy for the examples that we have reviewed, so why didn't all evolutionary biologists quickly follow suit? Perhaps this is because group selection had already been rejected by the time that Hamilton's and Lewontin's papers appeared. The debate was over and it only remained to avoid the egregious error of "group selection thinking." The task ahead was to develop the triumphant paradigm of individual selection. Against this background, the averaging fallacy made it easy to overlook the role of groups in so-called individual selection models.

We suspect that Simpson's paradox—or, rather, the lack of awareness of this phenomenon—was an additional reason the individualist paradigm was so powerful. If altruists by definition do worse than selfish individuals in the same group, it seems impossible for altruism to evolve. Talk of group selection seems futile in this context—a denial of the inevitable. Even if groups "compete" with each other, the fact remains that each group is subject to subversion from within. This line of reasoning cannot withstand criticism. Yet, on its face, it seems compelling. As a consequence, earlier, informal discussion of what is good for the group seemed naive, while equally informal discussion of what is good for the individual seemed perspicuous.

The examples of sex ratio and disease virulence as evidence for group selection should make it clear that knowledge of history is not just an idle pastime; it is required to understand the nature of the group selection controversy and to find the core of normal science that lies within. In addition, the controversy can be used as a case study to help historians, philosophers, and scientists understand the factors that make science so much more complicated in real life than it is in some models of *the* scientific method. Group selection theory and the averaging approach are a bit like the paradigms envisioned by Kuhn and others, but they also differ in important respects. The two paradigms clash in some ways (the interpretation of altruism) but not others (predicting the end product of what evolves). It is easy to translate between the two paradigms, once one appreciates their differences; Kuhnian talk of incommensurability is not appropriate in this instance. Both paradigms even deserve to coexist if, because they approach the same subject from different angles, they lead to different insights. Pluralism—the coexistence of multiple perspectives that "see" the same world in different ways—has a place in science, but only if the perspectives are properly related to each other (Dugatkin and Reeve 1994). A pluralistic scientist should be able to state clearly which perspective is being employed and how it can be translated into others. In the case of group selection, the pluralistic scientist should be able to say "I am employing the averaging approach and therefore can say nothing about group selection and altruism. To talk about these subjects, it is necessary to examine fitness differences within and between groups." Failure to distinguish between paradigms in this way has led to an incredible amount of wasted effort and confusion about the presence of altruism in nature and the importance of group selection as an evolutionary force.

We could provide additional examples of group selection and altruism that have been revealed by normal science but obscured by the averaging fallacy and other factors. Unlike the perpetual motion machine that exists only in the minds of dilettantes, group selection and altruism have passed the scientific test and can be said to exist in nature. To fully appreciate the importance of group selection, however, we must do more than consider examples on a case-by-case basis. We must examine the conceptual structure of evolutionary biology that has developed over the last thirty years.

2

A Unified Evolutionary Theory of Social Behavior

The fall of group selection theory during the 1960s was accompanied by the rise of other theories that became the foundation for the evolutionary study of social behavior. One was Hamilton's (1963, 1964a–b) theory of inclusive fitness, relabeled kin selection by Maynard Smith (1964). Another was the theory of reciprocal altruism (Trivers 1971), which later merged with evolutionary game theory (Maynard Smith 1982; Axelrod and Hamilton 1981). A third was selfish gene theory (Dawkins 1976, 1982), which makes even individuals appear to dissolve into collections of genes whose only interest is to replicate themselves.

All of these theories were developed to explain the evolution of apparent altruism without resorting to group selection. Here is a typical description of group selection and kin selection as alternative theories (Frank 1988):[1]

> Group-selection models are the favored turf of biologists and others who feel that people are genuinely altruistic. Many biologists are skeptical of these models, which reject the central Darwinian assumption that selection occurs at the individual level. (p. 37)
>
> According to Hamilton, an individual will often be able to promote its own genetic future by making sacrifices on behalf of others who

> carry copies of its genes . . . The kin selection model fits comfortably within the Darwinian framework, and has clearly established predictive power . . . Viewed from one perspective, the behavior accounted for by kin selection is not really self-sacrificing behavior at all. When an individual helps a relative, it is merely helping that part of itself that is embodied in the relative's genes. (p. 39)

Frank's comments about the Darwinian framework are historically inaccurate, since Darwin was the first person to suggest that altruism evolves by group selection. However, the rhetorical point of Frank's comment is clear enough; for a modern Darwinist, the ultimate rejection of an idea is to call it non-Darwinian! Notice also that the altruism that evolves by group selection is portrayed as genuine, whereas the altruism that evolves by kin selection is merely a form of self-interest.

Similarly, here is a typical account of selfish gene theory (Parker 1996):

> [The selfish gene] was designed to banish an infuriatingly widespread popular misconception about evolution. The misconception was that Darwinian selection worked at the level of the group or the species, that it had something to do with the balance of nature. How else can one understand, for example, the evolution of apparent "altruism" in nature? . . . Once one understands that evolution works at the level of the gene—a process of gene survival, taking place (as Dawkins developed it) in bodies that the gene occupies and then discards—the problem of altruism begins to disappear.

Inclusive fitness, game theory, and the selfish gene—all these ideas have led to great insights. We do not deny the advances in our understanding of social behavior that have taken place over the last thirty years. Nevertheless, a troubling vagueness appears when one tries to relate the theories to each other. Selfish genes are used to argue against groups as adaptive units, but somehow they allow individuals to be interpreted as adaptive units in many respects. Kin selection theory leads us to expect that helping behavior should be restricted to genetic relatives, yet nonrelated individuals can evolve to be very nice to each other through the mechanisms of evolutionary game theory.

Whatever evolves is a form of individual self-interest—or is it genetic self-interest?

In this chapter we will show that the evolutionary study of social behavior during the last thirty years has reflected a massive confusion between alternative theories that invoke different processes, on the one hand, and alternative perspectives that view the same process in different ways, on the other. The confusion has not entirely obscured the insights that emerge from each perspective, which is why we can acknowledge the genuine progress that has been made in the understanding of social behavior. Nevertheless, the confusion has occluded many additional insights and perpetuated falsehoods, especially in the understanding of altruism. Scientific progress could have occurred faster in the past, and it can proceed faster in the future. The key is to achieve a legitimate pluralism in which different processes are distinguished from different ways of viewing the same process.

We believe that a legitimate pluralism is possible and that it will lead to a unified evolutionary theory of social behavior. The theories that have been celebrated as alternatives to group selection are nothing of the sort. They are different ways of viewing evolution in multigroup populations. In most cases, the behaviors that are called "apparently altruistic" entail lower relative fitness for individuals within groups and evolve only because they increase the fitness of some groups relative to others. However, the theories are formulated in a way that obscures the role of group selection. These different formulations may lead to fresh insights about what evolves by natural selection—that is the advantage of multiple perspectives—but they cannot be used to argue against the *process* of group selection. When processes are distinguished from perspectives, a unified theory emerges that predicts the frequent evolution of genuine altruism and other group-advantageous behaviors.

In this chapter we will show how group selection is included in the major theories that were proposed as alternatives to group selection. We will also deal with the central issue of how to define groups. A growing number of evolutionary biologists have already achieved a legitimate pluralism (e.g., Boyd and Richerson 1980; Bourke and Franks 1995; Breden and Wade, 1989; Dugatkin and Reeve 1994; Frank 1986, 1994, 1995a–b, 1996a–c, 1997; Goodnight, Schwartz, and Stevens 1992; Griffing 1977; Hamilton 1975; Heisler and Damuth 1987; Kelly 1992; Michod 1982, 1996, 1997a–b; Peck 1992; Pollock 1983; Price 1970, 1972; Queller 1991, 1992a–b; Taylor 1988, 1992; Uyenoyama and Feldman 1980; Wade 1985). Acceptance of the unified theory is by no means complete, however. The view that group selection was authoritatively refuted decades ago and today exists only in the minds of a few heretics does not die easily. For many readers, including many evolutionary biologists, the material in this chapter will stand in stark contrast to what they have previously learned.

If science was nothing more than an exercise in hypothesis formation and testing, it could be described without much reference to history. It should be obvious by now that the scientific study of altruism is more complicated than what Kuhn called "normal science." What we have already shown for the subjects of sex ratio and disease virulence applies with even more force to the broad theoretical developments that we are about to review. The theory of group selection could have been accepted decades ago; to understand why it remained controversial, we must study not only the concepts but the history of the scientific discourse in which the concepts are embedded.

Kin Selection

W. D. Hamilton (1963, 1964a–b) is usually regarded as the father of modern kin selection theory. Six years earlier, however, G. C. Williams and D. C. Williams published a paper in the journal *Evolution* entitled "Natural Selection of Individually Harmful Social Adaptations among Sibs with Special Reference to Social Insects" (Williams and Williams 1957).[2] The starting point of their analysis was Sewall Wright's (1945) model of group selection.

It is worth describing Wright's model in detail (Box 2.1) because it highlights the importance that he placed on *relative* fitness. Wright imagined a trait that benefits the entire group, including the individuals expressing the trait. However, the trait has an individual cost that is not shared. This kind of altruism is exemplified by the story of the little red hen who shares the fruits of her labors but alone pays the cost. Wright made this idea mathematically precise with the equations in Box 2.1, in which g represents group benefit and s represents individual cost. It is easy to see that the freeloading genotype (aa)

Box 2.1. Sewall Wright's model of altruism

Wright (1945) imagined altruism as a behavior that benefits the entire group at a cost to the individual who performs the behavior. He stated this idea mathematically in the following model of natural selection of a gene at a single locus. The *W*s refer to the fitness of the three genotypes in a single group.

$$W_{aa} = (1 + pg) \tag{2.1}$$

$$W_{Aa} = (1 + pg)(1 - s) \tag{2.2}$$

$$W_{AA} = (1 + pg)(1 - 2s) \tag{2.3}$$

The A allele codes for the altruistic behavior, the a allele for selfish behavior. The group-level benefit of altruism is reflected by the term $(1 + pg)$, which increases with the proportion of altruists in the group (p). Heterozygotes (Aa), with one dose of the altruistic allele, have an individual cost of $(1 - s)$. And those homozygotes with two doses of the altruistic allele (AA) have an individual cost of $(1 - 2s)$. The group-level benefit is shared by everyone in the group and therefore cancels out of the equations when fitnesses are compared. The value of g is irrelevant and the A gene will always be selected against within groups when s is greater than zero. Group selection is required for A to evolve, but Wright only speculated about the process of group selection and did not provide an explicit model.

Wright's model of altruism needs to be distinguished from his more general "shifting balance" theory of evolution. Wright thought that natural selection cannot work well in single large populations because complex genetic interactions prevent the best combinations of genes from evolving. Just as plant and animal breeders develop new strains by selecting among inbred lines, Wright argued, the process of natural selection would work better in populations that were broken up into isolated groups. Different combinations of genes would be selected in each group, and the best of these groups could then replace the other groups. Wright developed his theory to explain the evolution of individual traits, such as coat color in guinea pigs, rather than social traits, such as altruism. The individual traits must be considered in a group context because they have a complex genetic basis and the fitness of the component genes depends on what other genes are common in the group. Unlike altruism, which is selected against within every group, the traits determined by the genetic combinations that evolve in Wright's shifting balance theory are all stable within their respective groups. In Chapter 4 we will discuss a model of group selection among multiple stable social systems that is close to Wright's shifting balance theory for individual traits. See Provine (1986) and Coyne, Barton, and Turelli (1997) for a modern assessment of the shifting balance theory.

always has the highest relative fitness, which is the same conclusion we reached for the model described in Box 1.1. In that case, altruists were not recipients of their own altruism. In Wright's model, the altruists benefit themselves along with everyone else in the group, but that does not alter their evolutionary fate. Natural selection within groups is totally insensitive to the g term because it does not affect *relative* fitness. Mathematically, the g term cancels out when fitnesses are compared, leaving the s term as the only relevant factor. Thus, a trait that greatly increases the *absolute* fitness of the individual expressing the trait (if g is large and s is small) will nevertheless fail to evolve because it decreases *relative* fitness (whenever $s > 0$). Natural selection within groups sees only the relative size of the slice, not the absolute size of the pie.

Wright's equations made the disadvantage of altruism within groups precise but did not describe the process of group selection that favors altruism. Instead, Wright merely speculated about how group selection might operate. He imagined that groups were spatially isolated populations connected by a trickle of dispersers, as shown in Figure 2.1. Wright knew that evolution is influenced by random processes in addition to the forces of natural selection. If two traits are equally fit, then their frequencies will vary without any particular pattern until one of them goes extinct, a process known as genetic drift. Similarly, if one trait is more fit than another, it is *more likely* to evolve but in some cases it may nevertheless go extinct by the same

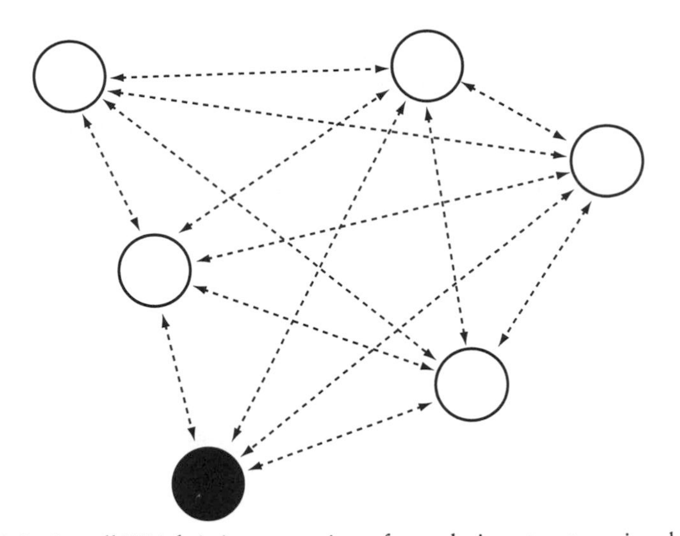

Figure 2.1. Sewall Wright's interpretation of population structure, in which groups are geographically isolated except for a trickle of dispersers. Altruists are supposed to evolve to fixation by genetic drift in a group (represented by the black circle) and to outcompete other groups by persisting longer and producing more dispersers.

drifting process. Wright therefore reasoned that altruism could become established by genetic drift in a few groups, despite its selective disadvantage. These altruistic groups might then outcompete the selfish groups by persisting longer and producing more dispersers that establish new groups.

Subsequent efforts to model Wright's scenario have cast doubt on the effectiveness of the process that he imagined (reviewed by Wade 1978; Wilson 1983). Unless the groups are very isolated from each other, immigrants from selfish groups will invade and take over the altruistic groups. If the groups are *that* isolated, however, it is hard to see how altruistic groups will reach other regions of the landscape. Even if we can fine tune the model to make it work, it fails to describe the vast majority of groups that exist in nature. Wright himself may have had little faith in his own scenario, since he ended his discussion with the following tepid statement: "It is indeed difficult to see how socially advantageous but individually disadvantageous mutations can be fixed without some form of intergroup selection" (p. 417).

Something like this must be going on, he seems to be saying, even if this particular version is unconvincing.

Williams and Williams (1957) used Wright's fitness equations but imagined a different kind of population structure. They considered a single, large, randomly mating population in which every female produces a clutch of offspring who interact only with each other during a period of their life cycle. A good example would be the interactions among baby birds in their nest. Williams and Williams realized that sibling groups are completely isolated from each other, as far as the behaviors expressed among siblings are concerned. An altruistic nestling will benefit the other birds in its nest, but will have no effect on the birds in other nests. The single, randomly mating population is not only a population of individuals but also a population of isolated sibling groups that form and dissolve every generation. Furthermore, variation among groups, which is required for altruism to evolve in the model portrayed in Figure 1.1 and which requires genetic drift in Wright's scenario, happens automatically when the groups are composed of siblings. The easiest way to see this is to imagine a population composed entirely of selfish types (aa) except for a single altruistic mutant (Aa). If the Aa mutant survives to adulthood, it will mate with a selfish aa type and produce a clutch of offspring that is composed of roughly 50 percent altruists (Aa) and 50 percent freeloaders (aa). Right away, the altruists are concentrated in a group with 50 percent altruists, while the vast majority of selfish genotypes exist in groups with no altruists. Wright assumed that genetic drift was required for altruism to become common within groups. Williams and Williams realized that this fragile assumption is unnecessary in the case of sibling groups.

In Chapter 1 we showed that the global increase in the frequency of altruists will be transient unless groups periodically mix or otherwise compete in the formation of new groups. The altruistic groups must somehow export their members to other parts of the global population; otherwise, individual selection against altruism will ultimately run its course within every group. It is unclear how Wright's scenario can handle this problem, because the isolation that is required for altruism to drift to fixation also makes it difficult for the altruistic groups to export their productivity. In contrast, it is obvious how sibling groups compete in the formation of new groups. The individuals simply grow up and leave their sibling groups to become adults who produce sibling groups of their own. Groups with more altruists contribute disproportionately to the total population, exactly as in our two-group model in Figure 1.1. Thus, group selection is an evolutionary force to be reckoned with when groups are composed of siblings.

The full model of William and Williams, which includes both Wright's equations and the process of group selection that favors altruism, is shown pictorially in Figure 2.2. A large population of individuals first forms randomly into mating pairs of adults, which give rise to isolated sibling groups. These groups may range from pure altruists (the progeny of AA × AA matings), to mixed groups (progeny of Aa × AA, Aa × Aa, etc.) to purely selfish groups (the product

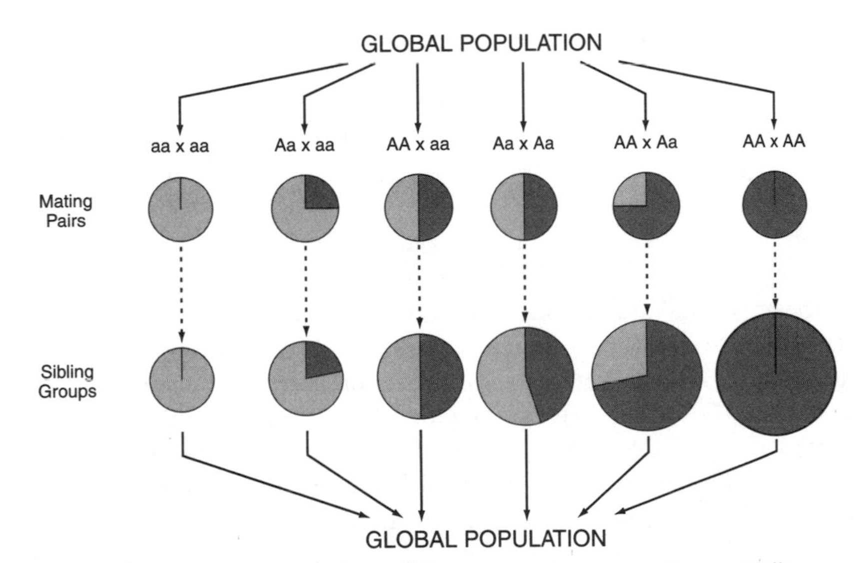

Figure 2.2. Population structure in sibling interactions, according to Williams and Williams (1957). Random mating in a large global population produces six types of mating pairs, which in turn produce six kinds of sibling groups that vary in the frequency of altruists. The frequency of the gene coding for altruism (A) in each group is shown in black. The altruistic gene declines in frequency within each mixed group (the black "slice" grows smaller), but each group benefits in direct proportion to the frequency of A (the circle gets bigger). The gene for altruism can evolve if the group-level benefit is sufficiently great, as compared with the individual-level cost.

of aa × aa matings). The altruists decline in frequency within all mixed groups (precisely as in our two-group model in Chapter 1), but groups contribute to the *global* population in direct proportion to their frequency of altruists. When both levels of selection are considered together, altruism evolves by group selection as long as the group benefit *(g)* is sufficiently large relative to the individual cost *(s)*.

The history of the group selection controversy contains many ironies, but this one is surely the greatest: George Williams, the future arch-critic of group selection theory, built the first model of group selection that really works. He clearly perceived the evolution of altruism among siblings as a group selection process. Now we can see why Williams later made family groups an exception to his sweeping claim that "group related adaptations do not, in fact, exist."

Hamilton's Alternative to Group Selection

The paper by Williams and Williams attracted little attention, at least compared with the reaction to a note published six years later by a British graduate student in the *American Naturalist* entitled "The Evolution of Altruistic Behavior" (Hamilton 1963) and the more extensive articles that soon followed in the *Journal of Theoretical Biology* (Hamilton 1964a–b). Hamilton was deeply influenced by the work of Ronald Fisher as an undergraduate student and had already decided to study the evolution of altruism when he applied to graduate school. He encountered not just indifference but hostility. This was the 1950s, and Hitler's "racial biology" had given British scientists a dread of studying anything connected with the genetics of behavior. Hamilton eventually found support at the London School of Economics sociology department, under the supervision of a professor who sympathized with heterodoxy but had no expertise in the subject (Hamilton 1996).

Shy and solitary, Hamilton never thought to ask for a desk. He worked in his room, in libraries, and in train stations when the libraries closed and he was too lonely to return to his room. Mathematics did not come easily to him, and at times he contemplated becoming a schoolteacher or a carpenter. Nevertheless, Hamilton eventually emerged from this intellectual wilderness with a theory to explain the evolution of altruistic behavior.

Hamilton began his note in the *American Naturalist* with the observation that group selection was the only existing explanation for the evolution of altruism but that it "must be treated with reserve so long as it remains unsupported by mathematical models." He then reviewed some of the problems with group selection models, concentrating on Haldane's (1932) treatment of the subject. Haldane had imagined a population structure in which new groups form by the fissioning of old groups (he had human "tribes" in mind). Haldane realized that altruism could momentarily increase in the global population if there was sufficient variation among groups, but he had difficulty seeing how the transient increase could be sustained. It seemed inevitable that natural selection within each group would eventually run its course. According to Hamilton, "The only escape from this conclusion (as Haldane hints) would be some kind of periodic reassortment of the tribes such that by chance or otherwise the altruists become re-concentrated in some of them."

Haldane's personal view of the world was that altruism *is* rare in humans and other animals, so he was content to drop the subject. Hamilton was dedicated to showing how altruism could evolve, but he also abandoned group selection as an explanation and developed what he saw as an alternative:

> As a simple but admittedly crude model we may imagine a pair of genes g and G such that G tends to cause some kind of altruistic behavior while g is null. Despite the principle of 'survival of the fittest' the ultimate criterion which determines whether G will spread is not whether the behavior is to the benefit of the behaver but whether it is to the benefit of the gene G; and this will be the case if the average net result of the behavior is to add to the gene-pool a handful of genes containing G in higher concentration than does the gene pool itself. With altruism this will happen only if the affected individual is a relative of the altruist, therefore having an increased chance of carrying the gene, and if the advantage is large enough compared to the personal disadvantage to offset the regression, or 'dilution,' of the altruist's genotype in the relative in question. (1963, pp. 354–355)

This is one of the first examples of adopting "the gene's eye view" that later would become the hallmark of selfish gene theory. We can examine Hamilton's logic by considering the same mutant altruistic allele that we used to analyze the Williams and Williams (1957) model. As before, the mutant Aa mates with an aa individual to form a sib-group composed of roughly half Aa and half aa. When an altruist benefits a sibling, the chance that the recipient also contains an altruistic allele is one-half. If the benefit to the recipient is greater than twice the cost to the altruist, there will be a net increase in the number of altruistic alleles as a result of the behavior. That is the basic point that Hamilton makes in the passage quoted above.[3] So far, however, we have calculated only the change in the number of altruistic genes. Natural selection works to increase *relative* fitness, and with group selection models we must examine relative fitness within and between groups. We already did this in our analysis of Williams and Williams (1957). We showed that the altruistic gene *declines* in frequency within the group and evolves only because it increases the fitness of its own group, relative to other groups.

In addition to adopting the gene's eye view, Hamilton wanted to explain his theory in a way that would make sense from the standpoint of an individual reasoning about how to behave (Hamilton 1996, p. 27). He imagined that an individual might try to maximize a quantity that includes its own fitness plus the fitness of others, weighted by the degree to which they were genetically related. This became the notion of inclusive fitness. An individual increases its inclusive fitness by behaving altruistically toward a sibling whenever the benefit to the sibling exceeds twice the cost to the individual. The concept of inclusive fitness succeeded in making Hamilton's theory intuitive, but it also allowed altruism to be reinterpreted. After all, is an individual behaving altruistically when it strives to maximize *its own* inclusive fitness? This ambiguity still exists in the evolutionary literature. Many regard the concept of inclusive fitness as an individualistic explanation of altruistic behaviors. Many others regard it as a kind of self-interest that is only apparently altruistic (e.g., the passage from Frank 1988 quoted at the beginning of this chapter). *Neither* interpretation is consistent with multilevel selection theory, which defines altruism in terms of fitness differences within and between groups. Multiple perspectives have led to multiple definitions of altruism and selfishness (Wilson and Dugatkin 1992; Wilson 1995; see Sober 1993b, pp. 100–112, for a simple derivation that relates Hamilton's rule to classical fitness).

Hamilton's model does not change any facts of the Williams and Williams model but merely calculates fitness in different ways, depending on whether the gene's eye view or the inclusive fitness view is employed. Both perspectives lose sight of the fact that natural selection operates against altruism (apparent or otherwise) in all mixed groups. The Williams and Williams model, which keeps track of fitness differences within and between groups, actually conforms nicely to Hamilton's own description of group selection, which he thought he was rejecting. The fitness of a sib-group is indeed proportional to its content of altruistic members, as Haldane supposed. The starting gene frequency that enables altruism to evolve is provided automatically by the mating process, which creates groups of 50 percent altruists. The frequency of altruists does indeed decline within groups, but this overall decline is halted by the periodic reassortment of the "tribes" (the sib-groups) and the reconcentration of altruists in new sib-groups. Unfortunately, Hamilton was apparently unaware of these connections; he cited Williams and Williams (1957) only in passing in his more extensive treatment (Hamilton 1964b, p. 35). From the very start, the distinction between group selection theory and what Hamilton called inclusive fitness theory was misconceived.

Maynard Smith's Haystack Model

John Maynard Smith was Haldane's last graduate student and by the 1960s had himself become one of Britain's premier evolutionary biologists. He quickly added his weight to the view that Hamilton's promising new theory was an alternative to group selection. Here is how Maynard Smith portrayed the two theories in a paper entitled "Group Selection and Kin Selection," published in 1964 in the journal *Nature:*

> By kin selection I mean the evolution of characteristics which favour the survival of close relatives of the affected individual, by processes that do not require any discontinuities in population breeding structure . . . There will be more opportunities for kin selection to be effective if relatives live together in family groups, particularly if the population is divided into partially isolated groups. But such partial isolation is not essential. In kin selection, improbable events are

> involved only to the extent that they are in all evolutionary change—in the origin of genetic differences by mutation. (p. 1145)

> If groups of relatives stay together, wholly or partially isolated from other members of the species, then the process of group selection can occur. If all members of a group acquire some characteristic which, although individually disadvantageous, increases the fitness of the group, then that group is more likely to split in two, and in this way bring about an increase in the proportion of individuals in the whole population with the characteristic in question. The unit on which selection is operating is the group and not the individual. The only difficulty is to explain how it comes about that all members of a group come to have the characteristic in the first place. (p. 1145)

Maynard Smith shared Wright's and Haldane's view of groups as spatially isolated multigenerational units. For group selection to work, altruism must first become established in a few groups by genetic drift—the "improbable event" that is not required for kin selection.[4] Then the altruistic groups must outcompete other groups by splitting at a greater rate, persisting longer, or producing more dispersers that found new groups. None of these unwieldy factors seemed to be required for Hamilton's theory.

To explore the likelihood of group selection, Maynard Smith built a fanciful model involving a species of mouse that lives entirely in haystacks (hence the model's name). Each haystack is founded by a single fertilized female, whose progeny breed among themselves for an unspecified number of generations. At the end of the year, the colonies break up and all the mice mate randomly in a single large population before repeating the cycle. The haystack model is shown pictorially in Figure 2.3. It begins in exactly the same way as the Williams and Williams (1957) model in Figure 2.2, with adults from a large population randomly mating and producing isolated groups of siblings. In the Williams and Williams model, the siblings disperse during the first generation after they have interacted but before they mate. In the haystack model, mating is entirely within haystacks and the groups persist for a number of generations before their members disperse. That is the *only* difference between the two models, as far as the population structure is concerned.

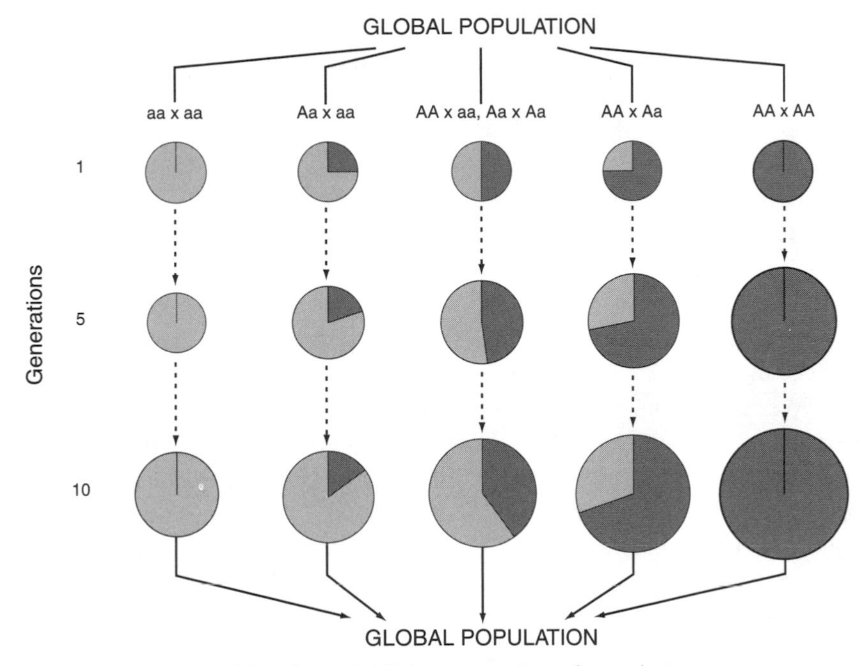

Figure 2.3. Maynard Smith's (1964) interpretation of population structure—the haystack model. The groups are initiated in the same way as the Williams and Williams (1957) model depicted in Figure 2.2, but they persist for a number of generations before blending back into the global population.

Even though the two models examine the same *process* (except for the duration of the groups), their authors had vastly different *perspectives*. For Williams and Williams, a group is a set of individuals that socially interact with each other. If an altruistic behavior is expressed only among siblings, then each set of siblings is an isolated group. Maynard Smith (and Hamilton) did not share this insight. For Maynard Smith, groups did not exist and the process of group selection did not begin until the haystacks became *breeding* populations.

Let us momentarily accept Maynard Smith's perspective and say that group selection begins with the second generation. The haystack model still affords an excellent opportunity to compare kin selection with group selection, as defined by Maynard Smith. We simply compare the degree of altruism that evolves when the mice disperse during the first generation (kin selection) with the degree of altruism that evolves if the mice stay together within haystacks for several generations between dispersal episodes (group selection).

Unfortunately, Maynard Smith did not do this. He never compared kin selection and group selection within a single mathematical framework but merely discussed kin selection verbally and presented the haystack model purely as a model of group selection. Furthermore, Maynard Smith added a simplifying assumption to the haystack model that amounts to a worst-case scenario for group selection. He assumed that the altruistic gene not only declines in frequency but goes completely extinct in all groups that are initially mixed! Altruists survive only in haystacks established by altruistic females who have mated with altruistic males. Given such powerful within-group selection, Maynard Smith concluded that altruism could not plausibly evolve by group selection. His subsequent efforts to model group selection (1976) also assume that altruists go completely extinct in all mixed groups between dispersal episodes. As Grafen (1984, p. 77) describes Maynard Smith's model, "mixtures of A_0 (selfish types) and A_1 (altruistic types) are considered too transient to matter."

For a fair comparison of kin selection with group selection in the haystack model, we first must specify an altruistic behavior with a cost to the altruist and benefits to recipients, as Wright, Williams and Williams, and Hamilton did. Then we must monitor the fate of the altruistic behavior in the haystack model as the number of generations between dispersal episodes is increased. Twenty-three years elapsed before this was done, long after group selection had entered its dark age (Wilson 1987). The results are shown pictorially in Figure 2.3. The altruists decline in frequency within all mixed groups for as long as the groups persist. However, even after ten generations they are far from extinct, so there is no warrant for assuming that within-group selection must completely run its course between dispersal episodes. In addition, groups with more altruists grow faster than groups with fewer altruists and this pattern also continues for as long as the groups persist. In other words, *both* within- *and* between-group selection intensify as the number of generations spent within haystacks is increased, so it is not obvious how the *balance* between levels of selection will be affected. Wilson (1987) showed that multigenerational groups are often *more* favorable for the evolution of altruism than single-generation groups in the haystack model, when the standard cost-benefit equations for altruism are used. Altruism can evolve by group selection even when the periodic reassortment of the "tribes" (the haystacks) and the reconcentration of altruists in new groups occurs at intervals of 10 or 15 generations.[5] In addition, genetic drift is no more required for multigeneration groups than for single generation groups. The same concentration process that allows altruism to evolve in the Williams and Williams model suffices for altruism to evolve in the multigeneration haystack model.

To summarize, Maynard Smith's distinction between kin selection and group selection can be faulted on two counts. First, he failed to recognize that a multigroup population structure exists during the first generation, when the original siblings interact with each other. The processes of within- and between-group selection that occur *after* the first generation in the haystack model are merely a continuation of the processes that occur *during* the first generation. Second, even if we accept Maynard Smith's definition of group selection, his pessimistic conclusion is based on the assumption that within-group selection is as strong as it can possibly be. If we use Hamilton's or Wright's equations for altruism in the haystack model, then altruism can evolve by group selection even as defined by Maynard Smith. As we saw in Chapter 1, female-biased sex ratios are a form of altruism that has evolved in population structures resembling the haystack model. If these problems had surfaced in 1964, the history of group selection theory might have been very different. Instead, the haystack model helped establish the view that group selection and kin selection were separate theories and that group selection models were too implausible to be taken seriously.

Hamilton Reconsiders

The next major event in the history of group selection theory was the appearance of an enigmatic American described by Hamilton (1996, p. 26) as "another solitary." George Price obtained his Ph.D. in chemistry in the 1940s and passed through an odd series of jobs; he participated in the Manhattan Project during World War II, conducted research in medicine, tried to write a book on foreign policy, and worked for IBM on computer design and the simulation of private enterprise market mechanisms. All of these jobs failed to satisfy what was evidently a deep need to understand and improve the human condition. Price therefore traveled to England in 1967 on his own savings to study evolutionary biology. His interest in evolution was partially stimulated by Hamilton's papers on altruism (Hamilton 1996, p. 172). This keen and restless intellect, without any biological training, proceeded to develop a totally fresh approach to the problem.

In a series of autobiographical essays that accompany his collected works, Hamilton (1996, pp. 172–173) described his first encounter with George Price and what has become known as the Price equation:

> A manuscript did eventually come from him but what I found set out was not any sort of new derivation or correction of my 'kin selection' but rather a strange new formalism that was applicable to every kind of natural selection. Central to Price's approach was a covariance formula the like of which I had never seen . . . Price had not like the rest of us looked up the work of the pioneers when he first became interested in selection; instead he had worked out everything for himself. In doing so he had found himself on a new road and amid startling landscapes . . . His voice was squeaky and condescending, rather guarded, on the phone . . . He spoke of his formula as 'surprising for me too—quite a miracle' . . . 'Have you seen how my formula works for group selection?' I told him, of course, no, and may have added something like: 'So you actually believe in that do you?' Up to this contact with Price, and indeed for some time after, I had regarded group selection as so ill defined, so woolly in the uses made by its proponents, and so generally powerless against selection at the individual and genic levels, that the idea might as well be omitted from the toolkit of a working evolutionist. If I did ask the question as to whether he believed in it I easily imagine George saying 'Oh, yes!', at once enthusiastic and imperious, followed probably by a long silence that would have been broken at last by some different topic.

Hamilton's feeling of awe for the Price equation is now shared by many theoretical biologists (reviewed by Frank 1995a). The equation represents the concept of selection in such a general and elegant way that it can be used to study all kinds of selection processes. Box 2.2 shows why the Price equation made group selection come to life for Hamilton. Imagine a single group, in which a gene that codes for a behavior is at a certain frequency (p). After the behavior has been expressed, the frequency of the gene within the group will have changed (from p to p') and the size of the group will have also changed (from n to n'). If the behavior is altruistic, then the gene will decline in frequency within the group ($p' < p$) but will increase the size of the group ($n' > n$). Furthermore, if there are many groups, we should expect a positive correlation between the frequency of altruists and the benefit to the group. The Price equation puts all of this together to calculate evolutionary change in the global population, which depends on the sum of two terms (see Equation 2.6 in Box 2.2). The first term is the average evolutionary change within single populations. The second term is the average evolutionary change caused by the differential fitness of groups. The two levels of individual and group selection seem to literally jump out of the Price equation. The same evolutionary forces are included in other mathematical models, including Hamilton's inclusive fitness theory, but they are not so clearly separated into within- and between-group components.[6]

Box 2.2. The Price equation

Price considered a global population divided into a large number of groups. Each group i begins with an initial size of n_i and an initial gene frequency of p_i. After selection, the size of the group changes to n_i' and the gene frequency changes to p_i'. If the gene codes for an altruistic behavior, then $n_i' > n_i$ and $p_i' < p_i$. The ratio $s_i = n_i'/n_i$ serves as a measure of group benefit.

To find the frequency P of the gene in the global population, we need to add the total number of altruistic genes across all groups and divide by the total number of genes in all groups. The frequency of the altruistic gene before selection is:

$$P = (\Sigma n_i p_i)/(\Sigma n_i) = p + \mathrm{cov}(n,p)/n \qquad (2.4)$$

The terms p and n denote the average gene frequency and the size of the average group. The covariance term, $\mathrm{cov}(n,p)$, measures the association between group size and gene frequency before selection. Suppose that 50 groups have size $n = 10$ and that each member possesses the gene for altruism ($p = 1$). Another 50 groups are of size 20 and contain no members with the gene for altruism ($p = 0$). The frequency of the gene in the global population of *individuals* is $P = 500/1500 = .333$, while the average gene frequency in a group is $.5(0) + .5(1) = 0.5$ (that is, half the *groups* have the gene). These two numbers don't match because of a negative correlation between group size and gene frequency. The covariance term accounts for this association, such that Equation 2.4 yields the right answer of .333 for the global frequency of the altruistic allele.

After selection, the global frequency of the altruistic allele is:

$$P' = (\Sigma n'_i p'_i)/(\Sigma n'_i) = p' + \text{cov}(n', p')/n' \qquad (2.5)$$

Price showed that the change in gene frequency in the global population, ($\Delta P = P' - P$), can be expressed by the following formula:

$$\Delta P = \text{ave}_{n'}(\Delta p) + \text{cov}_n(s, p)/\text{ave}_n s \qquad (2.6)$$

The first term is the change in gene frequency within the average group, weighted by the size of the group after selection (n'). It serves as a measure of within-group selection. The second term includes the covariance between group benefit *(s)* and the frequency of altruists in the group *(p)* before selection (when group size was n). It measures between-group selection. The more groups vary in the frequency of altruists, and the stronger the effect of altruists on group benefit, the more positive this term will become.

The left side of the Price equation (ΔP) gives the end product of what evolves in the global population. The right side shows the contribution of within- and between-group selection to that end product. When Hamilton reformulated inclusive fitness theory in terms of the Price equation, he discovered that the altruistic gene is selected against within every kin group (the first term is negative) but was favored by group selection (the second term is positive). The Price equation makes it obvious that both within- and between-group selection can influence the end product of what evolves.

Price understood all of this but stated it so briefly in his publications that few would appreciate its significance. In fact, Price abandoned his evolutionary studies after only a few years, just as he had abandoned his previous occupations. He became increasingly religious and devoted his life to analyzing the New Testament and helping London's homeless alcoholics. It was left to Hamilton to develop the implications of the Price equation for group selection theory. As Hamilton recalls:

> I am pleased to say that, amidst all else that I ought to have done and did not do, some months before he died I was on the phone telling him enthusiastically that through a 'group-level' extension of his formula I now had a far better understanding of group selection and was possessed of a far better tool for all forms of selection acting at one level or at many than I had ever had before. 'I thought you would see that', the squeaky laconic voice said, almost purring with approval for once. 'Then why aren't you working on it yourself, George? Why don't you publish it?', I asked. 'Oh, yes . . . But I have so many other things to do . . . population genetics is not my main work, as you know. But perhaps I should pray, see if I am mistaken.' (pp. 173–174)

Soon after, George Price committed suicide in an abandoned building where he had been living as a squatter after giving away all his belongings.

Hamilton (1975) presented his new views on group selection in a paper entitled "Innate Social Aptitudes in Man: An Approach from Evolutionary Genetics," which was published in an edited volume (Fox 1975). This important and neglected paper is notable for two reasons. First, Hamilton redescribed inclusive fitness theory as representing a multilevel selection process. Second, he freely speculated about the evolution of human behavior. Given the postwar intellectual climate in Britain and his own experience as a graduate student, the second foray seemed even more reckless and daring to Hamilton than the first.

The following passage from a section entitled "levels of selection" shows how strongly Hamilton had adopted the multilevel view, especially with respect to our own species:

> For example, as language becomes more sophisticated there is also more opportunity to pervert its use for selfish ends: fluency is an aid to persuasive lying as well as to conveying complex truths that are socially useful. Consider also the selective value of having a conscience. The more consciences are lacking in a group as a whole, the more energy the group will need to divert to enforcing otherwise tacit rules or else face dissolution. Thus considering one step (individual vs. group) in a hierarchical population structure, having a conscience is an "altruistic" character . . . As a more biological instance similar considerations apply to sex ratio, and here a considerable amount of data has accumulated for arthropods. (1975, pp. 135–136)

This passage is remarkably similar to Darwin's views on morality and group selection that we quoted in the Introduction (see also Hamilton 1987).

In his discussion of group selection, Hamilton emphasized that the periodic reassortment of groups is as necessary for the Price equation as for Haldane's model. If the groups are permanently isolated, then a global increase in altruism will be reflected in the Price equation, but it will be a transient phenomenon that cannot be sustained. The tribe-splitting scenario that Haldane envisioned was too difficult to analyze with the Price equation (it has since been modeled using computer simulations by Goodnight 1992), so Hamilton considered another kind of population structure, one that was more mathematically tractable:

> Therefore, noting hopeful auguries in Haldane's tribe-splitting no-migration idea, let us now turn to a model at the opposite extreme in which groups break up completely and re-form in each generation. Suppose that on reaching maturity the young animals take off to form a migrant pool, from which groups of *n* are randomly selected to be the groups of the next generation . . . (1975, p. 138)

Notice that these groups last only a fraction of a generation, as in the Williams and Williams (1957) model. Hamilton now saw the multigroup nature of social interactions, regardless of when and where breeding occurs. Social interactions among genetic relatives correspond to the nonrandom formation of groups. The significance of relatedness for the evolution of altruism is that it increases genetic variation among groups, thereby increasing the importance of group selection. Furthermore, *any* process that increases variation among groups will accomplish the same job. As Hamilton observed, "it obviously makes no difference if altruists settle with altruists because they are related . . . or because they recognize fellow altruists as such, or settle together because of some pleiotropic effect of the gene on habitat preference."[7] Genetic relatedness loses its status as the exclusive factor responsible for the evolution of altruism and becomes one of many factors that can promote group selection.

Hamilton (1975) left no doubt that the difference between inclusive fitness theory and group selection theory is a matter of *perspective,* not *process.* Earlier (1963) he had said that group selection should be treated with reserve so long as it remains unsupported by theoretical models. Now those models were available and the major so-called competing theory had vanished. The only process to explain the evolution of altruism was the one that Darwin identified long ago.

Reaction to the Price/Hamilton Model

In his biography of Price, S. A. Frank (1995a) remarked that "it must be unusual in the history of science for someone, without professional experience, to take up a field while in his forties and make significant contributions to the theoretical foundations of that field." It must be equally unusual for the inventor of a major new theory to announce cheerfully, several years later, that it is just another way of looking at the old theory after all.[8] In the Kuhnian version of scientific revolutions, the captain of a paradigm would rather go down with his ship than change his views. Yet Hamilton seemed to delight in his discovery and thought that others would share his enthusiasm. He seemed to relish the thought of his colleagues being torn between wanting to cite his paper for its powerful demonstration of group selection and not wanting to cite his paper for its bold exploration of evolution and human behavior (Hamilton 1996, p. 324).

Unfortunately, most of Hamilton's colleagues were not torn at all. His new views on group selection were accepted by a small sector of the evolutionary community, consisting primarily of theoretical biologists who could appreciate the beauty of the Price equation; otherwise, his positive view of group selection was ignored. The majority of evolutionary biologists not only continued to believe that the H.M.S. *Kin Selection* was afloat as an alternative to group selection, but they even imagined that Captain Hamilton was still at the helm! The passage by R. H. Frank (1988) that we quoted at the beginning of this chapter provides one of many examples of this view. Similarly, Richard Dawkins (1982) offers the following definition of group selection in the glossary of *The Extended Phenotype:* "A hypothetical process of natural selection among groups of organisms. Often invoked to explain the evolution of altruism. Sometimes confused with kin selection." In this book, Dawkins (1982, p. 6) cites Hamilton (1975) in a remarkable portrayal of the group selection controversy:

> The intervening years since Darwin have seen an astonishing retreat from his individual-centred stand, a lapse into sloppily unconscious group-selectionism, ably documented by Williams (1966), Ghiselin (1974) and others. As Hamilton (1975) put it, '. . . almost the whole field of biology stampeded in the direction where Darwin had gone circumspectly or not at all'. It is only in recent years, roughly coinciding with the belated rise to fashion of Hamilton's own ideas (Dawkins 1979), that the stampede has been halted and turned. We painfully struggled back, harassed by sniping from a Jesuitically sophisticated and dedicated neo-group-selectionist rearguard, until we finally regained Darwin's ground, the position that I am characterizing by the label 'the selfish organism', the position which, in its modern form, is dominated by the concept of inclusive fitness.

This passage is bizarre, given the actual content of Hamilton's 1975 paper. The group selection controversy includes a core of normal science, but it is often completely obscured by the paranormal.

Hamilton (1996, p. 324) himself suggests that his 1975 paper is still ahead of its time. It is rarely cited, and when it is—as in the Dawkins passage just quoted—references to it seldom reflect the paper's actual content.[9] An analysis of the journals included in *Science Citation Index* confirms Hamilton's suspicion: during 1994, the original version of Hamilton's theory (1964a–b) was cited 115 times, but the newer one (1975) was cited only 4 times. For much of the evolutionary community, the theory of kin selection was set in stone during the 1960s and thereafter lost its capacity for fundamental change—even at the hands of its own creator! If this is the situation *inside* the field of evolutionary biology, then it should come as no surprise that scholars from other disciplines who are interested in evolution see group selection primarily as a theory that died many years ago, along with the prospects for genuine altruism in nature.

Evolutionary Game Theory

Game theory was developed in the 1940s by economists and mathematicians to predict the behavior of people in conflict (Von Neumann and Morgenstern 1947). In standard game theory, people who are playing a "game" against each other decide what to do on the basis of rational choice. In evolutionary game theory, which was developed during the 1970s, the various behavioral options simply exist, as if they arose by mutation, and compete against each other in a Darwinian fashion. No assumption is made as to whether the behaviors are caused by minds (Sober 1985). Hamilton (1967, 1971a) was one of the first to use game theory to predict the "unbeatable strategy" for sex ratios and other traits. Trivers (1971) used game-theoretic ideas to develop his concept of reciprocal altruism. Maynard Smith and his colleagues (reviewed in Maynard Smith 1982) and Axelrod and Hamilton (1981) established game theory as a major theoretical framework for studying the evolution of social behavior. Evolutionary game theory is sometimes called ESS theory for "evolutionary stable strategy," a term coined by Maynard Smith.

One of the primary goals in the development of evolutionary game theory was to explain the evolution of cooperation among nonrelatives without invoking group selection. According to Dawkins (1980, p. 360), "there is a common misconception that cooperation within a group at a given level of organization must come about through selection between groups . . . ESS theory provides a more parsimonious alternative." This statement implies that cooperation can evolve within a single group and therefore does not require group selection. But do evolutionary game theory models really postulate only one group? The term *n-person game* clearly implies that social interactions occur among *n* individuals. Yet, game theory models that explore the evolution of cooperation do not assume a *single* group of size *n*, but rather *many* groups of size *n*. Although they are seldom explicitly described in this way, they assume the following population structure: Individuals in a very large population are distributed into groups of size n that are completely isolated from each other as far as the expression of behaviors is concerned. Just like baby birds in a nest, an individual is influenced by the members of its group, but not at all by the members of other groups. Usually the individuals are assumed to be distributed among the groups at random, but there is no impediment to changing this assumption in game theory models. After the behaviors that determine fitness have been expressed, the groups dissolve and the individuals blend back into the global population. Then the process is repeated. This population structure is identical to the one that Hamilton (1975) described and analyzed with the Price equation.

To examine fitness differences within and between groups, imagine that there are only two behaviors, altruistic (A) and selfish (S). Also imagine that groups consist of only two individuals, although the ideas may easily be extended to larger groups. The fitness of an individual depends on how it behaves and who its partner is. Let's say that altruists increase the fitness of their partner by 4 units at their own sacrifice of 1 unit. If both members of the group are altruists, then they trade benefits and get a net increase of 3 units. If it is a mixed group, then the altruist loses 1 unit and its selfish partner gains 4 units. If both members are selfish, then they receive no benefits and pay no cost. The possibilities are presented in a two-column, two-row table called the payoff matrix, shown in Box 2.3. These particular values correspond to a game called Prisoner's Dilemma, which was first described in a nonevolutionary setting. However, it should be obvious

Box 2.3. Evolutionary game theory and multilevel selection

Game theory models assume that social interactions take place in groups of size n. Within each group, an individual's fitness is determined by its behavior and the behavior of its social partners. When $n = 2$, the possibilities can be described in a 2×2 payoff matrix.

In the following game theory model of the evolution of altruism, altruists (A) give their partner 4 fitness units at a cost of 1 unit to themselves. An altruist gets a net benefit of 3 fitness units when paired with another altruist and suffers a loss of 1 fitness unit when paired with a selfish type (S). The selfish type gets 4 units when paired with an altruist and nothing when paired with another selfish type. A fitness of zero does not mean that the animal dies, but merely that its baseline fitness is unchanged by the social interaction. The payoff matrix looks like this:

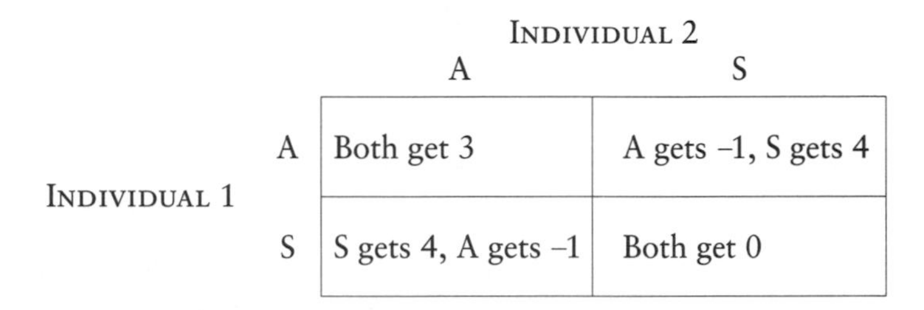

		INDIVIDUAL 2	
		A	S
INDIVIDUAL 1	A	Both get 3	A gets –1, S gets 4
	S	S gets 4, A gets –1	Both get 0

It is obvious that altruism has a low relative fitness within mixed groups (−1 vs. 4) but increases the fitness of the group (6 vs. 3 vs. 0). From the perspective of multilevel selection theory, this is a standard group selection model for the evolution of altruism in which the outcome depends on the amount of variation within and among groups. However, game theorists usually calculate the end product of what evolves by averaging the fitnesses of individuals across groups:

$$W_A = p_{AA}(3) + p_{AS}(-1) \tag{2.7}$$

$$W_S = p_{SA}(4) + p_{SS}(0) \tag{2.8}$$

The term p_{ij} is the proportion of i types that interact with j types (e.g., p_{AS} is the proportion of A types that interact with S types). When groups form at random and p is the frequency of A in the global population, then $p_{AA} = p_{SA} = p$ and $p_{SS} = p_{AS} = (1 - p)$. In the extreme case of no mixed groups, $p_{AA} = 1$, $p_{AS} = 0$, $p_{SA} = 0$, and $p_{SS} = 1$. Many game theorists conclude that A evolves by "individual selection" when $W_A > W_S$; this is an example of the averaging fallacy.

One of the most famous strategies in evolutionary game theory is "Tit-for-Tat," in which an individual (T) begins as an altruist and thereafter copies the previous behavior displayed by its partner. The payoff matrix for a population of T and S (which is selfish during every interaction) is shown below, assuming that an average of *I* interactions take place within each group.

		Individual 2: T	Individual 2: S
Individual 1	T	Both get 3*I*	T gets –1, S gets 4
	S	S gets 4, T gets –1	Both get 0

When T interacts with another T, they remain altruistic for all *I* interactions. When T interacts with S, it is exploited during the first interaction but then reverts to S-like behavior for the rest of the relationship. In contrast, A would have behaved altruistically during every interaction, resulting in a personal loss of $-1I$ and a gain of $4I$ for its selfish partner. From the standpoint of multilevel selection theory, T still has the lower relative fitness within mixed groups and requires group selection to evolve. However, the fitness differences within groups are small, in comparison with the fitness differences between groups. Random variation is sufficient for T to evolve by group selection, at least once T reaches a threshold frequency in the global population.

that the same numbers also represent the concepts of altruism and selfishness that we have been discussing all along.

It is easy to calculate relative fitnesses within and between groups from the payoff matrix. Altruists have a strong disadvantage within mixed groups (−1 vs. 4), but groups of altruists outperform mixed groups, which in turn outperform groups of selfish individuals (6 vs. 3 vs. 0). A fitness of 0 does not mean that the selfish groups go extinct. All individuals are assumed to have the same baseline fitness; the costs and benefits in the payoff matrix add to or subtract from that baseline. What evolves will depend on the relative strengths of within-group selection against altruism and between-group selection for altruism, which in turn depends on the amount of variation among groups. So far our model is no different from the one we presented in Chapter 1, except that the groups are smaller.

Is it really fair to call a pair of individuals a group, especially if they interact only briefly with each other before departing, never to meet again? It is important to remember that evolutionary game theory is a very general theoretical framework that can be used to model many kinds of groups that vary in both size and duration. In Hamilton's sex-ratio model, the groups persist until the progeny mate with each other and the payoffs are measured by counting the grandoffspring of the original females. Game theory has been used to model the evolution of virulence, in which the payoffs are computed by counting the descendants, produced over many generations, of the original colonists of a host (Bremermann and Pickering 1983, discussed in Chapter 1). We could easily rewrite Maynard Smith's haystack model, too, as a game theory model. For the sake of generality, terms must be defined on the basis of the entire theoretical framework and not on the basis of a special case. The duration of the group is left open in evolutionary game theory, which is sensitive only to the fitnesses of the "players" that result from their interactions within the group. We therefore cannot object to calling an ephemeral group of size $n = 2$ a group, even though it is a far cry from the groups that Wright and Haldane imagined. Williams and Williams (1957) and Hamilton (1975) already took a step in this direction when they contemplated groups that last for only a portion of the life cycle. We will examine the definition of groups in more detail at the end of this chapter.

So far we have shown that altruism in a game theory model is favored by group selection, but we have not shown whether it evolves. This question is usually answered within evolutionary game theory by averaging the fitnesses of individuals across groups. Depending on how the individuals are distributed into groups, a certain proportion of altruists will exist in AA groups while the remainder will be exploited in AS groups. Similarly, a certain proportion of selfish types enjoy the fruits of exploitation in AS groups while the rest put up with another grudger in SS groups. The net effect of all these forces determines the end product of what evolves. It is here that evolutionary game theory acquires its individualistic flavor. The behavior that confers the highest average fitness is assumed to evolve by individual selection, without requiring group selection. The presence of groups in the model is simply ignored. This is the averaging fallacy, pure and simple.

Let us bypass these semantic problems for the moment and focus on the question of whether A evolves. If we assume that individuals are randomly distributed into groups, then A does not evolve. Stated in the language of multilevel selection theory, within-group selection is stronger than between-group selection. Stated in the standard language of evolutionary game theory, S has the higher average fitness and evolves by individual selection. Now suppose that we increase the amount of variation among groups, perhaps by assuming that the pairs are genetically related (e.g., Wilson and Dugatkin 1991). Stated in the language of multilevel selection theory, between-group selection will eventually become stronger than within-group selection and altruism will evolve. Stated in the standard language of evolutionary game theory, A now has the higher average fitness and evolves by individual selection. According to the second approach, self-interest is defined as whatever evolves in the model, and altruism and group selection are defined out of existence.

One symptom of the individualistic perspective prevalent among game theorists is their use of the word *cooperation* rather than *altruism.* The equation that determines the behavior of A in our game theory model is identical to the one that Hamilton used to explore the evolution of altruism. Nevertheless, the word *cooperation* is used by evolutionary game theorists, presumably because it is easier to think of cooperation as a form of self-interest. The behavior is the same but it is labeled differently.

So far we have shown that altruism (as defined by multilevel selection theory) can evolve when there is enough variation in fitness among groups. However, altruism also can evolve when it entails only a modest disadvantage within groups. Consider the example of "Tit-for-Tat" (T), the most famous strategy in evolutionary game theory. Models involving the Tit-for-Tat strategy assume that members of a group interact repeatedly with each other before disbanding. T behaves altruistically during the first interaction and thereafter imitates its partner's previous behavior. This means that T remains altruistic toward other T's, but quickly switches to selfish behavior upon encountering an individual who repeatedly performs the S behavior. The second payoff matrix in Box 2.3 allows us to calculate relative fitnesses within and between groups. T still has the lower relative fitness within groups, because it loses the first interaction, but the magnitude of the disadvantage is much smaller than if T were unconditionally altruistic. As before, groups of altruists (TT) outperform mixed groups (TS), which in turn outperform groups of selfish individuals (SS). If we define altruism in terms of fitness differences within groups, Tit-for-Tat qualifies as an altruistic strategy. T requires less variation among groups to evolve than A, however, because it is a weaker form of altruism. As it turns out, even random variation among groups is sufficient for T to evolve by group selection, at least after it has surmounted a threshold frequency (Axelrod and Hamilton 1981).

For those who have become comfortable with the multilevel framework, it is child's play to see the groups in evolutionary game theory, calculate relative fitnesses within and between groups, and determine what evolves on the basis of the balance between levels of selection. This only makes it more remarkable that evolutionary game theory was developed as an alternative to group selection theory, as if it invoked an entirely different set of processes. The difference in perspective is so gripping that even today, many evolutionary game theorists would resist the interpretation of Tit-for-Tat as an example of altruism that evolves by group selection. However, there is at least one exception to this rule—the man who invented Tit-for-Tat.

The Tit-for-Tat strategy became famous as the winner of two computer game-theory tournaments that were sponsored by Robert Axelrod (1980a,b), a political scientist at the University of Michigan who had become interested in evolution. Axelrod asked his colleagues to submit strategies for behaving in the repeated Prisoner's Dilemma game, which he pitted against each other in Darwinian fashion. In every "generation" the strategies were played against each other and the average payoff for each strategy determined its frequency in the next generation. The Tit-for-Tat strategy was submitted by Anatol Rapoport of the University of Toronto and emerged as the winner against many other strategies, some of which were far more sophisticated. Here is Rapoport's (1991, pp. 92–93) comment on the widespread interpretation of Tit-for-Tat as an example of self-interest. In the following passage, C (for cooperate) refers to the act of behaving altruistically, and D (for defect) refers to the act of behaving selfishly:

> The most interesting and instructive result of those contests was the initial misconception about the reason for the success of Tit-for-Tat. This author, who submitted the program, was invited to give talks about the contest. In the discussions that followed, it became clear that many people had believed this program was "unbeatable." . . . The effects of ideological commitments on interpretations of evolutionary theories were never more conspicuous. To me, the most welcome result of Axelrod's experiments was the opportunity they provided for pointing out that "nice guys sometimes come in first" and for putting this homily into a scientific perspective . . .
>
> Altruism is naturally defined as a predisposition to act so as to benefit others at a cost to oneself. But this definition applies only to interactions between a pair of individuals. In a population, altruistic behavior by many may result in benefits to many, including the altruists.
>
> Returning to the Prisoner's dilemma contests, it is easy to see that Tit-for-Tat is anything but "unbeatable." In fact, it is eminently beatable. The only way a player of iterated Prisoner's Dilemma can get a higher score than the co-player is by playing more D's than the other, for only when one plays D while the other plays C does one get a larger payoff. But Tit-for-Tat can never get more D's than its partner in a sequential-play contest, because the only time it can play D is after the co-player has played D. Therefore, in every paired encounter, Tit-for-Tat must either draw or lose. It can never win a paired encounter.
>
> The reason Tit-for-Tat won both contests is because the "more aggressive" or "smarter" strategies beat each other . . In fact, the one unbeatable program is the one that prescribes unconditional D, since no other program can possibly play more D's than it. But in Axelrod's contests, two All-D programs playing each other would secure 1 point per play, whereas two Tit-for-Tat . . . would get 3 points per play.

As far as we know, Rapoport was not familiar with multilevel selection theory when he wrote this passage. Nevertheless, he found it easy to adopt a game theoretical perspective in which a behavior is called altruistic because it can be exploited within groups but nevertheless evolves because groups of altruists do better than groups of exploiters. The difference between multilevel selection theory and evolutionary game theory is a matter of perspective, not process.

The Selfish Gene

One of the concepts that Williams (1966) developed in *Adaptation and Natural Selection* was the gene as the fundamental unit of selection. According to Williams, to qualify as a unit of selection, an entity must be able to replicate itself accurately. Genes and asexually reproducing individuals have this property but sexually reproducing individuals and groups do not. This made genes seem somehow more fundamental than individuals, which now appeared as transient and irrelevant as groups. Richard Dawkins (1976, 1982) amplified and extended this concept into what is often called selfish gene theory. In Dawkins's fanciful language, individuals become lumbering robots controlled by genes whose only interest is to replicate themselves.

Selfish gene theory was widely regarded as a decisive argument against group selection, as the following quote from Alexander (1979, p. 26) attests:

> In 1966 Williams published a book criticizing what he called "some current evolutionary thought" and chastised biologists for invoking selection uncritically at whatever level seemed convenient. Williams' book was the first truly general argument that selection is hardly ever effective on anything but the heritable genetic units of "genetic replicators" (Dawkins, 1978) contained in the genotypes of individuals.

This kind of reasoning became one of the standard ways to raise and dismiss the issue of group selection in a few lines. Nevertheless, it suffers from a problem that is obvious in retrospect. Those naughty group selectionists thought that groups were like individual organisms in the harmony and coordination of their parts. If sexually reproducing individuals can become harmonious without being replicators, then why can't groups do the same? If the replicator concept can't say anything about individuals as adaptive units, how can it say anything about groups as adaptive units?

To explain adaptation at the level of individual organisms, Dawkins (1976) was forced to augment the concept of genes as replicators with another concept that he called "vehicles of selection." To use one of Dawkins's own metaphors, genes in a sexually reproducing individual are like the members of a rowing crew competing with other crews in a race. The only way to win the race is to cooperate fully with the other crew members. Similarly, genes are "trapped" in the same individual with other genes and usually can replicate only by causing the entire collective to survive and reproduce. It is this property of shared fate that causes selfish genes to coalesce into individual organisms that function as adaptive units.

The vehicle concept allows selfish gene theory to explain adaptation at the individual level, but it also opens the door to the possibility of groups as adaptive units. If individuals can be vehicles of selection, then why can't groups? Don't individuals in social groups sometimes find themselves in the same boat with respect to fitness? If so, why can't selfish gene theory explain the evolution of groups as the same kind of lumbering robots as individuals?

In short, the concept of genes as replicators, widely regarded as a decisive argument against group selection, is in fact *totally irrelevant* to the subject. Selfish gene theory does not invoke any processes that are different from the ones described in multilevel selection theory, but merely looks at the same processes in a different way. Those benighted group selectionists might be right in every detail; group selection could have evolved altruists that sacrifice themselves for the benefit of others, animals that regulate their numbers to avoid overexploiting their resources, and so on. Selfish gene theory calls the genes responsible for these behaviors "selfish" for the simple reason that they evolved and therefore replicated more successfully than other genes. Multilevel selection theory, on the other hand, is devoted to showing *how* these behaviors evolve. Fitness differences must exist somewhere in the biological hierarchy—between individuals within groups, between groups in the global population, and so on. Selfish gene theory can't even begin to explore these questions on the basis of the replicator concept alone. The vehicle concept is its way of groping toward the very issues that multilevel selection theory was developed to explain.

Let us see how selfish gene theory and multilevel selection theory offer different perspectives on the same set of processes. One of the beauties of multilevel selection theory is that it uses the same set of concepts to examine natural selection at every level of the biological hierarchy. It is easy to add a gene level to the individual and group levels that we have already considered. Natural selection occurs when genes differentially survive and reproduce within single individuals, when individuals differentially survive and reproduce within single groups, and when groups differentially survive and reproduce within a global population. Until recently, it has been unusual to think of natural selection taking place within single organisms. An organism has been regarded as a stable unit that retains the same genetic composition throughout its lifetime and faithfully transmits its genes in equal proportions when it sexually reproduces. As we shall see, this assumption is not entirely warranted, and the degree to which it *is* warranted requires an explanation.

In the standard model of selection at the individual level, a single genetic locus has two alleles (A,a), which combine to create three diploid genotypes (AA, Aa, aa). Each genotype has a fitness (W_{AA}, W_{Aa}, W_{aa}), which determines what evolves in the model. A multilevel selection theorist would say that natural selection takes place entirely at the individual level, or more precisely, among individuals within a single group. The Aa genotype is itself a mixed group of A and a genes that in principle could differ in their relative fitness. However, the rules of meiosis dictate that each allele is equally represented among the gametes. The only fitness differences that exist in the model occur among individuals within a single group. We therefore expect the evolution of genes that cause individuals to behave as organisms in the conventional sense of the word—beings that are built and designed to maximize their relative fitness in the single group.

Selfish gene theory doesn't change any facts about the model but merely places a special emphasis on the fact that it is genes that evolve. Even though the model says that fitness differences exist only among individuals, not within them, it is easy to average the fitnesses of genes across individuals to determine the end product of what evolves. The gene that evolves always has a higher average fitness than the genes that don't evolve, which accounts for the choice of the word *selfish*. A selfish gene theorist would insist that the gene is the fundamental unit of selection and that the individual is merely a vehicle of selection because the genes in an individual are in the same boat with respect to fitness.

In recent years, the view of individuals as stable genetic units has been challenged. Genes are increasingly viewed as social actors in their own right that can increase at the expense of other genes within the same individual. Some of the best-studied examples are genes that break the rule of meiosis by occurring in more than 50 percent of the gametes, a process known as meiotic drive (Crow 1979; Camacho et al. 1997; Lyttle 1991). These genes often decrease fitness at the individual level and sometimes are even lethal in their homozygote form. A multilevel selectionist would first look for fitness differences between genes within single individuals. In this case differences would be found, favoring the meiotic-drive gene. The next step would be to look for fitness differences between individuals within the group. These also would be found, since individuals who lack the meiotic-drive gene survive and reproduce better than individuals who have it. The multilevel selectionist would conclude that there is a conflict between levels of selection and would proceed to examine their relative strengths to determine what evolves. If gene-level selection is stronger than individual-level selection, then genes will evolve that make individuals *less* fit.

This example is identical to the models of altruism and selfishness that we presented earlier, except that we have frame-shifted down the biological hierarchy to consider genes interacting in individuals rather than individuals interacting in groups. We are accustomed to the idea that selfish individuals make groups less fit, so it makes sense that selfish genes make individuals less fit. Note that the word *selfish* in the previous sentence was used to refer not to whatever gene evolves, but to units that benefit at the expense of other units, within the next higher unit.

A selfish gene theorist would not change any facts about the model but would insist that everything that happens is gene-level selection. After all, genes are the replicators and when all is said and done, one gene will have a higher average fitness than the other. The fitness differences that exist within and between individuals will have to be described another way, by saying that the individual is no longer the only vehicle of selection. The word *selfish* has already been used to describe everything that evolves, so meiotic-drive genes and other genes that evolve by within-individual selection must be called something else, like *ultra-selfish* (Hurst, Atlan, and Bengtsson 1996).

Finally, we can frame-shift upward to consider a traditional group selection model. Individuals are stable genetic units, and altruists have a low relative fitness within groups but increase the fitness of their group. A multilevel selectionist would conclude that selfishness is favored by individual selection (or, more precisely, selection among individuals within groups) while altruism is favored by group selection (or, more precisely, selection among groups within the global population). What evolves depends on the relative strength of the opposing forces.

A selfish gene theorist would not change any facts about this model but would insist that natural selection is taking place at the gene level. After all, genes are the replicators and when all is said and done some genes will be more fit than others. The fitness differences within and between groups will need to be described another way, by saying that groups have become partial vehicles of selection. The selfish gene theorist would say that this is not an example of altruism but merely another example of selfish genes that only "care" about replicating themselves. Phrases such as "pseudo-altruism" (Pianka 1983), "mutual exploitation" (Dawkins 1982), or "self-interested refusal to be spiteful" (Grafen 1984) will be used in place of the word *altruism*.

At this point, some readers might be wondering why selfish gene theory should be regarded as a "theory" at all. Ever since genes were discovered, haven't they been thought to play a fundamental role in the evolutionary process? Isn't the standard textbook definition of natural selection "a change in gene frequency"? Doesn't the bulk of population genetics theory examine how fitness differences among individuals result in changes in gene frequency? Don't group selection models try to show how a gene for altruism can evolve by the differential fitness of groups? Hasn't it always been accepted that genes are the replicators? If so, then what is the point of a "theory" that relentlessly emphasizes that it is always, always, always, genes that evolve? Genes as replicators are a constant in virtually all models of biological evolution. A predictive theory needs to focus on the variables, which are the fitness differences that can occur anywhere in the biological hierarchy. Multilevel selection theory offers a precise framework for identifying these differences (among genes/within individuals, among individuals/within groups, among groups/within global populations, etc.) and for measuring their relative strengths. Selfish gene theory requires these same distinctions, but its central concept of genes as replicators offers no help. All the hard work is left for the tag-along concept of vehicles, which is not nearly as well developed as multilevel selection theory.

In retrospect it is hard to imagine how the focus on genes as replicators could have been advanced by Williams and Dawkins as an argument against group selection and how this argument could have become widely accepted. It can be dismissed in a single sentence: If sexually reproducing individuals and groups are not replicators, then the replicator concept cannot be used to argue that they are different from each other. Virtually everyone involved in the controversy now concedes that group selection is a question about vehicles, not replicators (e.g., Dawkins 1989, pp. 292–298; Cronin 1991, p. 290; Grafen 1984, p. 76; Williams 1985, p. 8). Unfortunately, past impressions die hard in science as in other walks of life. Many authors still use selfish genes as a quick and easy way to dispose of group selection theory (e.g., Parker 1996, quoted at the beginning of this chapter).

What Is a Group?

We have shown that all of the major theories proposed as alternatives to group selection—inclusive fitness theory, evolutionary game theory, and selfish gene theory—merely look at evolution in group-structured populations from different perspectives. In order to combine them into a single unified theory, however, we need a clear definition of groups. For Darwin (1871), groups were tribes that compete by direct conflict. For Haldane (1932), groups were tribes that compete by fissioning at different rates. For Wright (1945), groups were isolated populations that compete in the colonization of new groups. For Williams and Williams (1957), groups were sib-groups that last only a fraction of a generation. For Hamilton (1975), groups were any set of individuals that form for a period of the life cycle and influence each other's fitness.

Despite the diversity of these examples, they share a unifying theme that provides a simple definition of groups. In all cases, a group is defined as a set of individuals that influence each other's fitness with respect to a certain trait but not the fitness of those outside the group. Mathematically, the groups are represented by a frequency of a certain trait, and fitnesses are a function of this frequency. Any group that satisfies this criterion qualifies as a group in multilevel selection theory, regardless of how long it lasts or the specific manner in which groups compete with other groups.

This definition of groups is highly intuitive. After all, when we say that we are spending the evening with our bowling group, our bridge group, or our study group, we refer to the set of people with whom we will interact. If someone in your study group hasn't read the material, you will feel the consequences. If someone else in the library hasn't read the material, it will make no difference to you. That other person may be sitting at the same table, but he is not a part of your group. This is exactly how we must think of groups in the biological sense. The duration of the group and other details are determined by the nature of the interaction. Sibling groups last a fraction of a generation in the Williams and Williams (1957) model because the siblings disperse and stop interacting with each other. The haystacks in Maynard Smith's (1964) model are multigenerational because the siblings don't disperse; they and their descendants continue to interact with each other and their fitnesses are influenced by the initial genetic composition of the group. In multilevel selection theory, groups are defined *exclusively* in terms of fitness effects and everything else about groups, such as their duration and the manner in which they compete with other groups, follows from the nature of the interaction.

Although simple and highly intuitive, this definition of groups has a number of implications that must be kept in mind when thinking about multilevel selection. People often group organisms on the basis of spatial proximity. These groups *may* be appropriate for a multilevel selection model, but they may also be irrelevant. People in a library are obviously grouped around tables, but this does not mean that everyone at a table is a member of the same study group. Similarly, members of a study group may periodically disperse throughout the library to gather material, but they are still members of the same group. Individuals belong to the same group because of their interactions, not because they are elbow-to-elbow. We need to exercise the same care in identifying groups in nature as we would in identifying members of human groups. One of us coined the term "trait group" to emphasize the fact that groups must be defined on the basis of interactions with respect to particular traits (Wilson 1975).

Trait Groups

Defining groups on the basis of interactions resolves some of the confusion that exists concerning the relationship between group selection and kin selection. Imagine a species of butterfly that lays clutches of eggs on leaves. A leaf never receives more than one clutch and a caterpillar spends its entire juvenile stage on its natal leaf. Siblings interact with each other because they have no other choice. This population structure conforms to the Williams and Williams (1957) model, so we expect altruism to evolve by group selection. Now imagine another species of butterfly, whose offspring are more mobile. They quickly disperse from their natal leaf and mingle with caterpillars from other clutches. However, they have evolved the ability to recognize their siblings and also have evolved certain behaviors that they express *only* toward their siblings. Maynard Smith (1964) and Williams (1992) have suggested that kin selection might be regarded as a form of group selection in the first species but not in the second, because in the latter case there are no groups. If we define groups on the basis of interactions, however, we see that this distinction is misguided. As far as the evolution of sibling-only behaviors is concerned, the second species has the same population structure as the first species and can be modeled without a single change in the mathematical equations. In both cases, a mutant altruist will produce a sib-group with 50 percent altruists. Members of this sib-group will interact only with each other while the vast majority of nonaltruists will interact in sib-groups that contain no altruists. The fact that the groups are spatially circumscribed in the first species and behaviorally circumscribed in the second is irrelevant to the evolution of sibling-only behaviors.

A second implication of defining groups on the basis of interactions is that groups do not require discrete boundaries. Many species of plants (and some animals) have limited dispersal, such that offspring are deposited close to their parents. This kind of dispersal results in a genetically patchy population structure, despite the absence of discrete groups. In the language of kin selection theory, individuals are likely to interact with close relatives. In the language of multilevel selection theory, the genetic composition of the global population varies greatly over short distances. The patches are fuzzy and amorphous rather than discrete; over many generations, they dissolve and re-form, drifting around the physical environment like clouds in the sky. Nevertheless, it is still true that altruism is locally disadvantageous and requires the differential productivity of patches to evolve.

Multilevel selection theory recently led to an important discovery about plant-like population structures. Starting with Hamilton (1964a–b), limited dispersal has been thought to favor the evolution of altruism because it causes close relatives to interact with each other. According to multilevel selection theory, however, genetic variation among groups is not sufficient for the evolution of altruism. If groups remain isolated from each other, the global increase in the frequency of altruists will be transient and individual selection will ultimately run its course within each group. Altruistic groups must somehow export their progeny to other portions of the landscape for altruism to evolve. This is one reason why the isolated groups imagined by Wright (1945) are less favorable for the evolution of altruism than he hoped. It turns out that a similar problem exists for plant-like population structures. Limited dispersal creates patches of altruists and nonaltruists, but the many progeny produced by altruistic patches tend to fall back into the same patch and are not exported to other regions of the landscape. At the same time, the advantages of selfishness are local, allowing altruistic patches to be devoured by selfish invaders.

If we allow the plants to disperse a greater distance so that altruistic patches can export their productivity, the degree of patchiness declines. Limited dispersal has both a positive effect on group selection (by creating patches of altruists) and a negative effect (by limiting dispersal from those patches). These opposing forces exactly cancel, so that limited dispersal has no effect on the evolution of altruism (Queller 1992a; Taylor 1992; Wilson, Pollock, and Dugatkin 1992). Plant-like population structures are less favorable for the evolution of altruism than Hamilton hoped, for the same reason that isolated groups are less favorable for the evolution of altruism than Wright hoped. This fact was not discovered until 1992, when multilevel selection theory was applied to the problem. It had remained invisible to kin selection theory for almost thirty years, because kin selection treats genetic relatedness as the one and only relevant factor in the evolution of altruism. This example also illustrates how multilevel selection theory can identify situations in which altruism is *less* likely to evolve than previously thought, in addition to situations in which it is *more* likely. Our purpose is not to show that group selection and altruism are everywhere, but merely to see them where they exist.

A third implication of defining groups on the basis of interactions is that groups must be defined on a trait-by-trait basis. Continuing the example of our first butterfly species, imagine a mutant caterpillar that injects a chemical substance into the plant, weakening its defenses. Even though each sibling group remains on its own leaf, all caterpillars feeding on the plant will benefit from this particular trait, and so for this trait the group is defined as all caterpillars on the plant. Every trait has a sphere of influence that naturally defines the group.

Sometime the same set of individuals qualifies as a group with respect to many traits. An individual organism can be regarded as a group of genes that have a much more powerful effect on each other than on genes in other individuals with respect to most traits. If these traits evolve by individual selection, then the individual will become a highly adapted unit in many respects—foraging, mating, internal physiological processes, and so on. This is the way we usually think about organisms. Similarly, a bee hive is a group of bees that have a much more powerful effect on each other than on the bees in other hives with respect to most traits. If these traits evolve by group selection, then the hive will behave as a "superorganism" in most respects. It is important to realize, however, that adaptations are not always bundled together in the same group. The evolution of female-biased sex ratios requires groups that last long enough for the progeny of the original colonists to mate with each other before dispersing. Many species do not have this population structure and have evolved an even sex ratio. Group selection has been unimportant for this trait but may be highly important for other traits, such as predator defense, which have a completely different population structure. Even single organisms are not adaptive units in all respects. A trait such as meiotic drive that spreads by gene-level selection may be detrimental for the individual. The individual is not an adaptive unit with respect to that trait, however organismic it may be in other respects. Adaptation and multilevel selection must be evaluated on a trait-by-trait basis. Adaptations are sometimes bundled together in the same individual or group, but sometimes they are not.

These considerations help clarify the concept of vehicles in selfish gene theory. Dawkins (1976, 1982, 1989) states that individuals are exceptionally good vehicles of selection because the opportunities for within-individual selection are so limited. With only a few exceptions, the only way for a gene to increase its fitness is to increase the fitness of the entire genome. The implication has been drawn that most groups are not good vehicles of selection because they do not limit within-group selection to this degree (Sterelny 1996). This argument does not evaluate group selection on a trait-by-trait basis. In addition, it begs the question of how individuals became such good vehicles of selection in the first place. The mechanisms that currently limit within-individual selection are not a happy coincidence but are themselves adaptations that evolved by natural selection. Genomes that managed to limit internal conflict presumably were more fit than other genomes, so these mechanisms evolved by between-genome selection. Being a good vehicle as Dawkins defines it is not a *requirement* for individual selection—it is a *product* of individual selection. Similarly, groups do not have to be elaborately organized "superorganisms" to qualify as a unit of selection with respect to particular traits.

The fact that individuals are such good vehicles of selection, managing to suppress evolution within themselves as well as they do, suggests that population structure can itself evolve. A trait can alter the sphere of influence of other traits, variation among groups, the potential for fitness differences within groups, and other important parameters of multilevel selection theory. The coevolution of traits that influence population structure with traits that are favored by the new population structure can result in a feedback process that concentrates natural selection at one level of the biological hierarchy and bundles many adaptations into the same unit. Some evolutionary biologists have proposed that the history of life on earth has been marked by a number of major transitions in which previously autonomous units became integrated into higher-level units (Maynard Smith and Szathmary 1995). Molecules became organized into "hypercycles" during the origin of life itself (Eigen and Schuster 1977, 1978a–b; Michod 1983), genetic elements became neatly arranged into chromosomes, prokaryotic (bacterial) cells formed into elaborate communities that we call eukaryotic cells (Margulis 1970), and single-celled organisms built themselves into multi-celled organisms

(Buss 1987; Michod 1996, 1997a–b). The social insects are a more recent example of lower-level units coalescing into higher-level units (Seeley 1996). The transition is never complete and every unit, no matter how tightly integrated, has rogue elements that succeed at the expense of the unit. In addition, for every major coalescing event there must be thousands of other events in which the coalescence is only partial, with higher-level organization struggling to emerge from lower-level organization.

A Unified Theory and a Legitimate Pluralism

In science as in everyday life, it often helps to view complex problems from different perspectives. Inclusive fitness theory, evolutionary game theory, and selfish gene theory function this way in evolutionary biology. They are not regarded as competing theories that invoke different processes, such that one can be right and the others wrong. They are simply different ways of looking at the same world. When one theory achieves an insight by virtue of its perspective, the same insight can usually be explained in retrospect by the other theories. As long as the relationships among the theories are clearly understood, this kind of pluralism is a healthy part of science.

For many evolutionary biologists during the last three decades, one major theory has been excluded from this happy pluralistic family. Group selection has been treated as a theory that truly invokes different processes to explain the evolution of altruism and therefore can be proved to be wrong. Furthermore, the rejection of group selection during the 1960s was regarded as a watershed event that allowed evolutionary biologists to leave behind a pervasive and erroneous way of thinking, once and for all. Seldom has the death of a theory been proclaimed as loudly throughout the land as the death of group selection.

We have shown that the rejection of group selection reflects a massive confusion between process and perspective. The theories that were launched as alternatives to group selection are merely different ways of looking at evolution in group structured populations. This is not our own idiosyncratic interpretation but is becoming the concensus view among theoretical biologists and others who are most familiar with the conceptual foundations of evolutionary biology (e.g.,

Boyd and Richerson 1980; Bourke and Franks 1995; Breden and Wade 1989; Dugatkin and Reeve 1994; Frank 1986, 1994, 1995a–b, 1996a–c, 1997; Goodnight, Schwartz, and Stevens 1992; Goodnight and Stevens 1997; Griffing 1977; Hamilton 1975; Heisler and Damuth 1987; Kelly 1992; Michod 1982, 1996a–b, 1997; Peck 1992; Pollock 1983; Price 1970, 1972; Queller 1991, 1992a–b; Taylor 1988, 1992; Uyenoyama and Feldman 1980; Wade 1985). Indeed, we think that no other conclusion is possible, once the history of the group selection controversy is understood. That is why we have paid equal attention to concepts and history in this chapter.

Multilevel selection must be included in evolutionary biology's pluralistic family of theories. Does this mean that everyone has been correct and that all of the differences have been merely semantic? No; pluralism would be useless if it led to that kind of conclusion. To reach a more productive conclusion, we need to imagine how the debate would have turned out if the relationships among the theories had been understood from the beginning.

Darwin saw that individuals who help others will have fewer offspring than individuals in the same group who do not help. He also saw that groups of individuals who help others will have more offspring than groups of individuals who do not help. This was the first theory to explain the evolution of altruism. All subsequent theories have looked at the same process in different ways, but they have not proposed a different process. All perspectives should therefore agree that altruism—defined as behaviors that decrease relative fitness within groups but increase the fitness of groups—requires a process of group selection to evolve. Furthermore, all perspectives should agree that altruism *has* evolved and is an abundantly documented fact of nature. Assertions that altruism is only "apparent" and really an example of self-interest almost always involve a change in the definition of altruism, usually motivated by the averaging fallacy. Multiple definitions of evolutionary altruism may or may not be useful, but it is essential to be clear about which definition is being employed and how it relates to other definitions.

The group selection controversy concerns not only the evolution of altruism but also the interpretation of groups and other higher-level entities as organismic units. As Williams (1966) stated long ago, group-related adaptations can evolve only by a process of group-level selection. If Hamilton had presented his theory in the form of the Price equation in 1963, evolutionary biologists would have been forced to conclude that group selection is a significant evolutionary force that partially justifies the interpretation of groups as organismic units. The vagaries of history should not prevent us from reaching that same conclusion today. Neither should the existence of other frameworks that look at the process of multilevel selection in different ways. Nature exists as a nested hierarchy of units, and the process of natural selection operates at multiple levels of the hierarchy. Other frameworks may examine these facts from different perspectives but they are just plain wrong if they deny them as facts.

Unfortunately, the vagaries of history have erected a tremendous barrier to the acceptance of these fundamental conclusions about nature. The premature rejection of group selection and its consequences during the 1960s was such an important event that, for most evolutionary biologists, there was no turning back. Refinements of the theory that might have made a large difference when the initial consensus was being formed made almost no difference only a few years later. Scientific change that could have occurred rapidly was slowed to a glacial pace or frozen altogether.

We think that the group selection controversy has much to teach those who are interested in why science so often departs from the "normal" mode of hypothesis formation and testing. We hope that this book will serve as a starting point for more detailed historical and philosophical analyses. For our part, however, we will now leave these questions aside to explore the implications of multilevel selection theory, free from its tumultuous past.

3

Adaptation and Multilevel Selection

Adaptation is a powerful concept in part because it is relatively simple to employ. If one knows that something has been designed for a purpose, one can frequently predict its properties in minute detail: it must be structured in a certain way in order to perform its function. This way of thinking has revolutionized our understanding of organisms as entities that are partially designed by natural selection to survive and reproduce in their environments. Of course, it also is easy to misuse the concept of adaptation. If something has not been designed for a purpose, or if it is designed for another purpose, then the detailed predictions will be wrong and misleading. When evolutionary biologists disagree with each other, it is often about the use and misuse of adaptationism. Like fire and atomic power, the concept of adaptation must be handled with care.

In *Adaptation and Natural Selection,* Williams (1966) attempted to impose discipline on the study of adaptation. One of his most valuable insights was that adaptation at one level of the biological hierarchy does not automatically lead to adaptation at higher levels. Genes that spread at the expense of other genes within the same individual are often classified as diseases; to say that they proliferate like a cancer may be more than a metaphor. The design features that allow them to succeed *within* the individual are bad *for* the individual. Similarly, the behaviors that allow some individuals to succeed at the expense of others within the same group are often bad for the group. It is a familiar feature of human life that individual striving often leads to social chaos. Williams stressed that this outcome should frequently arise when natural selection acts at lower levels of the biological hierarchy. Nature should be full of higher-level units that behave maladaptively because their elements have been designed by natural selection to maximize relative fitness within those units. Adaptation should *never* be invoked at higher levels without identifying a corresponding process of natural selection at that level.

We do not agree with Williams about everything but we wholeheartedly agree on this point. Furthermore, it is a point that requires frequent repetition. As we stated in the Introduction, group-level functionalism is a major intellectual tradition that pervades everyday life and numerous academic disciplines. Many people axiomatically assume that societies, species, and ecosystems have evolved to function harmoniously. It is not easy to grasp the fragility of this assumption, and once grasped, the lesson learned is often troubling.

Seeing adaptations where they do not exist is a problem, but so also is failing to see adaptations where they do exist. As we have shown in Chapters 1 and 2, the averaging fallacy has made group selection invisible to many evolutionary biologists, who deny the existence of group-level adaptations as automatically as group-level functionalists accept them. The purpose of this chapter is to again impose discipline on the study of adaptation, for only by doing so may we employ the power and simplicity of functional thinking while avoiding its many pitfalls. We will outline a procedure for identifying adaptations where they exist, at all levels of the biological hierarchy. Multilevel selection theory provides new insights—ones that were not forthcoming from other perspectives; it also furnishes a framework for understanding the evolution of our own species.

Thinking about Adaptation at Multiple Levels

If natural selection can operate at multiple levels of the biological hierarchy, it is important to know what adaptations would look like at each level. Then we must know the relative strength of natural selection at each level to determine what actually evolves. We will outline a stepwise procedure for doing this, although we do not wish to imply that the steps must be taken in order or that all of them are required for the study of every trait. We will use two levels of selection (individual and group) to outline the procedure, although it is equally amenable to multilevel hierarchies.

Step 1: Determine What Would Evolve if Group Selection Were the Only Evolutionary Force

In this case, traits will evolve that maximize the fitness of groups, relative to other groups. If we are interested in predator defense, we might decide that the optimal group organization requires at least one individual to refrain from feeding and to scan the environment for predators at any one time. If we are interested in sex determination, we might decide that group fitness is maximized by a sex ratio in which females outnumber males. If we are interested in the effects of parasites on their hosts, we might predict an optimal degree of virulence that maximizes the number of emigrants that reach new hosts. The details of our prediction will depend on the biology of the particular species; our opening remark that adaptation is a relatively simple concept to employ does not mean that it can be used without empirical information. Notice that this exercise in group-level functionalism does not require one to believe that group selection *is* the only force at work; we merely want to know what to expect if it *were* the only force. We must be able to describe what group-level adaptations would be like, even if we are going to reject the hypothesis that they exist.

Step 2: Determine What Would Evolve if Individual Selection Were the Only Evolutionary Force

In this case, traits will evolve that maximize the fitness of individuals, relative to other individuals in the same group. If we are interested in predator defense, we might decide that individuals will attempt to position themselves so that other individuals are between them and the predator (Hamilton 1971b). For sex determination, we might decide that an equal proportion of sons and daughters is the "unbeatable strategy" within groups. For parasites, the best strategy might be to convert host tissue into parasite offspring at the maximum rate.

Some of these strategies may appear short-sighted, but that is just the point that Williams was trying to make when he stressed that well-adapted individuals do not automatically make a well-adapted group. It is only the relative size of the slice, and not the absolute size of the pie, that determines what is favored by natural selection within groups. As with step 1, this exercise in individual-level functionalism does not require one to believe that individual selection *is* the only force; we merely want to know what we should expect *if* it were the only force.

Steps 1 and 2 harness the power and simplicity of functional thinking without committing one to any particular position on adaptation or multilevel selection. If the traits are not well-adapted in any sense, then neither prediction will be accurate. To the extent that the traits under study have evolved by natural selection, steps 1 and 2 bracket the possibilities; the traits found in nature will lie somewhere between the extremes of what would evolve by pure within-group and pure between-group selection. The next step of the procedure is to determine the balance between levels of selection, but first we should acknowledge the progress that has already been made. Steps 1 and 2 avoid the averaging fallacy by forcing one to think separately about the relative fitness of individuals within groups and the relative fitness of groups in the global population. Furthermore, the predictions that emerge from steps 1 and 2 are often sufficiently different from each other to allow preliminary empirical tests of group vs. individual selection without even proceeding to step 3. In the case of sex determination, step 1 predicts a highly female-biased sex ratio and step 2 predicts an even sex ratio. A sex ratio of 75 percent daughters suggests a significant role for *both* individual and group selection, as Hamilton pointed out in 1967 (see Chapter 1).[1] If evolutionary biologists completed steps 1 and 2 for the long list of behaviors that they have been studying over the last three decades, they would see that real organisms frequently depart from the predictions of step 2 in the direction of step 1.

Step 3: Examine the Basic Ingredients of Natural Selection at Each Level

The process of natural selection requires three basic ingredients: (a) phenotypic variation among units, (b) heritability, and (c) differences in survival and reproduction that correlate with the phenotypic differences. To determine the balance between levels of selection, we need to examine these ingredients at each level.

Step 3a: Determine the pattern of phenotypic variation within and among groups. At all levels of the biological hierarchy, units must differ from each other before they can be sifted by natural selection. A population structure in which all the members of each group are identical and groups differ from each other is maximally conducive to group selection. Similarly, natural selection occurs only at the individual level when members of a group differ from each other but each group has exactly the same composition of members. Between these two extremes is a wide range of population structures in which phenotypic variation is partitioned into within- and among-group components.

Most evolutionary models presuppose a direct connection between genes and behavior. Genetic determinism is assumed, not because it is necessarily true but because it simplifies the models. In Chapter 1 we mentioned that mathematical models require simplifying assumptions, which nevertheless can be misleading when something important has been left out. Now it is time to question the assumption of genetic determinism in models of multilevel selection. Assuming a direct link between genes and phenotypic traits has important implications for the partitioning of phenotypic variation within and among groups. For example, the only way to get a phenotypically homogeneous group is to have a genetically homogeneous group. Altruism is favored in kin selection models because phenotypic similarity marches in lock step with genetic similarity. If groups are formed at random, then the amount of phenotypic variation among groups will decline as the size of the group increases. Randomly formed groups of size $n = 100$ should be much more similar than randomly formed groups of size $n = 2$. These conclusions, which are often taken for granted, are only as true as the simplifying assumption of genetic determinism.

Genetic determinism may be a serviceable idealization in some cases, but it is hardly the final word on phenotypic variation among groups. In fact, its inadequacy can be revealed by forming groups of real organisms in the laboratory and measuring the relationship between genetic and phenotypic variation empirically. In a classic experiment, Wade (1976, 1977) created groups of flour beetles by picking $n = 16$ individuals at random from a large laboratory population and placing them in small vials with flour to reproduce. Thirty-seven days later, Wade measured the total number of offspring produced by each group, which can be regarded as a group-level phenotypic trait. A standard way to think about this experiment would be to assume that there are genes for fecundity that vary among individuals. Some groups receive more fecundity genes by chance, which accounts for the differences among groups in the number of offspring produced. Phenotypic variation among groups should be low because each group was formed at random with a fairly large sample of unrelated individuals.

The actual results did not conform to these expectations. Phenotypic variation among groups was enormous, ranging from 365 progeny in the most productive group to only 118 progeny in the least productive group. In a series of follow-up experiments, Wade (1979) and McCauley and Wade (1980) showed that group productivity is not a simple sum of individual fecundities but reflects a complex interaction among a number of traits, including rates of development, cannibalism, and sensitivity to crowding.[2] Complex interactions turned small initial differences among groups in these primary traits into large differences in group productivity.

In retrospect, this result should not be surprising. At all levels of the biological hierarchy, genes code for a large number of products that interact with each other and the environment to produce the phenotypic traits upon which natural selection acts. It is well known that small genetic differences (sometimes only a single allele) can be magnified by the process of development to produce large phenotypic differences between individuals (Rollo 1994). In just the same way, small genetic differences among groups can be magnified by social and ecological interactions into large phenotypic differences. Both are examples of sensitive dependence on initial conditions, which also causes complex physical systems such as the weather to behave in odd and unpredictable ways (Gleick 1987; Wilson 1992). We still must ask if these differences are heritable, but as far as the *first* ingredient of natural selection is concerned, groups are usually more phenotypically variable than standard evolutionary models have led us to believe.

The relationship between genetic and phenotypic variation is even more complicated in human groups. Large groups of genetically unrelated people can behave in a uniform manner, especially when behaviors are regulated by social norms. Profound phenotypic differences can exist between human groups that have little or nothing to do with genetic differences. These forms of phenotypic variation among groups do not guarantee a role for group selection in human evolution, because the other ingredients of natural selection—heritability of the trait and fitness differences correlated with the trait—must also be present at the group level. Nevertheless, it should be obvious that the factors identified by standard evolutionary models are only a small part of the story as far as the *first* ingredient—phenotypic variation—is concerned. If group selection has been unimportant in human evolution, it is not for lack of phenotypic variation among groups.

Step3b: Determine the heritability of phenotypic differences. At all levels of the biological hierarchy, phenotypic differences must be heritable for natural selection to produce evolutionary change. Darwin defined inheritance as the tendency of offspring to resemble their parents; he offered this definition without knowing anything about the underlying mechanisms. Today we know a lot about the genetic mechanisms that cause offspring to resemble their parents, but the concept of heritability is not limited to genetic mechanisms. Cultural processes can also cause offspring to resemble their parents, and this kind of heritable variation can also serve as the raw material for natural selection (Boyd and Richerson 1985).

Parents and offspring obviously exist at the individual level, but what do these concepts mean when they are applied at the level of groups? Haldane imagined a population structure in which old groups fission to form new groups. The ancestry of groups can easily be traced in this example, but in other cases the groups simply dissolve into a global pool of dispersers from which new groups are formed. Maynard Smith (1987a,b) claimed that it is impossible to trace the ancestry of groups for this kind of population structure and concluded that heritability does not exist at the group level.

To deal with this issue, it will help to clarify the concept of heritability at the individual level with some simple examples. Imagine two alleles at a single locus that code for height. The AA genotype is tall, Aa is medium, and aa is short. To know if this trait is heritable, we need to relate the height of parents with the height of their offspring.

When AA mates with another AA genotype, all of the offspring are AA. When AA mates with Aa, the offspring are a mix of AA and Aa genotypes. When AA mates with aa, all of the offspring are Aa. Thus, AA parents produce a mix of AA and Aa offspring, depending on who their mates are. By the same reasoning, Aa parents produce a mix of all three genotypes and aa parents produce a mix of Aa and aa genotypes.

Next we must calculate the average height of offspring for each parental genotype. An example is shown in the top graph in Figure 3.1, which assumes that mating is random and that each allele is equally frequent in the population. Offspring of AA are shorter on average than their parents because they include both AA and Aa individuals. Similarly, offspring of aa are taller on average than their parents because they are a mix of aa and Aa. Nevertheless, there is still a positive correlation between parent height and offspring height. Offspring of AA are taller, on average, than the offspring of Aa, which in turn are taller than the offspring of aa. This means that the trait is *heritable*. If we select the tallest (or shortest) individuals to be the parents for the next generation, average height in the offspring generation will increase (or decrease).

Now let's change the example slightly, as shown in the bottom graph. Let's assume that the Aa genotype is tall while the AA and aa genotypes are short, as might be the case if heterozygotes have a physiological advantage that increases their growth rate. As before, AA parents produce a mix of AA and Aa offspring, Aa parents produce a mix of all three genotypes, and aa parents produce a mix of Aa and aa offspring. In this case, however, there is no correlation between the height of parents and the average height of their offspring. If we select the tallest individuals (Aa) to be the parents for the next generation, they will mate to produce the same mix of genotypes (AA, Aa, aa) that was present in the previous generation. If we select the shortest individuals (AA, aa), they too will mate randomly to produce the same mix of genotypes as before (mating within genotypes produces homozygotes, mating between genotypes produces heterozygotes). There is no correlation between parents and offspring, the trait is not heritable, and trait frequency will not evolve in response to natural selection.

Notice that phenotype is completely determined by genotype in both examples. This shows why the concept of heritability differs

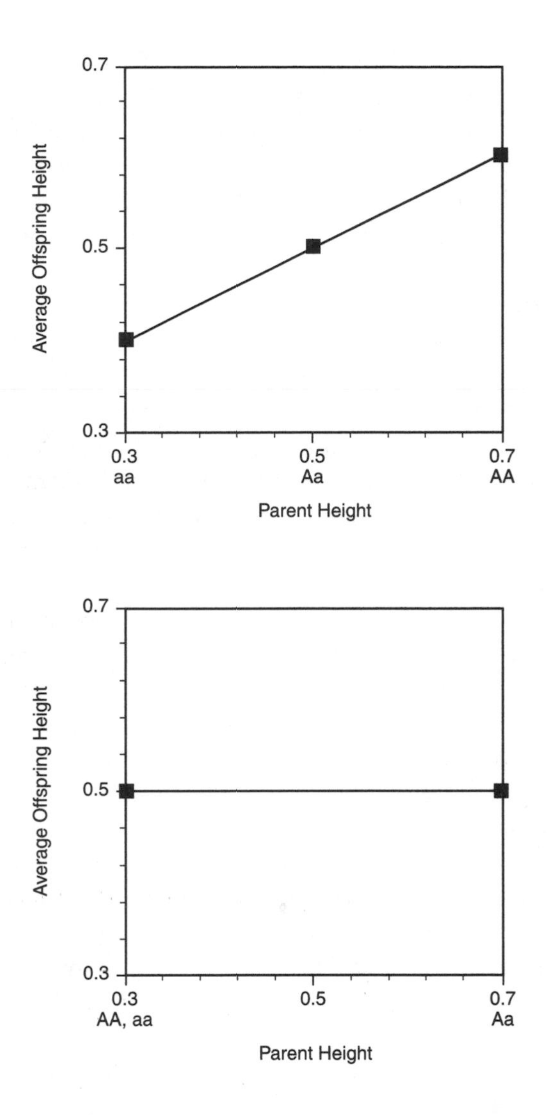

Figure 3.1. The concept of heritability in a simple genetic model. In the top graph, the AA genotype is tall, the Aa genotype is medium, and the aa genotype is short (numerical units are arbitrary). Sexual reproduction creates offspring that differ from their parents, but there is still a positive correlation between parent height and average offspring height. In the bottom graph, the Aa genotype is tall while the AA and aa genotypes are short. Now there is no parent-offspring correlation, even though height is still genetically determined. For both figures, the frequency of the A allele is 0.5 and mating is assumed to be random.

from the concept of genetic determination.[3] The reason that height is not heritable in the second example is that we have complicated the genotype-phenotype relationship. Shortness is caused by two genotypes (AA,aa) rather than one, and the genes in those genotypes, in different combinations, also cause tallness (Aa). Sexual reproduction therefore causes short parents to produce tall offspring, and vice versa.

Real genetic interactions are often so complex and poorly understood that the only way to determine heritability is to do an experiment. For example, we might measure wing length in a laboratory population of fruit flies and use the individuals with the longest wings as parents for the next generation. If their offspring have longer wings, on average, than the previous generation, then the trait is heritable. Response to selection is proof of heritability. What typically happens in experiments of this sort is that wing length (or some other trait) increases for a number of generations but then reaches a plateau at which it no longer responds to selection. Wing length may be genetically determined and variation in wing length may still exist, but once the plateau is reached there is no longer a correlation between parents and offspring. Presumably, the genetic interactions that cause wing length include both the simple pattern that is represented by our first example and the more complicated pattern that is represented by our second example. The simple genetic interactions allow the initial response to selection, but heritable variation is depleted as these genes evolve; what is left is nonheritable variation (Falconer, 1981).[4]

Against this background, we now can evaluate the concept of heritability at the group level. To keep the example as simple as possible, imagine an asexual population that consists of tall (A) and short (a) individuals in equal proportions. The individuals spend their juvenile stage in groups of size $n = 4$ but then disperse as adults and have offspring that thoroughly mix with each other before randomly settling into new groups. We are interested in the group-level phenotypic trait of average height. We measure the average height of many groups and obtain the distribution shown at the top in Figure 3.2. A few groups are very short (aaaa), a few are very tall (AAAA), while most are somewhere in between. Now we want to know if this group-level trait is heritable. Imagine that we can follow all of the adults that emerge from AAAA groups and trace their offspring into the next generation of groups. Because they settle at random, we will find them in Aaaa, AAaa, AAAa, and AAAA groups, but we will not find them in aaaa groups. We can regard all of these groups as the "offspring" of the AAAA parent groups and calculate their average height. If we repeat the exercise for the other parent groups (AAAa, AAaa, Aaaa, aaaa), we can construct a parent-offspring correlation at the group level, as shown in the middle graph in Figure 3.2. It is positive (taller parent groups produce taller offspring groups), demonstrating that the trait "average height" is heritable at the group level.

Of course, actually making these measurements may be tedious, but it is important to see that it is possible in principle to do so. Maynard Smith (1987a,b) didn't say that it was *difficult* to measure heritability at the group level; he claimed that the very concept of heritability doesn't apply here. Furthermore, there is a simpler way to measure group-level heritability. We simply perform a group selection experiment in the laboratory, identical in every way to the hundreds of artificial selection experiments that have been performed at the individual level. Continuing our example, let's say that we take all of the groups that are above average in height and use them as parents for the next generation of groups. Notice that we are selecting at the level of groups, not individuals. Many of the groups that we select will contain both A and a individuals, who will have the same number of offspring. Nevertheless, the fact that we selected the tallest groups will cause the distribution of offspring groups to shift in the direction of increased height, as shown at the bottom in Figure 3.2. Response to selection is proof of heritability, so the trait "average height" is heritable at the group level.

This example shows that the concept of heritability can be applied to all levels of the biological hierarchy. We may use different words for their beginnings and ends—groups "form" and "dissolve," individuals are "born" and "die"—but both groups and individuals take part in a cycle of generations. In both cases it is possible to trace ancestry, although it may be more difficult in the case of groups. In both cases it also is possible to measure heritability experimentally as a response to selection.

Multilevel selection experiments have been performed in the laboratory for over twenty years, starting with the classic experiments by Wade (1976, 1977) described above.[5] After forming groups and meas-

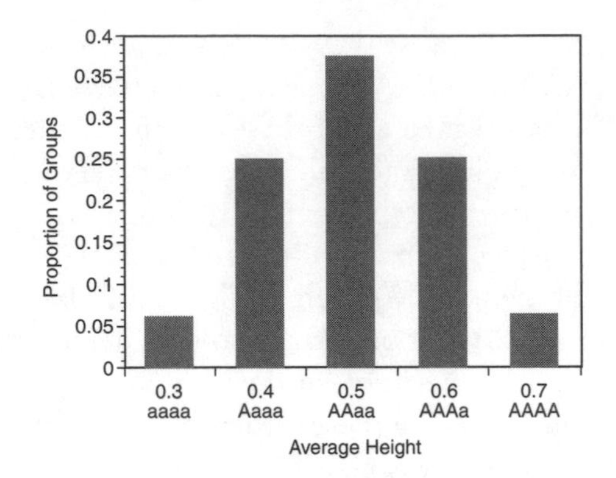
Proportion of Groups
0.4
0.35
0.3
0.25
0.2
0.15
0.1
0.05
0
0.3 aaaa
0.4 Aaaa
0.5 AAaa
0.6 AAAa
0.7 AAAA
Average Height

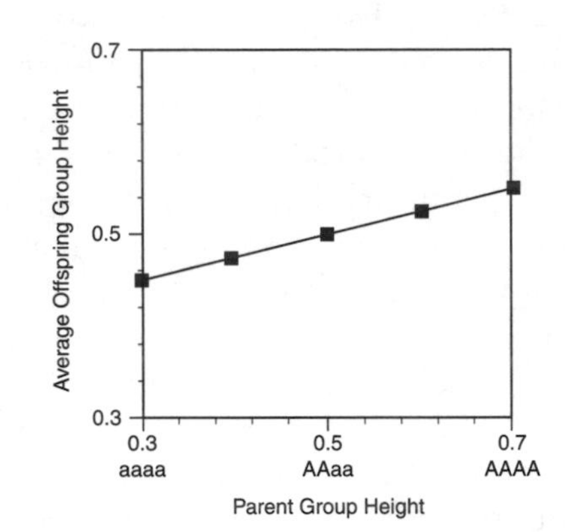
Average Offspring Group Height
0.7
0.5
0.3
0.3 aaaa
0.5 AAaa
0.7 AAAA
Parent Group Height

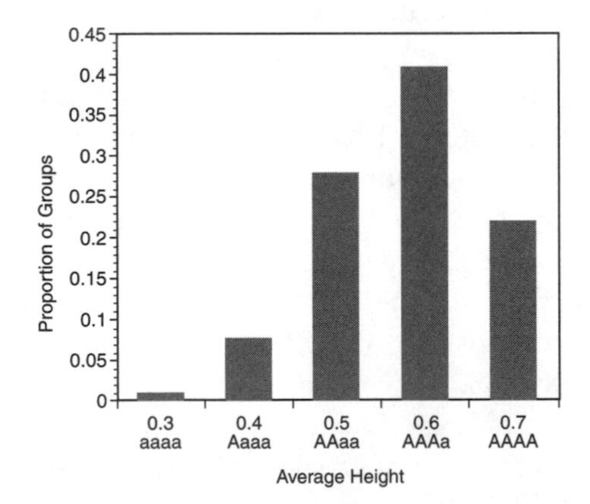
Proportion of Groups
0.45
0.4
0.35
0.3
0.25
0.2
0.15
0.1
0.05
0
0.3 aaaa
0.4 Aaaa
0.5 AAaa
0.6 AAAa
0.7 AAAA
Average Height

uring the trait "number of offspring after 37 days", Wade selected whole groups on the basis of their phenotypes. In one treatment, only the most productive groups were used as "parents" from which the next generation of groups was formed. In a second treatment, groups contributed to the next generation in proportion to their productivity. In a third treatment, the least productive groups were used as the "parents." Finally, in a fourth treatment, each group contributed the same number of progeny to the next generation regardless of its productivity. Individual selection occurred in all four treatments, favoring behaviors that maximize relative fitness within groups. Group selection favored high productivity in treatments 1 and 2 and low productivity in treatment 3; it didn't operate at all in treatment 4 because all groups were forced to be equally fit. The third treatment exemplified the scenario suggested by Wynne-Edwards (1962), in which the most productive groups go extinct, leaving more "prudent" groups to contribute to the next generation.

If heritability does not exist at the group level, then all four treatments should be dominated by within-group selection and so should remain similar to each other. Instead, the response to group selection was so great that after nine generations, *the average group from treatment 1 was over eight times more productive than the average group from treatment 3*. Furthermore, a response to group selection has been demonstrated in *every* laboratory experiment on multilevel selection that has been conducted since Wade's experiment, including

---

Figure 3.2. The concept of heritability at the group level. Asexual individuals are either tall (A) or short (a). Individuals are randomly distributed into groups of size $n = 4$ for a period of their life cycle. The bar graph at top shows the phenotypic distribution for the group-level trait "average height." The middle graph shows the correlation between a "parent" group and the average height of all the "offspring" groups that are derived from it. The positive slope indicates that height is heritable at the group level. Another way to demonstrate heritability is to perform an artificial selection experiment. If groups that are above average in height are used as "parents" for the next generation of groups, they produce "offspring" groups whose phenotypic distribution is shown in the bar graph at bottom. The response to selection indicates that the group-level trait "average height" is heritable.

cases where genetic differences among the groups were small (reviewed by Goodnight and Stevens 1997).

Critics of group selection sometimes dismiss these laboratory experiments as too artificial to address the issue of whether group selection occurs in nature. To evaluate this criticism, we need to distinguish the *intensity* of group selection from the *response* to group selection. Allowing only the most productive (or least productive) groups to contribute progeny to the next generation is indeed an extreme form of between-group selection that may or may not be common in nature (see next section). However, the response to group selection that is observed in laboratory experiments proves that phenotypic variation among groups is partially heritable. There is no reason to suppose that groups set up in the laboratory are different from natural groups in this respect.

Indeed, group-level traits are often more heritable in laboratory experiments than would be predicted on the basis of simple genetic models. It appears that the kinds of complex interactions that lead to *nonheritable* variation at the individual level can be *heritable* at the group level (Goodnight and Stevens 1997). Thus, a trait such as low fecundity might decrease relative fitness within groups and increase the fitness of groups. If we consider phenotypic variation alone, we might predict that individual selection is the stronger force and that the trait should fail to evolve. The experiments show that such traits can evolve because they are more heritable at the group level than at the individual level. This result was never anticipated by theoretical models that assume a simple genotype-phenotype relationship.[6]

The heritability of human behaviors at all levels is greatly complicated by cultural processes in addition to genetic processes (Boyd and Richerson 1985; Durham 1991; Cavalli-Sforza and Feldman 1981; Findlay 1992). The fact that a behavior is transmitted culturally should not be taken to mean that it is nonheritable. Cultural differences between human groups are often stable over long periods of time and are faithfully transmitted to descendant groups. They are heritable in the sense that offspring units resemble parent units, which is all that matters as far as the process of natural selection is concerned. We will consider cultural group selection in more detail in Chapters 4 and 5. For the moment, we conclude by saying that heritability must be evaluated at each level of the biological hierarchy.

There is no theoretical reason to doubt the existence of group-level heritability and the experimental evidence for it is surprisingly strong. Finally, inclusive fitness theory, evolutionary game theory, and selfish gene theory do not even remotely prepare one for the study of group heritability, the second fundamental ingredient of natural selection. Multilevel selection theory poses questions that were missed entirely by the other perspectives when they were regarded as alternatives to group selection.

Step 3c: Determine the fitness consequences of phenotypic variation within and among groups. If heritable variation exists, then the differential survival and reproduction of units will cause evolutionary change, leading the properties of the unit to "fit" the environment. This is the relation called adaptation. The rate of evolutionary change and the degree to which natural selection at a given level prevails against opposing forces depends on the intensity of selection. For example, Wade's (1976, 1977) experiment included two group-selection treatments for high productivity. The first treatment was a form of truncation selection in which only the most productive groups contributed to the next generation. The second treatment was a milder form of group selection in which every group contributed in proportion to its productivity. A response to group selection was observed in both treatments but was stronger in the first because the fitness differences between groups were greater.

As we mentioned in the previous section, the experiments of Wade and others are sometimes criticized for imposing a more extreme form of group selection than probably exists in nature. The best reply to this criticism is that a response to group selection has been observed for a *range* of selection intensities in laboratory experiments, from strong to weak. It is also worth pointing out, however, that group selection in nature is sometimes just as intense as the most extreme laboratory experiment. A desert leaf-cutting ant *(Acromyrmex versicolor)* provides an example of extreme group selection that is remarkably similar to Wade's experiment (Rissing, Pollock, Higgins, Hagen, and Smith 1989). After ants come together in highly synchronized mating swarms, new colonies form in clusters under the canopies of trees that provide shelter from the heat. The new colonies, which are founded by multiple females, are the groups in this example. Unlike the queens of most ant species, who raise the first generation of workers from fat reserves, *A. versicolor* queens must forage for leaves to initiate a new fungus garden. This task is not shared equally by the co-foundresses but rather is performed by a single queen who becomes a specialized forager. From the standpoint of the colony, it makes sense for one individual to forage because efficiency is gained through experience. However, above-ground foraging is a dangerous activity that substantially reduces the fitness of the specialized forager, relative to her colony mates. In short, specialized foraging benefits the group but is selected against within groups.

It is tempting to assume that the co-foundresses are genetically related and that specialized foraging evolves by kin selection, but studies have shown that founding females are *not* related. Nor is it true that one individual is forced to become a specialized forager by social dominance. The real answer appears to involve an especially strong form of truncation selection at the group level. As soon as the first generation of workers emerges, they raid the broods of other new colonies. Ultimately, only a single colony within a cluster will survive to grow to maturity. Specialized foraging hastens the production of new workers and gives the colony a crucial head start over other colonies.

This example from nature is remarkably similar to Wade's laboratory experiment. Instead of many groups of flour beetles in an incubator, we have many incipient ant colonies under the canopy of a tree. Instead of the trait "number of offspring after 37 days," we have "speed of worker emergence." Just as Wade selected the most productive groups in his first treatment, so natural selection favored the fastest colonies under the canopy. Specialized foraging is indeed disadvantageous within groups, but it is even more advantageous at the group level. In general, the intensity of natural selection at each level needs to be considered, along with phenotypic variation and heritability, in any investigation of what evolves.

A Hypothetical Investigation

Now that we have sketched a procedure for studying multilevel selection, let's see how it might be put to use in a hypothetical example. Imagine that you are studying a disease organism. You know enough about its biology to determine the level of reproduction that will maximize the number of disease progeny that reach new hosts (step 1). You have also identified a strain that reproduces much faster than this and would rapidly replace a more "prudent" strain within single hosts (step 2). You study the population structure of the disease and discover that most hosts are infected by numerous propagules, which provides ample opportunity for within-group selection. On the basis of this information, you work out the balance between levels of selection (step 3) and decide that the short-sighted virulent strain should replace the prudent strain. In other words, the population should lie close to the individual selection end of the continuum (step 2).

Now you discover a third strain with the following interesting property. Not only does it grow at a prudent rate itself, but it releases a chemical that causes all other strains within the same host to grow at the same rate. This changes all the details. The virulent strain no longer has a huge fitness advantage within mixed groups (a change in the intensity of selection; step 3c), but it still has the between-group disadvantage of quickly killing its host whenever it exists by itself. Upon further study, you discover that there is a small cost to producing the chemical, which makes the regulator strain slightly less fit than other strains within single groups. However, this disadvantage is trivial in comparison with the disadvantage of the first prudent strain that did not produce the chemical, and so allowed the virulent strain to reproduce unchecked. Armed with this new information, you work out the balance between levels of selection (step 3) and determine that the regulator strain should evolve because the group-level advantage of prudence far outweighs the within-group cost of producing the chemical. In other words, the population should lie close to the group selection end of the continuum (step 1).

The use of the procedure leads to a number of insights in this example. First, the discovery of the regulator strain radically changes the details, but the revised prediction still falls within the extremes of pure within-group and pure between-group selection that were established during steps 1 and 2. Thus, the first two steps bracket the possibilities and provide a disciplined way of examining the complicated details. Second, the factors usually associated with kin selection tell only part of the story. Extreme genetic variation among groups facilitates group selection but is not required for group selection to be an important evolutionary force. Third, notice that the first avirulent strain was more altruistic than the regulator strain that actually evolved. The first avirulent strain benefited the group at great expense to itself. It *could* have replaced the virulent strain if the population structure had been more favorable for group selection (e.g., if there had been more variation among groups). However, the regulator strain accomplished the same group-level effect at a smaller cost to itself within groups, so it replaced the more altruistic strain. Group selection favors traits that increase the relative fitness of groups, not altruism *per se*. Finally, notice that the regulator strain is slightly altruistic because there is a small cost to producing the chemical. Spectacular altruism has been eliminated from the population but muted altruism remains. All of these themes will appear again in our analysis of group selection and human evolution in Chapters 4 and 5.

New Insights from a New Perspective

In Chapter 2 we stressed that it often helps to view a complex process from different perspectives; this constitutes a legitimate form of pluralism. In practical terms, a perspective deserves to exist to the extent that it provides new insights, even if they can be accounted for in retrospect by other perspectives.

Multilevel selection theory allows us to see possibilities that have not been obvious from other perspectives. Indeed, focusing on the fundamental ingredients of natural selection at each level of the biological hierarchy affords so new a view of evolution that it is difficult to translate the basic concepts into the language of the other perspectives. How would truncation selection be studied from the standpoint of inclusive fitness theory? How would heritability at the group and individual levels be studied from the standpoint of game theory? In this section we will review two especially novel insights that multilevel selection theory provides but that are hard to see from the other perspectives.

Natural Selection at the Level of Multispecies Communities

When groups are defined as the set of individuals whose interactions affect each other's fitness, they often include members of different species in addition to members of the same species. Other perspectives on natural selection, such as inclusive fitness theory and evolutionary game theory, are almost invariably limited to interactions among members of the same species.[7] Multilevel selection theory, in contrast, can easily be extended to explain the evolution of communities of species as adaptive units. The best illustration is an experiment by Goodnight (1990a,b) that built upon Wade's group selection experiments that we have already described. Goodnight's experiments were similar to Wade's except that his groups were composed of two species of flour beetles *(Tribolium castaneum* and *T. confusum)* and therefore were miniature communities. Goodnight treated the density of *one* of the species as a community-level phenotypic character in an artificial selection experiment. In other words, whether a community was used as a "parent" to found the next generation of communities depended upon the number of *T. confusum,* regardless of the number of *T. castaneum.* However, for those communities that were selected as parents, both species in the community were used to create the "offspring" communities. In other treatments, the density of *T. castaneum* and emigration rates rather than density were used as community-level phenotypic characters.

Goodnight's communities responded to higher-level selection, but when he investigated the mechanistic basis of the response, he discovered that it had an interspecific ecological basis in addition to an intraspecific social basis. Because selection operated on the entire community as a unit, genes in *either* of the two species that caused an increase in the phenotypic character (density or emigration rate of one of the species) were selected. In fact, the experiment selected a particular *interaction* between the genes in the two species. When a species in one of the selected lines was raised by itself without the other species, the response to selection disappeared. When a species in one of the selected lines was raised with the other species from the original stock population, as opposed to the other species from the same selected line, the response to selection also disappeared. Selecting entire communities as a unit favored genes in both species that functioned as a single interacting system to produce the response to selection. The fact that the genes existed in separate gene pools was unimportant.

Examples of community-level selection also can be found in nature. Many species of insects have evolved to reproduce on patchy, ephemeral resources, such as carrion, dung, and wood. Other species that lack wings, such as mites, nematodes, and fungus, have also evolved to utilize these resources and travel from patch to patch by hitchhiking on the bodies of the more mobile insects. The use of one species for transport by another is termed "phoresy." In some species of carrion beetles, the average individual arriving at a carcass carries four or five species of mites, numbering over 500 individuals, tucked into the nooks and crannies of its exoskeleton (Wilson and Knollenberg 1987). These hitchhikers can potentially have positive or negative effects on their carrier species, both while on board and when they are reproducing on and around the carcass. The hitchhikers require the insect for dispersal, so those with a positive effect will cause their local community to contribute more to future resource patches, as surely as the flour beetle communities artificially selected by Goodnight. However, hitchhikers that benefit their carrier do not necessarily increase their relative abundance within their local community. Benign species that benefit their community are vulnerable to the same problems of freeloading as benign genes within individuals that benefit their group. Community-level selection is therefore required for the hitchhikers to evolve into an integrated ecological system that benefits the carrier. Some species of wood-feeding beetles cannot survive without their hitchhikers, who systematically disable the tree's defenses and convert the wood into a nutritious resource for the larval insects that will carry the community's progeny to future resource patches (e.g., Batra 1979). The multispecies community on the *outside* of the beetle has evolved to benefit the beetle, as surely as have the genes on the *inside* of the beetle! Other phoretic species appear to have a more neutral effect on their host, but when removed they are soon replaced by species from the surrounding environment that have a negative effect (Wilson and Knollenberg 1987). In other words, commensalism (a neutral effect) is not an absence of interactions but an evolved set of interactions that eliminates negative effects on the host that might otherwise occur. The same phoretic associates that do not harm their own hosts act as competitors and predators with respect to other insect species that they do not depend upon for transport (reviewed by Wilson 1980, pp. 119–126). Some phoretic associations have evolved a neutral effect, rather than a beneficial effect, in part because the carriers live in a benign environment that is not easily improved by the activities of the phoretic associates (Wilson and Knollenberg 1987). Removing their own negative effect is the best that these hitchhikers can do.

Communities of symbiotic organisms that live inside their hosts are subject to the same kinds of multilevel selection pressures as the phoretic communities that cling to the outside of their host. Even if a symbiotic community can potentially benefit its host, it may not evolve to do so if the population structure favors within-community processes over between-community processes. Some host species have evolved to manipulate the population structure of their symbionts; in a way, they play the role of the experimenter in an artificial selection experiment. Frank (1996b, 1997) has suggested that some hosts even divide their symbionts into a germ line, which remains latent and is transmitted to the host offspring, and a somatic line, which does the active work within the host. The advantage of this arrangement is that the freeloaders that evolve by within-community selection in the somatic line are not passed to the next generation of hosts. This is an excellent example of how a single phenomenon (the germ-soma distinction) can exist at the level of genetic interactions within individuals (Buss 1987; Frank 1991) and at the level of interactions among species in a multispecies community.

Practical Applications of Multilevel Selection Theory

Artificial selection at the individual level has been used as a tool to evolve domesticated strains of plants and animals since before Darwin. The simple idea is to use individuals with the desired properties as breeders for the next generation. Multilevel selection theory makes it obvious that the same methods can be used to evolve groups and communities that perform useful functions. There is nothing technically difficult about selecting at the level of groups and communities; it is merely a matter of *seeing* them as evolvable units whose properties can be shaped by natural and artificial selection.

Group selection was recently used to evolve a new strain of chicken that is likely to save the poultry industry millions of dollars (Muir 1995; Craig and Muir 1995). Chickens have been pushed to their genetic limits as individual egg-laying machines by countless generations of artificial selection. Chickens do not lay eggs as individuals, however, but as members of social groups. In former times, the groups were flocks that roamed outside or in indoor floor pens. In today's highly automated poultry industry, chickens are often crowded into multiple-hen cages that are efficient from the standpoint of feeding and the collection of eggs but inhumane to the birds. Stress and aggression are so high that even after their beaks are trimmed (itself a painful operation), annual mortality can approach 50 percent.

It is interesting that selecting for increased egg production at the individual level can actually reduce the productivity of individuals. Suppose that we record the number of eggs laid by all the hens in a flock and choose the most productive hens as breeders for the next generation. These hens may have been most productive because they are the best egg-laying machines. Or, they may have been most productive because they are the nastiest hens in the flock and were able to suppress the production of the other hens. Using the most productive hens as breeders would then result in a flock of nasty hens with a lower egg production than before. This outcome, which makes it seem as if the trait is negatively heritable, has actually occurred in chicken-breeding programs (Craig et al. 1975; Bhagwat and Craig 1977) and in laboratory studies of multilevel selection (reviewed by Goodnight and Stevens 1997). It seems paradoxical until we realize that it is the essence of what Williams meant when he said that maximizing relative fitness *within* the group often *decreases* the fitness *of* the group.

To select for chickens as *social* egg-laying machines, it is necessary to select at the level of whole chicken societies. Muir (1995) repeated Wade's experiment using chickens instead of flour beetles. Groups of hens (those housed in the same cages) were scored for egg production, and hens from the most productive groups were used as breeders for the next generation of groups. The response to group selection was as dramatic for chickens as it was for flour beetles. Annual egg production increased 160 percent in only 6 generations. Muir (1995, p. 454) calls this improvement "astonishing," since efforts to increase egg production by selecting at the individual level are virtually never this effective. The increase in egg production was caused by numerous factors, including more eggs laid per day and lower mortality. In fact, mortality due to stress and aggression was so low in the group-selected strain that beak trimming became unnecessary. If this strain becomes widely used in the poultry industry, the projected annual savings will far exceed the money spent by the U.S. government for basic research in evolutionary biology.

This example indicates the tremendous practical benefits that can be gained from selecting at the level of whole groups. There are no technical barriers to achieving the benefits; it is merely a matter of seeing groups as evolvable units that can be shaped by the methods of artificial selection. The same is true of multispecies communities. In human life, the most difficult problems can be tackled only by teams of specialists interacting in a coordinated fashion. It would be useful to develop coordinated teams of species to tackle difficult problems, such as breaking down toxic compounds in the soil. Multilevel selection theory suggests a procedure: Place sterilized soil containing the toxic compound in a large number of flasks. Inoculate each flask with a sample from a natural soil community that includes bacteria, fungi, protozoa, and myriad other organisms. Community composition will vary among flasks, if only at random, and small differences will become magnified by complex interactions as the communities develop. After a period of time, select the communities that were most successful at degrading the toxic compound as the "parents" for new generation of communities. This procedure could result in the evolution of multispecies communities that function as coordinated teams to solve important practical problems. Despite its simplicity, to our knowledge it has never been attempted. Goodnight (1990a,b) is the first person in the history of biology to treat multispecies communities as evolvable units.

The Issue of Parsimony

We have described conceptual and empirical procedures for determining the level(s) at which a given trait has evolved by natural or artificial selection. We now need to consider the issue of parsimony, which seems to offer another set of guidelines for answering the same questions. Rather than examining the empirical details, some biologists have suggested that group selection hypotheses should be rejected because they are unparsimonious. This appeal to parsimony is a leitmotif in Williams's *Adaptation and Natural Selection* and was repeated by Dawkins in *The Selfish Gene*. Parsimony has become part of the litany repeated by those who casually reject group selection as an important evolutionary force. It is therefore important to re-examine the issues in the light of modern multilevel selection theory. We believe that three different parsimony arguments have been made by Williams, Dawkins and others. The first two utterly fail but they do so for different reasons; the third contains a kernel of truth, but its significance has been widely misinterpreted.

The first argument is simply a variant of the averaging fallacy. Dawkins (1976) argues that the gene is the unit of selection by way of a suggestive analogy. He imagines a crew coach who wants to put together a rowing team composed of eight oarsmen. The coach makes the candidate oarsmen form up in different groups of eight and has these groups race against each other. The rowers are then assorted into new groups and a second competition is held. After repeated trials of this type, the coach consults the results to decide which eight oarsmen will make the team. Dawkins asks whether the coach is selecting individual rowers or is selecting a group of rowers. He says that the former description is more parsimonious. The elongated racing shells are analogs for chromosomes; individual oarsmen are genes. Just as it is supposedly more parsimonious to think of the crew coach as choosing individual oarsmen, so it is supposed to be more parsimonious to view natural selection as choosing individual genes.

There is a confusion in this argument between the process of natural selection and its product. If the criterion for a population's evolving is that the genes in the population change frequency, then natural selection produces evolution only when some genes are more successful than others in achieving representation in successive generations. However, this fact about the result of natural selection leaves open what type of selection process has taken place. For example, if group selection causes altruism to evolve, then genes for altruism increase in frequency and genes for selfishness decline. Since natural selection can always be "represented" in terms of genes and what happens to them, this fact cuts no ice when the question is whether group selection or some other type of natural selection was the process responsible. In fact, the metaphor of the rowing crew is intended to describe the standard process of individual selection, in which genes are always in the same boat with other genes and can succeed only by enhancing the collective effort. To say that it is more parsimonious to think in terms of genes for this example is to obscure the distinction between standard individual selection and intragenomic conflict, in which some genes succeed at the expense of other genes in the same boat. We dub this criticism of group selection the "representation argument" (Wimsatt 1980; Sober 1984, 1993b). It is a fallacy; in particular, it is the averaging fallacy with the word *parsimony* thrown in for good measure.

The second parsimony argument states that group selection is inherently improbable. For example, Wright's (1945) model of group selection seems to require an unlikely combination of events. Altruism must drift to fixation against the current of natural selection in at least one group. Then the group must persist long enough and send out a sufficient number of dispersers to counterbalance the disadvantage of altruism in other groups. At the same time, the altruistic group must be protected from invasion by selfish individuals from other groups. The prospect of all these events co-occuring is like the probability of a randomly thrown dart hitting a very small portion of a dart board (Sober 1990).

Wright's (1945) model is indeed implausible, but this conclusion can hardly be extended to the entire subject of group selection. In the Williams and Williams (1957) model, groups are isolated for only a portion of the life cycle, the altruistic gene need not drift to fixation within groups, and altruistic groups export their productivity when their members disperse. These assumptions are eminently reasonable. As a *general* theory, group selection passed the plausibility test long ago. Populations that are subdivided into groups are a biological fact of life for many species, including human beings. Evolutionary models must take this population structure into account, whether they go by the name of group selection, inclusive fitness theory, game theory, or selfish gene theory. When a population is subdivided into groups, the within- and between-group components of selection must be estimated empirically, not resolved *a priori* by insisting that group selection is "unparsimonious."

Although we think that neither the representation argument nor the dartboard argument holds water, we do think that parsimony has a role to play in the controversy over the units of selection, and in inference generally. Conventional scientific practice embodies a preference for simpler theories. The simplest hypothesis is usually identified as the null hypothesis and is considered innocent until proven guilty. If two hypotheses explain the observations equally well, the null hypothesis is chosen on the basis of its simplicity. However, this point should not obscure the fact that data relating to the two hypotheses must be obtained and that the more complex hypothesis is preferred when it explains the observations better than the null hypothesis. It would be an abuse of the principle of parsimony to adopt the null hypothesis without looking at any data at all. Parsimony is a tool for interpreting observations, not for doing without them (Forster and Sober 1994).

Williams (1966, p. 18) says that a model that postulates multilevel selection is more complex than a model that postulates selection operating solely within a single population. We agree, and if both models explain the observations equally well, then the simpler should be chosen. However, this does not allow one to reject a multilevel selection hypothesis without consulting data. Still less does it allow the rejection of the entire theory. Multilevel selection hypotheses must be evaluated empirically on a case-by-case basis, not *a priori* on the basis of a spurious global principle. We have provided numerous examples in which the simpler hypothesis (based on pure within-group selection) and the more complex hypothesis (based on multilevel selection) lead to different predictions, with the weight of evidence supporting multilevel selection. A legitimate concern for parsimonious hypotheses, upon which conventional scientific procedure is based, poses no special problem for multilevel selection theory.

Preparing for the Study of Humans

Multilevel selection theory can be used to study an enormous range of subjects, from the origin of life (Michod 1983) to the major transitions of life that created individual organisms as we know them (Maynard Smith and Szathmary 1995), to the structure of single-species societies and multispecies communities. We have touched upon many of these subjects, but now it is time to focus on the evolution of our own species. In Chapters 4 and 5, we will argue that our lineage is a recent example of an evolutionary transition in which lower-level units (individuals) coalesce into functionally integrated higher-level units (groups). However, human beings have an impressive ability to change their behaviors over the short term in response to their immediate environment. To proceed further we must understand facultative behaviors—behaviors that are expressed under some conditions but not others—in terms of multilevel selection theory.

Most animals can alter their behavior in response to environmental change, either innately, by learning, or by cultural transmission. Let's see how facultative behavior is conventionally studied before relating it to multilevel selection theory. Imagine a dangerous aquatic habitat with a predator lurking around every corner and a safer habitat from which predators are excluded. A fish species that lives entirely in the dangerous habitat will evolve cautious behaviors to avoid being eaten. A second fish species that lives entirely in the safe habitat will evolve to behave more boldly. If we place individuals of the cautious species in the safe habitat, they will not necessarily adjust their behavior. Safety is something that they have never experienced in their evolutionary past, so there is no reason to expect them to behave adaptively. Similarly, members of the bold species that are placed in the dangerous habitat should simply disappear down the gullets of predators; they are not likely to adjust their behavior. Species that have evolved on isolated islands without predators are like this, and will approach human travelers utterly without fear.

Now imagine a third fish species, whose environment includes both a safe and a dangerous habitat. If individuals cannot detect which habitat they are in, they should evolve to be somewhat intermediate in behavior between the bold and cautious species. If individuals can detect which habitat they occupy, then they should evolve to behave like the cautious species in the risky habitat and like the bold species in the safe habitat. In short, facultative behaviors should evolve whenever the environment is variable and this variation can be detected by the organism. The *facultative* behavior that evolves in a particular environment is similar to the *fixed* behavior that would evolve in a species inhabiting *only* that environment. This is the standard way of thinking about the evolution of facultative behaviors.[8]

Now let's modify this example to include multilevel selection. Consider a species of wasp that lays eggs on hosts, as modeled by Hamilton (1967). Hosts are always parasitized by a single female, and as a result the population has a highly female-biased sex ratio. In a second wasp species, hosts are always parasitized by exactly ten females, favoring a more even sex ratio. In a third wasp species, the number of females parasitizing a given host varies from one to ten. If females cannot detect the number of other females that have parasitized a given host, we expect a sex ratio to evolve that is intermediate between the first two species. However, if females can perceive this important environmental variable, we might expect them to produce a highly female-biased sex ratio on hosts that they colonize by themselves and a more even sex ratio on hosts that have been parasitized by many other females. This prediction was tested on *Nasonia vitripennis,* a tiny wasp that parasitizes fly pupae (Orzack, Parker, and Gladstone 1991). Amazingly, females are able to assess the state of their environment and alter the sex ratio of their offspring accordingly.[9]

This example shows that the balance between levels of selection must be evaluated on a trait-by-trait basis for facultative behaviors as well as for fixed behaviors. If a wasp can perceive that she is the only one to colonize a host, behaviors can evolve that are adapted to this particular population structure, which is characterized by extreme genetic variation among groups. The facultative behavior that evolves therefore lies close to step 1 of our procedure—assessing what would evolve if only group selection were responsible for evolutionary change. Hosts colonized by many parasites represent another population structure, one in which most of the genetic variation exists within groups. The facultative behavior that evolves therefore lies close to step 2 of our procedure. Whenever facultative behaviors evolve that are restricted to a given situation, the balance between levels of selection must be evaluated on the basis of the population structure relevant to that situation.

Perhaps the most dramatic example of a facultative behavior that reflects a shift in population structure occurs in honey bees. When the queen dies and cannot be replaced, the only opportunity for reproduction is for the workers to lay unfertilized eggs, which will become males. All workers can compete with each other in this activity, so the death of the queen causes a sudden change in population structure. The hive, which previously was a marvel of group-level functional organization, deteriorates into a barroom brawl as the workers scramble to consume the hive's resources, killing the eggs of their sisters in order to lay their own eggs (Seeley 1985). Thus, even bees in the same hive—a paradigm of the superorganism—are innately prepared to respond facultatively to shifts in population structure that have recurred throughout their evolutionary past.

The way that facultative behaviors are studied in multilevel selection theory follows directly from the way that they are traditionally studied. Nevertheless, it leads to some unfamiliar conclusions about human behavior. Imagine, for example, that you are interacting with total strangers whom you will never meet again. Now imagine interacting with good friends whom you have known for years. It is obvious that most people would behave differently in these two social situations. However, many people would be reluctant to say that either set of behaviors is altruistic. In the first place, even a psychological egoist would behave differently toward friends and strangers. In the second place, each set of behaviors seems to make good sense from the standpoint of individual advantage. For example, it appears to be smart to mistrust and perhaps even to exploit total strangers while remaining trustworthy to good friends. If you profit from interacting with your friends, how can you be an altruist?

This way of thinking about human behavior feels as comfortable as an old shoe, but it fails to define altruism or identify the level of adaptation responsible for the behavior according to multilevel selection theory. The fact that a psychological egoist would behave differently toward friends and strangers is irrelevant. We didn't ask what the wasps and bees thought when they facultatively changed their behaviors; why should we make an exception for humans? We will focus on psychological mechanisms later, but for now it is important to ignore them entirely and stick to definitions that are based on fitness effects. The fact that the individual profits by facultatively changing its behavior is more relevant, if we equate "profit" with "fitness." Recognizing this fact tells us that the behaviors are adaptive but it does not identify the level at which they succeed. Adaptation is not the same thing as selfishness! To identify the adaptive unit, we must follow the first two steps of the procedure outlined at the beginning of this chapter. Does the individual's behavior increase the fitness of all the social partners in a group (step 1), or does it increase the fitness of the individual, relative to its social partners (step 2)? If the answer is close to that provided by step 2 for behaviors expressed toward strangers, then we can conclude that the individual is behaving selfishly when interacting with strangers. If the answer is close to that provided by step 1 for behaviors expressed toward friends, then we can conclude that the individual is behaving for the good of its group when interacting with friends. The fact that the individual profits along with everyone else in the group does not alter our categorization. Genes that benefit the entire genome (including themselves) evolve at the individual level; by the same token, individuals who benefit the entire group (including themselves) evolve at the group level. If the individual causes his friends to benefit more than himself, he is altruistic. The fact that the individual can easily toggle between the two sets of behaviors is irrelevant. Each facultative behavior must be analyzed on its own terms, against the background of the relevant population structure. That is the procedure we followed when examining the behaviors of fish, wasps, and bees, and humans should be no exception.

When we view our own species through the lens of multilevel selection theory, we discover that human behavior cannot be placed, in its entirety, at any one point on the continuum from pure group selection to pure individual selection. As the most facultative species on earth, we span the entire continuum. Like bees, human beings may be innately prepared to claw their way to the top of highly dysfunctional groups *and* to participate in group-level superorganisms, depending on the population structure that they naturally encounter or build for themselves.

Adaptation: Handle with Care

Adaptation is a powerful concept but it must be handled with care. In *Adaptation and Natural Selection,* Williams (1966) attempted to impose discipline on the subject. He exposed many examples of sloppy thinking, and some aspects of his analysis are still valid. Others must be reconsidered, however, and his pessimistic conclusion that "group-level adaptations do not, in fact, exist" has not withstood the test of time. Unfortunately, the study of adaptation has again become undisciplined, and many evolutionary biologists have lost the ability to see higher-level adaptations when they exist. This chapter therefore has the same purpose as Williams's book—to provide a guide for using adaptationist ideas as a source of insights while avoiding their many pitfalls.

The procedure that we have outlined preserves the core of Williams's own analysis. We agree with Williams that group selection is required to explain group-level adaptations. Williams consistently defined individual selection in terms of relative fitness within groups and seldom committed the averaging fallacy. His analysis of sex ratio and parasite virulence follows the procedure that we have outlined, in which traits that evolve entirely by group selection (step 1) and traits that evolve entirely by individual selection (step 2) are imagined as extreme possibilities. However, other aspects of Williams' analysis need to be rejected. His focus on genes as replicators cannot be used to argue against groups as adaptive units, and his use of a parsimony criterion to reject group selection is no longer appropriate.

Our procedure may seem complex in some of its details, but it may be the simplest way to study what is, after all, a complex process. Steps 1 and 2 harness the power and simplicity of functional thinking without forcing one to believe that everything is adaptive. They bracket the possibilities and avoid the averaging fallacy by requiring one to imagine what would evolve by pure within-group and pure between-group selection. The task of determining the balance between levels of selection can become complex (step 3), but all of the details can be understood in terms of the fundamental ingredients of natural selection—phenotypic variation, heritability, and fitness consequences. If the predictions that emerge from steps 1 and 2 are sufficiently different from each other, then empirical tests are possible even in the absence of detailed knowledge about step 3. The procedure that we suggest is not the only way to study natural selection, but it enables one to recognize adaptations where they exist, at all levels of the biological hierarchy.

4

Group Selection and Human Behavior

We have shown that natural selection is a multilevel process that sometimes molds groups into adaptive units. In this chapter we will argue that group selection may have been an especially important force in human evolution. Human social groups may have the potential to be as functionally integrated as bee hives and coral colonies.

How have different intellectual traditions sought to understand the nature of human society? A recurrent theme in Western philosophy and religion has been the analogy between the social macrocosm and the individual microcosm. In the *Republic,* Plato compared harmoniously interacting social classes to the organs of a healthy person. Joseph Butler (1726) claimed that "it is as manifest that we were made for society and to promote the happiness of it, as that we were intended to take care of our own life and health and private good." Religions frequently employ the superorganism metaphor, as in the following quotation from the Hutterites (Ehrenpreis 1650, p. 11): "True love means growth for the whole organism, whose members are all interdependent and serve each other. That is the outward form of the inner working of the Spirit, the organism of the Body governed by Christ. We see the same thing among the bees, who all work with equal zeal gathering honey . . ." In everyday life, people frequently invoke the good of society in addition to their own self-interest to explain their actions. Moral behavior and social policy are usually defined and justified in terms of the common good.

These group-centered views were amply represented in the human sciences during the first half of the twentieth century. According to the psychologist D. M. Wegner (1986, p. 185), "Social commentators once found it very useful to analyze the behavior of groups by the same expedient used in analyzing the behavior of individuals. The group, like the person, was assumed to be sentient, to have a form of mental activity that guides action. Rousseau (1767) and Hegel (1807) were the early architects of this form of analysis, and it became so widely used in the 19th and early 20th centuries that almost every early social theorist we now recognize as a contributor to modern social psychology held a similar view."

More recently, the group-level perspective has fallen on hard times, although to different degrees in different disciplines. Social groups are seen not as adaptive units in their own right but as by-products of individual self-interest. According to the psychologist D. T. Campbell (1994, p. 23), "Methodological individualism dominates our neighboring field of economics, much of sociology, and all of psychology's excursions into organizational theory. This is the dogma that all human social group processes are to be explained by laws of individual behavior—that groups and social organizations have no ontological reality—that where used, references to organizations, etc. are but convenient summaries of individual behavior."

The individualistic perspective is perhaps most extreme among evolutionary biologists interested in human behavior, as illustrated by the passage from Alexander (1987) that we quoted in the Introduction. They reject the concept of human groups as superorganisms for a simple reason—the genetic structure of human populations is very different from what is found in other ultrasocial species. When groups are formed by asexual reproduction from a single individual, as in coral colonies, their members are genetically identical (except for mutations), so it is not surprising that they act as a single organism. And when social insect colonies are founded by a single diploid female carrying the sperm of a single haploid male, the members of these groups are highly related genetically, even if they are not identical. We might therefore expect them to behave largely, although not entirely, like a single organism. More recently, it has been discovered that some social insect colonies are founded by multiple queens or by queens who have mated with multiple males. We might expect an even lower degree of altruism and group-level functional organization in these species, a point that we will return to below.

In contrast to these ultrasocial species, most human social groups consist of a mix of nonrelatives and genetic relatives of varying degrees. If modern hunter-gatherer societies can be used as a guide, average genealogical relatedness among group members was probably above zero during most of human evolution but not nearly as extreme as is true for social insects and clonal organisms. The conclusion seems to follow that we should not expect a human social group to act much like a single organism.

The problem with this reasoning is that it treats genealogical relatedness as the only important variable in the evolution of group-level functional organization. It is a literal interpretation of kin selection theory, in which the degree of altruism that evolves depends just on the degree of relatedness between the donor and the recipient of the altruistic behavior. As we have seen, however, kin selection is a special case of a more general theory—a point that Hamilton (1975, 1987) was among the first to appreciate. In his own words, "it obviously makes no difference if altruists settle with altruists because they are related . . . or because they recognize fellow altruists as such, or settle together because of some pleiotropic effect of the gene on habitat preference." We therefore need to evaluate human social behavior in terms of the general theory of multilevel selection, not the special case of kin selection. When we do this, we may discover that humans, bees, and corals are all highly group-selected, but for different reasons.

In this chapter we will present a series of models that show why genealogical relatedness is not required for group selection to be a strong evolutionary force. These models may apply to many species but they are especially relevant for humans, because the mechanisms that substitute for genealogical relatedness often require sophisticated cognitive abilities and, in some cases, the ability to transmit culture. Thus, these models may explain not only why humans are group-selected, but why they have experienced a unique variety of group selection.

Assortative Interactions

Group selection requires variation among groups—the more variation the better—but the variation does not need to be due to genealogical relatedness. Imagine a very large population of unrelated individuals who vary in their altruistic tendencies. Suppose that each individual's degree of altruism is observable and that membership in a social group requires the consent of all parties. Each individual wants to associate with the most altruistic partners it can find. Thus, if the groups are of size *n,* the *n* most altruistic individuals in the population will form one group, the *n* next most altruistic individuals will form another group, and so on down to the *n* least altruistic individuals, who associate not by choice but by default (alternatively, the least altruistic individuals could choose to remain solitary; Kitcher 1993). If the total population is extremely large, then the members of each group will be virtually identical and almost all the variation in altruism will be concentrated at the between-group level. Groups formed in this way would be almost as favorable for the evolution of altruism as groups formed by clonal reproduction.

Admittedly, this scenario is unrealistic in several respects. The altruistic tendencies of an individual may not be readily observable (Dawkins 1976), one cannot expect everyone to know everyone else in a very large population, membership in a group may not require the consent of all parties, and so on. Nevertheless, even if we relax some of these extreme assumptions, it seems likely that partial information and partial control over membership will still create highly nonrandom groupings. Here, then, is a mechanism for the evolution of altruism that is potentially as important as kin selection.

The idea that choosing associates favors the evolution of altruism is deeply intuitive, but it has received little attention from evolutionary biologists. One reason for this neglect is that initial efforts to model assortative interactions encountered a difficulty that can be called *the problem of origination* (Wilson and Dugatkin 1997). In mathematical models, it is convenient to assume that behaviors exist as discrete traits that arise by mutation and therefore initially exist at a very low frequency in the population. If we model discriminating altruists in this way, we discover that they do poorly at a very low frequency because they seldom encounter other altruists with whom to associate.

Altruism can easily evolve after the discriminating altruists exceed a threshold frequency, but a model that requires altruists to exist at a frequency of 20 percent (for example) before they can be favored by natural selection fails to address a fundamental problem about how altruism can evolve.

Kin selection models do not suffer from the problem of origination because an initial mutant altruist (Aa) who survives long enough to mate with a nonaltruist (aa) immediately produces a group of offspring of whom 50 percent are altruists. Perhaps for this reason, kin selection models became enormously influential while models of assortative interactions had minimal impact. Nevertheless, the problem of origination may not be as serious as it first appears. It may be convenient to model behaviors as discrete traits that enter the population at mutation frequency, but real behaviors are often continuous in nature and have a mean and a variance. For example, imagine a group of minnows who suddenly discover that a large and ominous fish is lurking in the vicinity. Is it a predator? If so, is it hungry? The only way to find out is for at least one of the minnows to approach the fish and try to provoke a reaction. If the inspector discovers that the large fish is not a threat, then all the minnows can resume foraging. Of course, the inspector is risking its life to obtain information that everyone will share, so it has the lowest relative fitness within the group and qualifies as an altruist. One way to model the evolution of inspection behavior is to imagine a single mutant inspector that arises in a population of noninspectors. Another way is to imagine that individuals vary in a continuous fashion, with some approaching the potential predator more closely than others. Natural selection may favor either a greater or a lesser tendency to inspect, but whatever the outcome, there will always be variation around the mean. Most behaviors, including predator inspection behavior in fish, are continuous, not discrete.

Behaviors and other traits tend to be continuous when they are influenced by genes at many loci. Models exist for studying the evolution of continuous traits, but they are more complex than models of discrete traits and have seldom been used to study the evolution of altruism (Boyd and Richerson 1980). When altruism is modeled as a continuous trait, the problem of origination largely disappears (Wilson and Dugatkin 1997). It may be hard for a mutant altruist to find another altruist to interact with, but it is easy for an individual who is above average to find another individual who is above average. This is true even when the *average* degree of altruism is very low. Of course, we still need to relax the extreme assumptions of the initial scenario presented above, but at least the problem of origination can be removed as an initial stumbling block when behaviors are continuous in nature.

A recent model by Wilson and Dugatkin (1997) allows assortative interactions to be compared directly with kin selection as a mechanism for the evolution of altruism. The individuals in a large population vary with respect to a single continuous character, such as response to a potential predator. Within-group selection favors one value of this character (e.g., staying at least 1 meter away from the potential predator), while between-group selection favors another value (e.g., approaching to within 10 centimeters). If within-group selection is the only evolutionary force operating on the population, then the distribution will become centered on the within-group optimum. In other words, the average individual in the population will be maximally fit from the standpoint of within-group selection, some individuals will deviate in a way that is good for the group, and others will deviate in a way that is bad for the group. Maladaptive variation around an adaptive mean often occurs when the behavior is influenced by genes at many loci. Every generation the maladaptive individuals on both sides of the distribution are removed by natural selection, but then they are restored by mutation and recombination.[1]

Starting from this worst-case scenario in which the population is centered on the within-group optimum, Wilson and Dugatkin (1997) added a population structure in which groups of size *n* are formed in a variety of ways. In one set of computer simulations, groups were formed by randomly selecting *n* individuals from the global population. Random sampling creates a modest amount of variation among groups, depending on the value of *n*. Thus, the average group is centered on the within-group optimum (as is the global population), but some groups will be more altruistic than others by chance and these will contribute more individuals to the global population. Random grouping introduces a modest component of group selection into the process, which pulls the phenotypic distribution of the population toward the group optimum. The balance between levels of selection with random variation among groups is shown for one set of parameter values in Figure 4.1.

In a second set of simulations, kin groups were formed by first selecting two individuals at random from the population and having them mate to form offspring groups of size *n*. An offspring group also varies in a continuous fashion, but its distribution is centered on the average of its parents' phenotypes. For example, if an individual who approaches within 90 cm and an individual who approaches within 80 cm of the potential predator mate with each other, their offspring will approach to within 85 cm on average but there will be some variation around this mean. Once again, this pattern is anticipated when behaviors are influenced by genes at many loci. As might be expected, kin groups vary much more than randomly formed groups. To create a highly altruistic kin group, we need only pick two highly altruistic individuals at random from the population. To create a highly altruistic random group, we must pick *n* highly altruistic individuals at random from the population. Group selection is therefore a more powerful evolutionary force when groups are composed of full siblings, pulling the population further in the direction of the

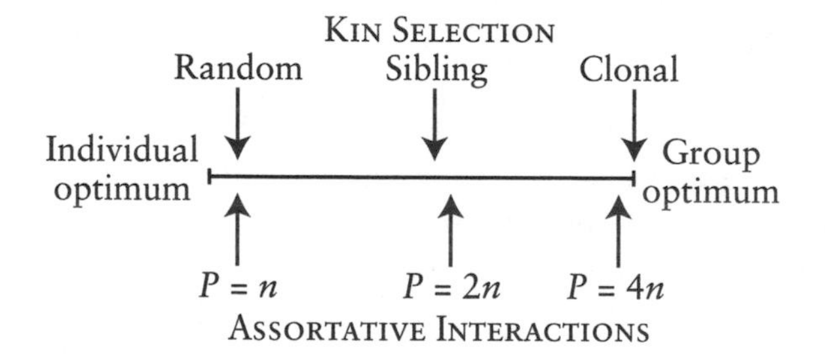

Figure 4.1. A comparison of kin selection and assortative interactions as mechanisms for the evolution of altruism. For kin selection, groups are formed from the offspring of two randomly mating individuals (in which case group members are siblings) or from one asexual individual (in which case group members are clones of each other). For assortative interactions, groups are formed by drawing P individuals at random from the population, ranking them with respect to altruism and splitting them into groups of size n. This mechanism creates nonrandom groupings that are comparable to groups of full siblings when $P = 2n$ and clonal groups when $P = 4n$.

between-group optimum. The new balance between levels of selection is shown for one set of parameter values in Figure 4.1.

In a third set of simulations, clonal groups were formed by selecting single individuals at random from the global population and having them reproduce asexually to form groups of size n. Genetic uniformity within groups leaves group selection as the only evolutionary force, pulling the population all the way to the between-group optimum. So far the model duplicates the insights of kin selection theory, in which the degree of altruism that evolves is proportional to the degree of genealogical relatedness.

Against this background, we now can evaluate assortative interaction as a mechanism for the evolution of altruism. Imagine that individuals do not know the altruistic tendencies of every other individual in the population and must select their associates from among a much smaller subset. To be specific, suppose that individuals exist in randomly formed neighborhoods of size P, whose members learn each other's altruistic tendencies and sort into groups of size n as originally described. If $P = n$, then there is no opportunity for assortative interactions and the groups form at random. If we double the size of the neighborhood so that $P = 2n$, then groups are formed by drawing $2n$ individuals at random from the global population (the neighborhood), ranking their altruistic tendencies, and splitting them down the middle to form two groups. These groups will obviously be nonrandom samples of the global population but will still contain substantial variation within groups. If $P = 4n$, then groups are formed by randomly drawing $4n$ individuals, ranking them, and splitting them into four groups, and so on. Variation between groups obviously increases with P, favoring between-group selection. The question is how large P must be for assortative interactions to rival kin selection as a mechanism for the evolution of altruism.

The answer is that P can be remarkably small (Figure 4.1). *When* $P = 2n$, *the altruism that evolves among genealogically unrelated individuals is comparable to the altruism that evolves among full siblings by kin selection. When* $P = 4n$, *genealogically unrelated individuals are almost as altruistic as genetically identical members of clonal groups.* These results may seem extraordinary to readers who think of kin selection as the one and only mechanism for the evolution of altruism, but they follow directly from viewing kin selection as just one of several sorting processes. Sampling two unrelated individuals at random and creating groups of size *n* from their progeny sorts altruists and nonaltruists into separate groups about as well as sampling 2*n* unrelated individuals at random, ranking them with respect to altruism, and splitting them into two groups at the midpoint.

Of course, our model may be unrealistic in other respects. Not all groups are formed by individual choice and it may be difficult to discover the altruistic tendencies of others, especially if individuals have evolved to conceal their selfish tendencies. These problems must be taken seriously, but it should be obvious that assortative interactions are potentially a powerful mechanism for the evolution of altruism. It certainly would be premature to conclude that human social groups cannot be altruistic or functionally organized on the grounds that their members do not have a high degree of genealogical relatedness.

Assortative interactions require that organisms have a certain amount of cognitive sophistication to discern the altruistic tendencies of others. However, there is evidence that even animals as simple as the guppy *(Poecilia reticulata)* can choose associates on the basis of prior interactions. The example of predator inspection that we described earlier is not hypothetical but is actually practiced by guppies and many other species of fish. Dugatkin and Alfieri (1991a) showed that individual guppies differ in their propensity to approach predators. Behavioral variation was continuous, satisfying one of the major assumptions of the model. In a separate experiment, Dugatkin and Alfieri (1991b) allowed three guppies to inspect predators in an aquarium divided into three lanes by transparent panels. The guppies were then placed in an apparatus that allowed the fish that occupied the center lane to indicate a preference for one of the two side fish by swimming over to join it as a companion. The side fish that went closer to the predator was consistently chosen as a future associate, and the preference remained stable over a period of at least four hours. It is possible that preference was not based on predator inspection *per se* but on other characters that correlate with predator inspection. This possibility was tested by repeating the experiment and allowing a fourth guppy, who did not observe the predator inspection, to choose among the two side fish. The more altruistic side fish was not preferred in this experiment, demonstrating that preference was based on the actual act of inspection. Guppies inspect each other at the same time that they inspect the predator.

These experiments indicate that guppies can perceive the altruistic behaviors of their neighbors and attempt to associate with the most desirable partners, satisfying most of the assumptions of our model. However, it has *not* yet been demonstrated that preference results in assortative interactions. In the Dugatkin and Alfieri (1991b) experiment, guppies in the center lane preferred the more altruistic side fish regardless of their own altruistic tendencies. As expected, everyone loves an altruist. An extra step is required to show that altruists can form their own groups and avoid the company of nonaltruists. This experiment can easily be performed by measuring the predator-inspection tendencies of unrelated individuals and allowing them to form groups in an environment that includes predators. If the groups are nonrandom with respect to predator-inspection behavior, then this result will provide strong evidence that above-random variation among groups can be created by assortative interactions, rather than genealogical relatedness, in an animal as cognitively simple as the guppy.

If guppies can accomplish nonrandom groupings with their meager cognitive abilities, then imagine what might be achieved by our own species! Humans have a fantastic ability to acquire information about others based on personal interaction, direct observations, and cultural transmission. This information is often used to seek out trustworthy individuals and avoid cheaters in social interactions. The ability to detect cheaters is likely to select for the ability of cheaters to avoid detection (Trivers 1971), but deception is unlikely to be completely successful (Frank 1988; Wilson, Near, and Miller 1996). Even if initial intentions can be disguised, behavioral tendencies can be reliably discerned from actual behaviors, whenever the history of past interactions is known. It is hard to fake inspecting a predator; a guppy either approaches or fails to approach, and its behavior can be observed by all. Similarly, it is hard to fake altruism or conceal selfishness in human social groups whose members have a long history of interaction. The large anonymous societies of today are recent inventions; human evolution took place in small groups whose members had extensive opportunities to observe and talk about each other. Information about a single antisocial act can quickly spread through a social network and spoil a person's reputation, with grave consequences for future social interactions (e.g., Boehm 1993; MacDonald 1994). Human social interactions among unrelated individuals are anything but random, and our ability to learn and to change our behavior according to what we learn provide a powerful mechanism for the evolution of altruism and other group-advantageous behaviors.

Rewards, Punishments, and the Amplification of Altruism

Imagine that you are an anthropologist studying a group of hunter-gatherers whose lives closely approximate the ancestral human condition. You know that meat is a resource that is critically important for fitness. You have also been schooled in the theory of individual selection, which makes you expect that hunters will attempt to keep the game that they catch for themselves and their kin. If they share with nonrelatives at all, it must be in a way that will be reciprocated by those individuals in the future.

What you discover is not at all what you expected. The sharing of meat is scrupulously fair. Even after you weigh with portable scales all the meat that is captured over a period of weeks, there is no statistically significant difference between the hunter's share and the shares of other members of the group. Neither is meat shared in a way that is likely to be reciprocated in the future. Some men are better hunters than others and spend more time at it, yet they share equally with those who provide less and are unlikely to provide more in the future.

Your explanatory framework has failed you completely. Then, to your relief, you discover that there are some individual benefits to hunting after all. It turns out that women think that good hunters are sexy and have more children with them, both in and out of marriage. Good hunters also enjoy a high status among men, which leads to additional reproductive benefits. Finally, individuals do not share meat the way that Mr. Rogers or Barney the Dinosaur would, out of the goodness of their hearts. Refusing to share is a serious breach of etiquette that provokes punishment. In this way, sharing merges with taking. These new discoveries make you feel better, because the apparently altruistic behavior of sharing meat that would have been difficult to explain now seems to fit comfortably within the framework of individual selection theory.

This scenario informally describes the current conceptualization of meat sharing in hunter-gatherer societies (e.g., Blurton Jones 1984, 1987; Hawkes 1993; Kaplan and Hill 1985a,b; Kaplan, Hill, and Hurtado 1984; reviewed by Wilson 1998). A critical resource, which has had a strong effect on fitness throughout human evolution, is usually shared among members of social groups without respect to genealogical relatedness or expectation of reciprocity. The most direct effects of hunting increase the fitness of the group and decrease the relative fitness of the hunter within the group, since the hunter expends time and energy and takes risks to advance a common good. If this were the end of the story, we would call the hunter an altruist. Yet, the hunter enjoys other rewards and avoids punishment in a way that makes us want to reclassify him as an egoist. Those who provide the rewards and punishment seem like egoists too, since they get meat for their efforts. We will accept these observations as facts for the purpose of our analysis. Our point is that the facts don't fit comfortably within the framework of individual selection theory. The system of rewards and punishments that cause meat to be shared may look and even be selfish in the *psychological* sense of the word, but our evolutionary definitions must be based on fitness effects. As evolutionists schooled in multilevel selection theory, we need to understand rewards and punishments in terms of selection within and between groups.

Let's call hunting and sharing the *primary* behavior and the rewards and punishments that others confer on hunters the *secondary* behavior (Ellickson 1991). By itself, the primary behavior increases the fitness of the group and decreases the relative fitness of the hunter within the group. If it evolved without the complications of rewards and punishments, there should be little controversy about calling it altruistic and saying that it evolved by group selection. That is why it seems so problematic from the standpoint of individual selection. The evolution of the secondary behavior (the rewards and punishments) can be analyzed in exactly the same way as the evolution of the primary behavior. By causing another individual to perform an altruistic primary behavior such as hunting and sharing, the secondary behavior indirectly increases the fitness of the entire group. At the same time, the secondary behavior is likely to require at least some time, energy, or risk for the individual who performs it. The secondary behavior therefore requires group selection to evolve, just as the altruistic primary behavior would if it evolved without the secondary behavior being present. Economists call this a second-order public goods problem: Any behavior that promotes a public good is itself a public good (Hawkes 1993; Heckathorn 1990, 1993).

Although the altruistic primary behavior (by itself) and the secondary behavior both require group selection to evolve, there is an important difference between them. The individual costs of the primary behavior are substantial. There is simply no way to hunt game without investing large amounts of time and effort and perhaps also taking risks. That is why the primary behavior intuitively seems altruistic. The individual cost of the secondary behavior might be substantial, but it might also be very small. It is an important fact of human life that individuals can often greatly increase (reward) or decrease (punish) the fitness of others at trivial costs to themselves. It is the low cost of imposing rewards and punishments and the fact that they elicit benefits that make secondary behaviors appear psychologically egoistic rather than altruistic. From the evolutionary standpoint, however, the fact that the cost is trivial does not alter the level at which the behavior evolves. Secondary behaviors evolve *more easily* by group selection than primary behaviors because they are less strongly opposed by within-group selection, but they still evolve by group selection. The package of primary and secondary behaviors therefore remains a group-level adaptation.

An example from Jewish society in thirteenth-century Spain will help to clarify these ideas. A law was then in place that a man who did not pay his taxes would have a blot placed upon his genealogy (MacDonald 1994). Reproduction was highly regulated in this society, and a marriage could not take place unless the genealogies of both the bride and the groom were consulted and approved. A blot on a person's genealogy could prevent a marriage. Thus, failure to pay taxes could have severe costs, not only for the individual but also for his descendants. Similar laws discouraged individuals from converting to other religions or even associating with members of other groups.

Once again, we have a primary behavior (giving money to the group) that by itself would be considered highly altruistic. The primary behavior is reinforced by a secondary behavior (the law) that makes alternatives to the primary behavior even more costly. Giving money to the group no longer appears altruistic, but we still must explain the law that makes it selfish. Is the *law* selfish? The purpose of the law and its enforcement is to promote the welfare of the group, which is the same function as the primary behavior. It is the *costs* of the primary and secondary behaviors that are different. It costs an individual a great deal to fork over a large fraction of his income. It costs the keeper of the genealogy almost nothing to make a blot.

Our argument about rewards and punishments can be summarized in a more general form as follows:

1. For a group to behave as an integrated unit (as an organism), many activities must be performed that would be altruistic if they occurred in the absence of associated rewards and punishments. The altruistic nature of these activities is imposed by the external environment. It is just a fact of life that resource acquisition, defense, aggression, and the like on behalf of the group require time, energy, and risk on the part of the individuals who perform the behaviors. We have termed these activities "primary behaviors."
2. For altruistic primary behaviors to evolve in the absence of associated rewards and punishments, a population structure is required that is highly favorable for group selection. Thus, these behaviors might be seen in groups of close relatives or groups whose members have been winnowed by assortative interactions, but not in randomly composed groups.
3. A variety of rewards and punishments can be used to make primary behaviors less costly or even selfish from the standpoint of within-group selection. We call these "secondary behaviors" and their evolution must be analyzed in the same way as any other behavior.
4. When a secondary behavior is used to promote a primary behavior that would otherwise be altruistic, the secondary behavior is also altruistic from the evolutionary perspective. In other words, the secondary behavior increases the fitness of the group by causing the primary behavior to be expressed and decreases relative fitness within the group, if there is any time, effort, or risk associated with performing the secondary behavior. Even if *no* individual costs were

involved, secondary behaviors would merely be neutral from the standpoint of within-group selection.

5. In comparison with altruistic primary behaviors, which by their nature are costly if performed without associated rewards and punishments, secondary behaviors can often be performed at a low cost to the actor. Therefore, secondary altruism can evolve in population structures that are less favorable for group selection than the ones required for primary altruism. At the extreme, secondary behaviors that involve *no* cost can evolve if there is *any* heritable variation among groups.

The use of secondary behaviors to promote altruistic primary behaviors can be called *the amplification of altruism*. The population structure of many human groups may not be sufficient for altruistic primary behaviors to evolve by themselves, but may be sufficient for primary and secondary behaviors to evolve as a package. Since the secondary behaviors cause the primary behaviors, behaviors that evolve in human groups can be similar to those that evolve in species with more extreme population structures, such as clonal organisms and social insect colonies.

An analogy with natural selection at the individual level might help clarify these ideas. In standard population genetics models, beneficial genes are assumed to increase the fitness of the individual without being selected against within the individual. Genetic variation among individuals is required for the gene to evolve, but the variation created by random mating is sufficient. Complex and elaborate adaptations therefore evolve at the individual level without the issue of altruism ever arising. Now consider a mutant gene that uniformly increases the fitness of everyone in a social group, including the individual who carries the gene. It would be selectively neutral within the group but would cause its group to outperform other groups whose members lack the gene. More than one group and genetic variation among groups would be required for the gene to evolve, but random variation would be sufficient. If many genes of this sort existed, we would expect complex and elaborate adaptations to evolve at the group level without the issue of altruism ever arising. This is not the way that group selection has been traditionally viewed, but why is that? Presumably, because it is difficult to imagine behaviors that benefit whole groups at no cost to the individuals (and their genes) that perform them. As we have seen, there is a measure of truth to this argument. The activities that allow groups to function as adaptive units require time, energy, and risk on the part of individuals. However, the imposition of rewards and punishments may constitute an important class of behaviors that *do* benefit whole groups at little or no cost to the individual. Furthermore, merely by the proper application of rewards and punishments, virtually *any* high-cost group-beneficial behavior can be promoted; under these circumstances, ultrasociality could evolve without extreme genetic variation among groups.

The analogy between the gene-individual relationship and the individual-group relationship can be taken a step further. As we described in Chapter 2, the traditional view of genes benefiting the individual at no expense to themselves has been challenged. An individual organism is increasingly being viewed as a diverse genetic and cellular community in which it is possible for some elements to succeed at the expense of others. The community can evolve as an adaptive unit only to the extent that the potential for evolution within the community can be suppressed. Thus, the rules of meiosis ensure that all autosomal genes are fairly represented in the gametes (Crow 1979). A single-cell stage in the life cycle and the sequestering of reproductive cells into a germ line minimizes the potential for within-individual selection (Buss 1987; Michod 1996a,b).

In short, even single individuals require rules and regulations to become the adaptive units that we call organisms. The entire language of social interactions among individuals in groups has been borrowed to describe genetic interactions within individuals; "outlaw" genes, "sheriff" genes, "parliaments" of genes, and so on (Alexander and Borgia 1978; Leigh 1977). If communities of genes and cells can evolve a system of rules that allow them to function as adaptive units, then why can't communities of individuals do the same? If they do, then *groups will be like individuals,* which is the proposition that we are seeking to establish.

Before we return to human groups, it will be interesting to revisit the social insects. As mentioned before, genetic uniformity within social insect colonies is sometimes not as extreme as originally thought because colonies are founded by more than one female or by a female that had mated with more than one male. If group selection depended entirely on genetic uniformity within groups and variation among them, we might expect these colonies to be less well adapted at the group level than colonies founded by a single female that had mated with a single male. In general, this prediction has *not* been confirmed. Honey bee queens usually mate with many males, yet the functional organization of bee colonies remains truly superorganismic (Seeley 1996). For example, a gene that causes honey bee workers to lay unfertilized eggs (which develop into males) would be favored by within-hive selection, even if it disrupted the well-being of the hive. A gene that causes workers to refrain from laying unfertilized eggs therefore counts as altruistic and requires substantial variation among groups to evolve. Despite an extensive search, honey bee workers have been observed to lay unfertilized eggs only rarely, in part because the egg-layers are attacked by other workers and their eggs are eaten (reviewed by Seeley 1996). Ratnieks (1988; Ratnieks and Visscher 1989) calls these behaviors "policing," borrowing yet another term from human social interactions. In addition to the primary behavior ("refrain from laying eggs"), we therefore have secondary behaviors to consider ("attack workers who lay eggs"). The population structure of honey bee colonies is sufficient for "refrain/attack" to evolve as a package, even if it is not sufficient for "refrain" to evolve by itself. To summarize, extreme genetic variation among groups may be insufficient to explain all instances of ultrasociality, even in the social insects. The amplification of altruism by rewards and punishments can help primary behaviors evolve in such circumstances.

Our analysis of rewards and punishment leaves many gaps. What are the actual costs involved in rewarding and punishing others? What are the cognitive skills required for individuals to use rewards and punishments to modify the behaviors of their social partners? Is cheating a problem for the low-cost altruism associated with secondary behaviors, as it is for high-cost altruism? These are important questions for the future, but our present goal is simply to bring the subject into proper focus. A proverb states that the hardest thing for fish to see is water. Rewards and punishments are part of the water of human experience, and they are often interpreted as broadly compatible with the concept of self-interest. It may turn out that they *are* compatible with the *psychological* concept of self-interest, a question that we will address in Part II. But this does not mean that they can be explained as the product of within-group selection in multilevel selection theory. Just as group selection is required to explain altruistic primary behaviors when they evolve by themselves, so also is group selection required to explain the system of secondary behaviors that causes primary behaviors to be expressed. Seeing rewards and punishments as products of group selection goes a long way toward explaining how human social groups can be organismic even though they do not have the same population structure as clonal organisms or social insect colonies.

Social Norms and Cultural Evolution

The first ingredient of natural selection at any level is phenotypic variation among the units. As we discussed in Chapter 3, most evolutionary models assume that phenotypic variation is directly related to genetic variation. The degree of altruism in the group is directly proportional to the frequency of altruistic genes in the group. The experiments of Wade (1976, 1977), Goodnight (1989; Goodnight and Stevens 1997), and others showed that these assumptions need to be questioned. Large phenotypic differences can be caused by small genetic differences, at both the group and individual levels.

In human social groups, the relationship between genes and phenotypes is even less direct. Human groups can behave very differently without being genetically different. Behavioral differences can also develop over extremely short time periods, as when a group makes a collective decision to change its behavior or a new religion is founded that quickly attracts a large following. Since phenotypic differences are the only thing that natural selection "sees," we cannot ignore the psychological and cultural processes that intervene between human behavior and human genes.

The second ingredient of natural selection at any level is heritable variation. It might seem that cultural differences between groups are not heritable by definition, but this interpretation of heritability is too narrow. Heritability means resemblance between parents and offspring. New individuals must resemble the old individuals from which they are derived, and new groups must resemble the old groups from which they are derived. No more is required for multilevel selection to operate, regardless of whether the differences are mediated by genetic or cultural processes. We can take many steps away from genetic determinism and still talk about the fundamental ingredients of natural selection: phenotypic variation, heritability, and fitness consequences.

Viewed from this perspective, human behavioral evolution appears to be more conducive to group selection than simple genetic models would suggest (Boehm 1997a,b; Boyd and Richerson 1985, 1990a). Human social groups are never genetically uniform, but they are often quite uniform behaviorally, especially when the behaviors are reinforced by social norms. A new behavior does not necessarily spend a long time at low frequency—it can quickly become the *only* behavior that is practiced by a social group. Heritability can be even greater for groups than for individuals when cultural processes are involved. A social group can maintain its distinctive behavioral characteristics even though the members of the group are constantly changing.

The rules of the human evolutionary game are clearly different from what simple genetic models would suggest, but we must be specific about the nature of these rules before we can make further progress. Cultural evolution *per se* is not more favorable than genetic evolution for group selection. The rule of cultural transmission—"Imitate the most successful individual in your group"—would be disastrous for group selection because it would be sensitive only to relative fitness within groups. If group selection is a strong force in human cultural evolution, a special set of transmission rules must have evolved that make it so.

To proceed with our analysis, we will make a specific claim about human cultural evolution: *In most human social groups, cultural transmission is guided by a set of norms that identify what counts as acceptable behavior.* People who violate the norms are subject to punishment or exclusion from the group. The specific behaviors that are sanctioned by social norms vary widely from group to group, but the existence of norms that are reinforced by punishment and/or exclusion is common to most human social groups.

We will evaluate the generality of our empirical claim in the next chapter (see also Boehm 1993, 1996, 1997a,b). Our present task is to analyze the effects of social norms on multilevel selection. It should be obvious that many social norms promote behavioral uniformity within groups, thereby reducing the potential for within-group selection. Social norms also promote phenotypic stability and heritability at the group level. Consider two imaginary cultures, the squibs and the squabs. The squibs follow the social norm "Be altruistic to fellow squibs, punish those who don't, and punish those who fail to punish." The squabs have the norm "Solve your own problems." They freely exploit fellow squabs, who may attempt to retaliate as individuals but without the backing of their tribe's moral outrage. The altruistic squibs will outperform the quarrelsome squabs in all situations that involve between-group processes, such as direct conflict, foraging for a common resource, founding new groups, and so on. The problem of cheaters and freeloaders within groups, which is so often used to argue against the evolution of altruism, is not a problem for the squibs because cheaters and freeloaders are severely punished. Of course, this forces us to examine the evolution of punishment. What does the punisher get from enforcing the norms? Well, failure to punish is itself punished. That is part of the squibs' creed. Then what does the person who punishes the person who fails to punish get from enforcing the norms? At the end of this inquiry is a behavior that benefits the group at some expense to the individual who performs the behavior. As we argued in the previous section, this is not a problem if the individual costs are sufficiently small. Thus, our present analysis of human social norms relies on our previous analysis of low-cost rewards and punishments.

Now suppose that there is some movement of individuals between groups. Every year, a fraction of squibs and squabs intermarry, and the new couples join one or the other of the two groups. Squibs who become squabs are mercilessly exploited and are reduced to nothing unless they change their ways. Squabs who become squibs are mercilessly punished or sent away unless they change their ways. People are good at changing their ways, so the squibs and squabs remain behaviorally uniform, despite the movement of individuals between groups (Eibl-Eibesfeldt 1982; Simon 1990).[2] The group differences are more stable than the individual differences.

If the world consisted only of squabs, altruism would be an impossibility. If the world consisted only of squibs, selfishness would be an impossibility. It seems that the question of altruism vs. selfishness has become arbitrary, a matter of which social norms happen to become established in the population. As we shall see below and in the next chapter, there is a measure of truth to this statement. However, in a world that includes both squibs and squabs, who compete with each other in the formation of new groups, the squibs will replace the squabs. We therefore note the following asymmetry: *Within-group selection can favor any behavior, depending on the social norm of the group. Between-group selection favors only social norms that lead to functionally adaptive groups.*

Our imaginary example of the squibs and squabs informally describes a theory of cultural group selection that has been developed by Boyd and Richerson (1985, 1990a,b, 1992). The key innovation of their theory is the concept of *group selection among multiple stable equilibria*. In standard group selection models, the traits that benefit the group are evolutionarily unstable (or at best neutral) within the group. One of the most important insights to emerge from evolutionary game theory is that social interactions can result in multiple stable equilibria. Because the fitness of each strategy depends on the other strategies expressed within the group, there are often majority effects; a strategy can persist in a group in which it is common but be unable to invade a group in which other strategies are common. The most familiar example is the interaction between all-defect and tit-for-tat in repeated prisoner's dilemma games; each trait can resist invasion by the other.

Multiple stable equilibria create a form of variation among groups that is not opposed by within-group selection. The behaviors that are favored by different equilibria may well differ in group fitness, but all of them are stable within groups by definition.[3] Even small differences in group fitness may therefore be sufficient to cause one behavior to replace another over a sufficiently long period of time.

Social norms and cultural evolution are not required for group selection among alternative stable equilibria, but they vastly increase its potential. Virtually *any* behavior can become stable within a social group if it is sufficiently buttressed by social norms (Boyd and Richerson 1992). The costs and benefits that are naturally associated with the behavior are simply overwhelmed by the rewards and punishments that are attached by the social norms. With a variety of groups deploying different social norms, group selection is free to sift among a vast number of alternative primary behaviors, each of which is internally stable within the group in which it is normatively sanctioned.

So far we have painted a picture in which human social groups vary in ways that are internally stable. The variation affects performance at the group level, which causes some traits to replace others by the process of group selection. Now we need to ask how variation among groups originates. One possibility is rational thought. Suppose that the squibs originated from an individual squab who awoke one morning with a brilliant idea. She would form a society in which everyone is nice, punishes those who aren't, and punishes those who fail to punish. She convinces some others of the wisdom of her idea and they travel to another valley to found their utopia. The rest is history, as the helpful squibs prosper and ultimately replace the quarrelsome squabs.

Rational thought may account for many phenotypic differences among groups. Among anthropologists, it has not been accorded the attention it deserves as an agent of cultural change (Boehm 1996). In the next chapter, examples will be given of new social norms that are created by highly rational and collective decision-making processes, which specify not only how everyone in the group should behave but also how the norms will be enforced. Nevertheless, it would be a mistake to conclude that *all* adaptive features of social norms are consciously and rationally designed, as Boehm (1996) is careful to stress. The human mind is too feeble to comprehend the consequences of behaviors in all their ramifications (Simon 1983), and irrational beliefs may arise and persist for a variety of reasons. Many behaviors have adaptive consequences at the group level without anyone knowing why or even if they are adaptive, as Adam Smith emphasized with his metaphor of the invisible hand. A mechanism of blind variation and selective retention (Campbell 1974) is required for these behaviors to be incorporated into social norms. To modify our previous example, suppose that the squab who arose one morning with a brilliant idea wasn't consciously thinking about how to design a utopian society. Actually, she had been smoking a powerful weed and a vision appeared that she couldn't explain. Nevertheless, she convinces others to have faith in her and off they go to the next valley. What happens next depends on how the new system of social norms functions, relative to the old system. Group-level functional design will decide the outcome, regardless of whether the social norms are rationally designed or irrationally inspired.

So far, we have imagined that variation among groups originates as a saltational event, starting with an individual who quickly attracts a following. Continuous variation in social norms among groups is also possible. Modifying our example yet again, suppose that the next valley is colonized not by a prophet and her flock but by a segment of the squab tribe who has decided to emigrate. The new group has every intention of following the old ways, but faulty memory alone will cause at least some variation in social norms, which can be selected at the group level if it is functionally significant. This kind of small-scale cultural variation has been amply documented by anthropologists (e.g., Barth 1989, Hudson 1980; Jorgenson 1967).

To summarize, variation among groups can arise by conscious design or by arbitrary processes. It can originate as a major saltational jump or in imperceptible steps. These are important distinctions, with major implications for the evolutionary process, but all of them provide the raw material of heritable phenotypic variation for natural selection to operate upon at the group level. Once again, an analogy with natural selection at the individual level might help to clarify these ideas. In genetic evolution, a new mutant can be radically different from the wild type (a macromutation) or only slightly different (a micromutation). Mutations are usually thought to occur at random, but processes that approximate Lamarckism are at least conceivable, in which mutations appear to anticipate environmental change (e.g., Steele 1979). Micromutations, macromutations, arbitrary mutations, and directional mutations have different implications for the evolutionary process, yet all of them can provide the raw materials for individual selection, resulting in individual organisms that are functionally designed to survive and reproduce in their environments. Similarly, all of the mechanisms for generating group-level variation that we have described above can result in social groups that are functionally designed to survive and reproduce in their environments, relative to other social groups. By this process, groups can evolve into adaptive units.

Group Selection and Cultural Diversity

Boyd and Richerson's theory of group selection goes a long way toward explaining how large groups of unrelated individuals can evolve into functionally organized units. In addition, their theory has implications for the nature of cultural diversity. In standard group selection models, within-group selection promotes selfish behaviors. In Boyd and Richerson's theory, within-group selection can promote virtually *any* set of primary behaviors, depending on the social norms. When these primary behaviors are examined in the absence of the social norms that promote them, they could appear to be altruistic, selfish, or just plain stupid. As the title that Boyd and Richerson used for their 1992 article puts it, "Punishment allows the evolution of cooperation (or anything else) in sizable groups." The costs and benefits of the primary behaviors simply become irrelevant when they are overwhelmed by the rewards and punishments associated with social norms.

Evolutionary biologists interested in human behavior occasionally emphasize the importance of social norms but generally assume that individuals can freely choose their own primary behaviors, whose costs and benefits are unmodified by secondary behaviors. This assumption greatly reduces the diversity of behaviors that can be favored within groups. The behaviors that do evolve can then be interpreted in narrow functionalist terms. Critics of adaptationism frequently emphasize the diversity of behaviors across cultures that seem to make no functional sense outside the context of cultural norms. The difference between these two positions is often perceived as so great that members of each side barely speak to each other. Boyd and Richerson's analysis of social norms may allow the positions to be reconciled.

The diversity of behaviors that can be favored by within-group selection is trimmed by between-group selection, leaving a subset that is adaptive at the group level. For a number of reasons, however, we should still expect to see behavioral diversity among human social groups that can be explained only within the context of the social norms. In the first place, group selection may be a strong force in cultural evolution, but it is not the only force. Return to the squab prophet who founded squib society after smoking a strong weed. Her vision of a new society may include elements of altruism that account for the squibs' success, but it also may include other elements that are meaningless or even dysfunctional for the group. Since all of the elements are favored within groups by the squib social norms, they can evolve only as a package that includes good and bad elements alike. This is similar to the concept of hitchhiking in population genetics theory, in which bad and neutral genes evolve by being associated with good genes. Cultural hitchhiking is especially likely when individuals are unaware of what counts as adaptive behavior (or don't care about this consideration to the exclusion of others) and merely follow tradition.

In the second place, behaviors that are adaptive at the group level can be produced by more than one set of norms. Once again, an analogy with evolution at the individual level will help clarify this idea. Imagine an artificial selection experiment for wing length that is performed on ten isolated populations of fruit flies. Every generation, only those individuals with the longest wings are used as parents for the next generation. After many generations, long wings have evolved in all ten populations. Now look at the proximate genetic and developmental mechanisms that have evolved in each population to make long wings. These need not be the same across populations. The process of artificial selection "sees" only long wings, not the proximate mechanisms that make long wings. We therefore expect the populations to be similar at the phenotypic level, but not necessarily at the level of the proximate mechanisms that create the phenotype. This kind of experiment has actually been performed in the laboratory (Cohan 1984), but it has been performed on an even grander scale in natural populations, where it is known as "convergent evolution." The natural world is full of species that have evolved to do the same things in different ways.

By the same token, even if group selection has caused human social groups to evolve into adaptive units, the cultural traits that promote group welfare may be highly diverse. Prosocial behaviors might be accomplished by watchfulness and public shaming in some cultures and internalization of values at an early age in others. Group selection "sees" only the behavioral product, not the process that creates the product. In the next chapter we will present evidence suggesting that the mechanisms producing prosocial behaviors in human groups are more diverse than the behaviors themselves.

Consolidating Our Gains

We have covered a lot of ground in four chapters and it will be useful to review our progress before we continue. We began with Darwin's original conjecture that human moral virtues afford "little or no advantage" within groups but can evolve by natural selection at the group level. We then introduced the standard model of altruism that captures Darwin's idea in a mathematical form. We showed that the fundamental ingredients of group selection are the same as the fundamental ingredients of individual selection: a population of units, which vary in a heritable fashion, with corresponding variation in fitness.

We also showed that when altruism evolves by group selection, it can always be made to appear as a form of selfishness by averaging the fitnesses of individuals across groups. We called this the "averaging fallacy" because it vacuously defines selfishness as "anything that evolves." No one has tried to defend the averaging fallacy in its general form, but the error has been committed in many specific cases and has led to the erroneous conclusion that group selection is not involved in the evolution of the traits under study. The entire history of evolutionary biology during the last three decades, in which group selection was rejected while other frameworks became the foundation for the study of social behavior, would have been different if the averaging fallacy had been avoided. Ironically, Hamilton (1975) was among the first to see the proper relationships among the theories, but even he could not alter the Dr. Jekyll and Mr. Hyde conception of group selection and its so-called alternatives.

In Chapter 3 we discussed how to think about adaptation from the standpoint of multilevel selection theory. We suggested a procedure in which groups are first imagined as adaptive units, functionally designed to maximize their fitness relative to other groups. As a second step, individuals are imagined as adaptive units that maximize their fitness relative to other individuals in the same group. These two steps are relatively simple because they involve thinking in terms of functional design, which is highly intuitive. To the extent that a trait has evolved by natural selection, steps 1 and 2 bracket the possibilities, with real populations lying somewhere in between. The third step involves examining the fundamental ingredients of natural selection at each level to see where the population is likely to lie between the two extremes of pure between-group and pure within-group selection.

This three-step procedure makes it clear that group selection can be a significant evolutionary force in ways that do not always require groups of genetic relatives. For all its insights, kin selection theory has led to the constricted view that genealogical relatedness is the one and only mechanism for the evolution of altruism; because of this theory's widespread acceptance, altruism has eclipsed adaptation as the central question of sociobiology. Multilevel selection theory expands the view by focusing on *adaptation* as the central question and examining the fundamental ingredients of natural selection—phenotypic variation, heritability, fitness consequences—that are required for adaptations to evolve at all levels of the biological hierarchy. Altruism can be understood only in the context of this broader framework.

This framework is especially relevant to the evolution of human behavior. Human social groups *seem* highly organized and have been interpreted as superorganisms for centuries. Many evolutionary biologists have rejected this interpretation because human groups do not have the same genetic structure as bee hives and coral colonies. Multilevel selection theory casts doubt on this objection and demands that human groups be evaluated in terms of the fundamental ingredients of natural selection. By these criteria, it is plausible that group selection has been a very strong force throughout human evolution. The mechanisms that substitute for genealogical relatedness probably operate in many species, but they do so especially in human populations, because they require sophisticated cognitive abilities and (in some cases) the cultural transmission of behavior. Thus, multilevel selection theory has the potential to explain not only why humans are ultrasocial, but why they have experienced a unique variety of group selection.

➤ 5 ≺

Human Groups as Adaptive Units

We have portrayed human groups as potentially adaptive units that are comparable to bee hives, coral colonies, and single individuals in their functional organization. This picture is in harmony with some intellectual traditions, since the superorganism metaphor has been used to describe human societies for centuries. The same picture clashes with other intellectual traditions, including methodological individualism in the human sciences and the individualistic perspective in evolutionary biology.

So far, our picture has been based on theoretical models, which show how even large groups of unrelated individuals can evolve into adaptive units. In this chapter we will examine how well the models describe real human societies. One way to evaluate the models would be to pick a specific aspect of human behavior and follow the three-step procedure that we outlined in Chapter 3. For example, decision making is a cognitive process that is centrally related to fitness. Making a decision involves imagining a number of possible solutions to a problem, gathering relevant information, and evaluating the alternatives. All of these steps might benefit from the combined action, division of labor, parallel processing, and other advantages of groups interacting in a coordinated fashion. It is conceivable that human groups have biologically and culturally evolved to function as integrated decision-making units (step 1 of the procedure outlined in

Chapter 3). It is also possible that individuals function as self-contained decision makers who attempt to maximize their relative fitness within groups (step 2). Wilson (1997) reviewed the psychological literature on decision making to evaluate where human groups are located between these two extreme possibilities. The vast scientific literature on human behavior allows at least a preliminary evaluation of many other subjects along the same lines.

In this chapter we take a somewhat different approach by conducting a survey of human cultures around the world. Our goal is not to concentrate on a specific aspect of human behavior, but rather to evaluate the major factors identified by the models that would make group selection a significant force in human biological and cultural evolution. These include social norms that can be enforced at low cost, the formation of groups by assortative interactions, and between-group replacement processes. Our analysis will be based on a survey of twenty-five cultures, randomly selected from several hundred described in the Human Relations Area Files (HRAF), an anthropological data base designed to facilitate cross-cultural comparisons (Murdock 1967). We will then elaborate some of our most important points by reviewing more detailed studies of selected cultures.

A Survey of Randomly Selected Cultures

Human cultures are sometimes claimed to be so diverse that attempts at generalization must inevitably fail. This assessment is too pessimistic, but it is true that a minefield of biases must be avoided. The most common bias is to pick and choose among cultures to find those that illustrate one's own point of view. Just as any argument can be supported by the selective use of statistics, virtually any statement about human nature can be supported by the selective use of anecdotes from the anthropological literature. The way to avoid this problem is to study a *random* sample of world cultures. To facilitate this enterprise, anthropologists developed a data base known as the Human Relations Area File (HRAF). The HRAF consists of ethnographies of hundreds of cultures from around the world that have been read, coded, and indexed according to a large number of categories.[1]

One of the categories is "norms" (code #183), which is defined as follows:

> Native and scientific definitions of custom (e.g., as ideal patterns, as ranges of variation within limits, as statistical inductions from observed behavior); positive and negative norms (e.g., folkways, taboos); verbalized and covert norms; investment of norms with affect and symbolic value (e.g., mores, idealization); discrepancies between ideals and behaviors; configurations of norms (e.g., culture complexes, institutions).

To see what has been written on the norms of any culture in the HRAF, merely look up code #183 for that culture. What you find are microfilmed pages of the original ethnographies with the number 183 handwritten in the margins, next to the passages that are relevant to norms. Code numbers for hundreds of other subjects are also written in the margins, for many thousands of pages of ethnographies; compiling the HRAF was a truly herculean task, especially before the age of computers.

For our survey, we first selected twenty-five cultures at random from the 354 cultures in the version of the HRAF available to us.[2] We then collected a minimum of five and a maximum of ten pages marked with code #183 for each culture. Cultures with fewer than five pages were discarded and replaced with another randomly chosen culture. For cultures with more than ten pages, the first ten pages were collected without regard to specific content. The cultures that are included in our random sample are shown in Table 5.1. They range from hunter-gatherer groups that approximate the ancestral human condition,[3] to highly stratified societies such as the Indian caste system and the Native Americans of the Pacific Northwest. The tremendous cultural diversity that is the hallmark of our species is amply represented in our sample, and any general conclusions that emerge from our survey should apply to most of the other cultures in the HRAF data base. We cannot be accused of picking and choosing our cultures to support the points that we are trying to make.[4]

Even with a random sample of cultures, it is possible to pick and choose among specific passages to support a given point of view.

Table 5.1 A sample of twenty-five cultures randomly selected from the Human Relation Area Files. Code refers to the HRAF identification number for each culture.

Culture	*Code*	*Region*	*Subsistence*
Senoi	AN6	Asia	Hunting, gathering, swidden agriculture
Afghanistan	AU1		Pastoralist, agriculture
Tamil	AW16		Agriculture
Bhil	AW25		Agriculture, formerly hunting, gathering
Mbuti	FO4	Africa	Hunting, gathering
Kpelle	FD6		Agriculture
Tallensi	FE11		Agriculture
Somali	MO4	Middle East	Nomadic pastoralism
Amhara	MP5		Agriculture, animal husbandry
Fellahin	MR13		Agriculture
Nootka	NE11	North America	Fishing
Amish	NM6		Agriculture
Navaho	NT13		Agriculture, raiding
Mescalero	NT25		Hunting, gathering, raiding
Papago	NU28		Desert agriculture
Manus	OM6	Oceania	Fishing
New Ireland	OM10		Agriculture
Truk	OR19		Fishing, agriculture
Yap	OR22		Fishing, coconut raising
Samoa	OU8		Fishing, agriculture
Gilyak	RX2	Russia	Fishing, sea mammal hunting
Chukchee	RY2		Hunting, reindeer herding
Toba	SI12	South America	Hunting, gathering, agriculture
Cuna	SB5		Fishing
Paez	SC15		Agriculture

Methods have been developed to avoid this bias and also to turn verbal accounts of human behavior into a numerical form that can be analyzed statistically. For example, a method known as the Q-sort consists of sorting a stack of cards with statements printed on them (Block, 1978; Tetlock, Peterson, McGuire, Chang, and Feld 1992). A person who has read a verbal account sorts the cards into a number of categories, ranging from highly appropriate to highly inappropriate, depending on how well the statement on each card fits the verbal account. Thus, a statement such as "individual behavior is tightly regulated by social norms" might be ranked as appropriate or inappropriate, on the basis of the entire account rather than a specific passage.

Quantitative methods such as the Q-sort do not eliminate the bias of the reader. Different people who read the same material may still sort the cards in different ways. In addition, the original accounts were written by anthropologists who were not perfectly objective and sometimes had strong ideological biases. A single culture might be described very differently by a Marxist, a libertarian, a liberal, or a feminist. These biases can be partially overcome by having multiple readers perform the Q-sort on multiple accounts of the same culture. An ambitious survey would confine itself to cultures that have been studied by at least three different anthropologists. Each ethnography would be read by three separate readers, who are kept ignorant about the purposes of the survey. This design would allow genuine differences between cultures to be distinguished from differences between ethnographers of the same culture and differences between readers.[5]

Even with all of these precautions, additional biases can creep in that lead to false conclusions. For example, we have been told that most early ethnographies (which contribute disproportionately to the HRAF) overemphasize the importance of social norms in tribal societies. Further research on the same societies often reveals more flexible and individualistic aspects of behavior. It may be true that anthropologists have changed the way that they interpret other cultures over the decades, which could be verified by comparing Q-sorts of early and more recent ethnographies for a sample of cultures. However, it is important to avoid the assumption that knowledge always advances and that modern ethnographers are invariably more enlightened than their predecessors. Recent anthropological studies that have been inspired by evolutionary theory tend to portray people in all cultures as free agents who shrewdly calculate their options to maximize their inclusive fitness (e.g., Betzig, Borgerhoff Mulder, and Turke 1988; Betzig 1997). This perspective did not emerge from ever-better ethnographies that are converging on the truth, but from a conceptual framework that was imported from evolutionary biology and acts as a lens through which information is viewed. We should be wary of biases in ourselves, as well as of biases in our predecessors. As we shall see, many of the ethnographers in our survey appreciate both the importance of social norms and the latitude for improvisation within those norms.

We are in the process of developing a Q-sort that addresses the major issues raised by the theoretical models, allowing us to conduct the ambitious survey that we have outlined above. In this chapter, however, we will describe the results of a less formal survey, citing passages that illustrate the general trends that seem to emerge. Thus, we have taken the important first step of choosing a random sample of cultures, but we have *not* yet taken the subsequent steps to avoid other biases, including selective quoting of the survey material. Of course, we have not intentionally distorted the results of our survey, but the whole purpose of scientific methodology is to eliminate unintended biases. Our conclusions therefore should be regarded as preliminary until the more ambitious survey is completed.

Now that we have discussed our own biases, we also need to consider possible biases in the reader. We stated in the Introduction that our readers are likely to come from three broad intellectual traditions: individual-level functionalism, group-level functionalism, and anti-functionalism. Each reader is likely to find some of our conclusions plausible—even self-evident—while remaining skeptical about others. Unfortunately, readers from the three traditions will probably disagree on which conclusions are plausible! Individual-level functionalism has been the dominant tradition in evolutionary biology and the human sciences for the last three decades. Our primary goal is therefore to correct its bias by showing that human groups can and often do function as adaptive units. For group-level functionalists, however, we need to show that human groups do not *invariably* function as adaptive units. As the most facultative species on earth, people are able to increase their relative fitness within groups *and* to merge themselves into a group-level organism. The outcome depends on the population structure that people naturally encounter or build for themselves. For anti-functionalists, we need to say that they are partially right. Many human behaviors cannot be explained in narrow functionalist terms, outside the context of the cultural system. Admitting this is not a rejection of evolutionary theory, however; we stand by the claim that human behavior can be studied productively within the Darwinian framework. Multilevel selection theory includes elements of all three intellectual traditions.

With these caveats in mind, we now can ask how verbal accounts of human cultures from around the world fit the assumptions and predictions of the theoretical models.

The Importance of Social Norms

Social norms can favor the evolution of group-level adaptations by altering the costs and benefits of altruistic primary traits and by creating a diversity of internally stable social systems that can be sifted by group selection. Social norms exist in all cultures, but their strength and the degree to which they regulate behavior within groups is more open to question. The current evolutionary view of human behavior tends to portray individuals as free agents who can employ any strategy they want to maximize their inclusive fitness. This view does not deny the existence of social norms, but it does accord them a minor role in the regulation of behavior.

In contrast, our survey suggests that human behavior is very tightly regulated by social norms in most cultures around the world. Passages such as the following could be provided for most of the cultures in our sample: "Any infringement of the socially accepted way or value of life may be a crime . . . , however small it may be . . . They have a large body of civil laws, a system of rights and obligations in all spheres of life, economic, social and religious, which are fulfilled very scrupulously" (in reference to Bhil culture; Naik 1956, p. 223); "All more or less important acts of social life, even including sacrificing one's life in a battle of vengeance for one's clansman, are categorical imperatives of a religious world-outlook which neither allow hesitation or require compulsion" (Gilyak; Shternberg 1933, p. 116); "Even the most insignificant and routine action in the daily life of the family is potentially of major concern to the band as a whole . . . It is important that there should be a pattern of behavior that is generally accepted, and which covers every conceivable activity" (Mbuti; Turnbull 1965, p. 118).

Some observers of the Papago people of the American Southwest summed up the general dynamic of social control in small tribal societies:

> Public opinion censures those who deviate from accepted standards of conduct. In any community so small that all the members and

> their affairs are known to one another, this is a powerful force. Among the Papago there are almost no conventionalized variations from the normal behavior pattern; the guiding principles apply to everyone. No man can ignore them because of his wealth, position as village leader, or special skill . . . The facts that failure to perform in a socially approved manner will inevitably become common knowledge and that general censure is sure to follow constitute a powerful deterrent to breaking the mores. (Joseph et al. 1949, p. 166)

We strongly suspect that these statements will be regarded by some readers as self-evident and by other readers as the folly of anthropologists who didn't understand the true nature of the cultures they were studying. It is a massive folly if the statements are incorrect, since they exist for the vast majority of cultures in our random sample, and therefore (in all probability) for the majority of cultures in the entire HRAF data base. If the statements are roughly correct, then we can conclude that human behavior is strongly regulated by social norms in most cultures around the world and presumably throughout human history. This conclusion may seem obvious to some people on the basis of common experience, but it is more profound for those who are trying to explain human behavior in formal scientific terms. In particular, it provides evidence for a mechanism that substitutes for genealogical relatedness, making group selection a strong force in human evolution.

The Cost of Enforcing Social Norms

In Chapter 4 we claimed that individuals can often greatly increase or decrease the fitness of others at trivial cost to themselves. Low-cost rewards and punishments are essential for the maintenance of social norms and the amplification of altruism in groups of unrelated individuals. Our survey provides many examples of how even the strongest biological urges can be overridden by low-cost rewards and punishments. One ethnographer who spent many years among the Gilyak of eastern Siberia recorded only three clear violations of prohibitive norms (Shternberg 1933, p. 184). The first involved an old man who purchased a new wife the same age as his own son. When the father died, his son lived for a time (and presumably had sexual relations) with his stepmother. The son's attraction to a woman his own age may appear natural to us, but it evidently shocked the entire community, who "spoke of him as some kind of a monster." The second incident involved a man who married a woman from the clan that took wives from his own clan. This violated the marriage norms of the culture, even though it would not constitute incest in our terms. The third incident involved an older brother who married the widow of his younger brother. The normal procedure was to determine the remarriage of widows by a gathering of clansmen. According to the rules of Gilyak society, older brothers had "no right whatsoever" to the widows of their younger brothers.[6]

All three of these incidents involved reproduction, the stuff of biological fitness. The offenders might appear to merely be doing what comes naturally by competing for mates that were perfectly suitable from the biological standpoint. The desire to mate is one of the strongest biological urges, and competition among men for women is one of the most disruptive aspects of social life (Daly and Wilson 1988). How can such powerful urges be inhibited by social norms, and what are the costs for those who impose the norms? According to Shternberg (1933, p. 184), "In the first two cases the violators had to go into voluntary exile, i.e. settle outside the settlement and lead a lonely existence, with all the deprivations of clan blessings associated with ostracism."[7] The marvelous phrase "had to go into voluntary exile" suggests the potentially low costs of imposing rewards and punishments that we emphasized in Chapter 4. Enforcers of the social norm do not have to fight the deviant or otherwise spend much time, energy, or risk. They merely have to *decide* that the deviant must go into exile and the deed is done. The balance of power so obviously favors the group over any particular individual that an actual contest does not take place.

These incidents are also remarkable for their rarity. In fact, two of the three offenders are described as "Russified Gilyaks" who took a supercilious view of the customs of their own people. Shternberg describes them as "exceptional cases bound up with psyches changed under the influence of an alien culture." According to Shternberg, Gilyaks normally react to the idea of marriage between forbidden categories with the same kind of visceral disgust that many people in our own society reserve for incest or homosexuality. The norms are so internalized that they do not require enforcement.

Another example of low-cost enforcement comes from a Melanesian island society, the Lesu (Powdermaker 1933 p. 323). One man's pig had broken into another man's garden and eaten his crop of taro. The offended person was not angry but the transgressor was annoyed by all the talk the incident occasioned, which was damaging his prestige. The transgressor therefore tried to give a pig to the offended person to "stop the talk." The offended person replied that this would be foolish and stopped the talk by declaring that the incident should be forgotten. This is one of many examples in our survey in which gossip appears to function as a powerful form of social control. What is the cost of enforcing norms in this example? The person who was wronged did nothing except forgive the transgressor and refuse payment. The people who gossiped would otherwise have talked about something else. The prestige that the transgressor was so anxious to recover is like a magical substance that can be given and taken away at will.

Social Norms Are Not All-Powerful

Despite the powerful social norms that exist in most cultures, our survey makes it clear that individuals do attempt to violate norms for their own advantage, surreptitiously when they are in a position to be punished and openly when they are not. Strong social norms would be unnecessary if there were not equally strong impulses to break them. We are not claiming that human groups always function as adaptive units, and the results of our survey would not support such a claim. Many of the anthropologists whose ethnographies comprise the HRAF were clearly not starry-eyed romantics; they were quick to notice self-serving behaviors that undermine the welfare of the group. Passages such as the following are reasonably frequent in our survey: "Despite the kindly interest of the supernaturals, the beneficial influence of ceremony and tradition, and the virtuous professions of most individuals, the Apache is faced with a world in which sorcery, deceit, ingratitude, and misconduct are not uncommon occurrences" (Hoijer and Opler 1938, p. 215); "The talking chiefs whose duty it is to act as custodians of the political arrangements, are open to bribery and manipulation. The holders of smaller titles who wish to advance their position are ready and willing to bribe" (Samoa; Mead 1930, p. 21); "The natives' keen sense of the importance of social obligations hides a tendency to aggressive self-assertion . . . Firstly, they say, unscrupulous people never hesitate to break a customary rule if it suits their purpose" (Tallensi; Fortes 1945, p. 9); "Superior access to food arises through rupturing the norms of the system, through uninhibited demands and through the exercise of emotional coercion, by screaming" (Toba; Henry 1951, p. 218).

The eternal conflict between sanctioned and unsanctioned behavior is well described by an observer of the Kpelle in central Liberia:

> There are two broad avenues to these goals, hard work and efficient management of one's resources and a reputation for fairness in dealing with others, or one can use sorcery, witchcraft, thievery and exploit one's advantages over others. The latter course is expedient and dangerous, but most ambitious individuals use a combination of sanctioned and unsanctioned means to achieve power. Finally, overambition is checked through various forces of retribution. There are the moot and town-chief's courts which penalize the overly-ambitious through fines and public ridicule. There is the power of the secret societies and Zos ("medicine-men") to exact swift and often deadly punishment. Lastly, there is the force of public opinion vented through gossip and other forms of harassment to keep an individual in check. (Lancy 1975, p. 29)

The "two broad avenues" discussed in this passage correspond to behaviors that benefit (or do not disrupt) the entire group and behaviors that profit the individual at the expense of others in the same group. The language of the ethnographer naturally fits the important categories of multilevel selection theory.

Social Norms and Cultural Diversity

Now that we have established the importance of social norms, we turn to the primary behaviors they sanction. To a large extent, the content of a norm depends on the culture that is being examined. Among the Fellahin of Egypt, "members of each sex dress more or less alike, and the observing of this convention is a very important factor in social equality. Their maxim in this connection is 'Eat what you yourself like, but dress according to what others like'" (Ammar 1954, p. 40). Yet in the Pacific island of Samoa, there is "no restriction on new ideas, and anything fresh that is likely to look well is eagerly adopted" (Grattan 1948, p. 117). For the Amish of North America, "the marriage norm is not love, but respect" (Hostetler 1980, p. 156). Yet for the Paez, of highland Colombia, "a wife will laboriously make the multicolored *keutand yahas* which she will give as tokens of love to her husband" (Bernal Villa 1953, p. 188). The Nootkan personality is described as "nonaggressive, rather amiable, disliking and disapproving violence in conflict situations" (Drucker 1951, p. 456). In contrast, an Amharan male "considers a pitiless regime of fasting to be the only way he can keep his hostile impulses subdued, and believes that children must not be spared the rod lest they be rude and aggressive" (Levine 1965, p. 85).

Some social norms appear downright dysfunctional at all levels, as the following example from Ethiopia attests:

> A number of other, more specific, norms have had the effect of discouraging inventiveness among the Amhara. Experimentation with matter was inhibited by the disdain for puttering about with one's hands—doing anything, that is, similar to the activities of the socially dejected artisans and slaves. Hence the peasant retains the same rudimentary tools for wresting a subsistence from nature that he has used for millennia, and searches about the woods for a properly shaped piece of wood rather than improve his art of carpentry. (Levine 1965, p. 87)

This example contrasts with many other cultures in our sample, in which crafts are highly developed and innovation is encouraged within bounds.

Behavioral diversity, the hallmark of our species, is amply represented in our survey. The fact of diversity is obvious but its causes are not. Some of it can probably be explained in functional terms, but functionalism seems inadequate to explain all of it. Is it really functional for the Nootkans but not the Amhara to have a sense of humor, for the Paez but not the Amish to express affection in marriage, for the Samoans but not the Fellahin to have a sense of fashion? Those who are wary of evolutionary approaches to human behavior often claim that functionalism (including evolutionary functionalism) is inadequate to explain all details of human behavioral diversity. We agree, but the problem is to provide other explanations that go beyond vague appeals to "culture." We suggest that low-cost secondary behaviors play a crucial role in the creation and maintenance of diversity in primary behaviors, which are nonfunctional outside the context of the cultural system. In other words, if we could eliminate the ability to reward and punish primary behaviors at low cost but retain all other aspects of human cultural processes, we might witness a dramatic collapse of behavioral diversity. Low-cost secondary behaviors may therefore provide a better explanation of nonfunctional behaviors than vague appeals to cultural processes.

Two kinds of explanation are required for a full understanding of the evolution of nonfunctional behaviors. First, we need to know how a behavior that has no function can evolve in a population. Once we understand this process, we need to know why some functionless behaviors have evolved rather than others. Following Boyd and Richerson (1985, 1990a,b, 1992), we have provided the first kind of explanation by showing how powerful secondary behaviors can promote functionless primary behaviors. If humans are more apt to impose and respond to secondary behaviors than other species, we can predict that human primary behaviors should defy functional explanation more often than primary behaviors in other species. This is an important prediction that supports the anti-functionalist's claim. However, our ability to furnish the second kind of explanation is more limited. We can say that functionless behavior should be more common in humans than other species, but we cannot explain why a particular functionless behavior has evolved in a particular culture. That kind of understanding probably requires detailed historical knowledge of the culture, and it may turn out that some behaviors evolved mainly by chance.

Social Norms and Group-Level Functional Organization

Despite the cultural diversity that we have emphasized above, which does not necessarily require a functional explanation, there is one sense in which the twenty-five cultures in our sample do not seem to vary. In every case, many of the social norms appear designed to forge groups of individuals into well-functioning units. This conclusion emerges so strongly from the ethnographies and seems so embedded in the minds of the people themselves that a functional interpretation appears warranted. In culture after culture, individuals are expected to avoid conflict and practice benevolence and generosity toward all members of a socially defined group, as the following quotes attest:

> The basic axiom of everyone is that all his existence and welfare are completely in the hands of the gods, in particular the clan gods, who shower their beneficence not on one but on all. Any attempt to monopolize the gifts of the gods must inevitably incur just punishment by the common benefactors of the clan. (Gilyak; Shternberg 1933, pp. 115–116)

> Loyalty is proclaimed all around. Conviviality is the norm among peers, though the authoritarian character of the Amhara family inhibits camaraderie between those of greatly differing ages. Unlimited succor in time of sickness and death and profuse commensality in happier hours are important values to the Amhara peasant and ones about which he is self-conscious and articulate. (Levine 1965, p. 83)

> The extended family was not a self-sufficient economic unit, even though this was possible in theory. The family group ordinarily formed part of a larger encampment, the members of which are linked together by ties of kinship, friendship, and propinquity. Within the camp, the norms requiring the sharing of food were so pronounced that the entire community could be considered a single production-distribution-consumption unit. (Apache; Basehart 1974 p. 139)

> The actions of the members of the society were oriented to a normative order, to accepted values and beliefs, and to the correctness of certain sanctions for inducing conformity . . . There was agreement on certain patterns of cooperation within groups, structured with great flexibility around kinship and affinal ties, for economic subsistence, co-residence, sexual satisfaction, and the raising of children. There was agreement on . . . the value of acquiring property, of hard work, of reciprocity, of generosity. Disputes should be settled

> through compromise and arbitration. Force should be used only against witches and aliens. Conformity should be secured through respect, praise, cooperation. Deviance should be punished through disrespect, ridicule and withdrawal of cooperation. (Navaho; Shepardson 1963 p. 48)

> A primary function of an economic system like that of the Pilaga is to convert the product of the individual into a social form. The individual's catch of fish or load of forest fruit must be given a social meaning, and must be transmuted from private into public property. (Toba; Henry 1951, p. 218)

Similar passages could be cited for the other cultures in our random sample, and therefore most of the cultures in the HRAF. Thus, we can conclude in general that *social norms function largely (although not entirely) to make human groups function as adaptive units, even when their members are not closely related.* Some readers may find this statement self-evident, but its obviousness does not diminish its significance within the framework of multilevel selection theory.

Evidence for Between-Group Replacement Processes

Boyd and Richerson (1990b, 1992) have shown that social norms do not automatically promote behaviors that benefit the group. If social norms consistently have this property, they must have been winnowed by a between-group selection process.[8] Group selection might be an ongoing process, or it might have occurred in the past and resulted in psychological and cultural mechanisms that anticipate the outcome of the winnowing process (see the section on the evolution of facultative behaviors in Chapter 3). A surprising number of ethnographers in our survey not only describe the social norms of a culture, but also suggest that the norms are maintained by an ongoing process of between-group selection. In other words, it is obvious to the ethnographers and often to the people themselves that social norms function to keep the group together and that if the norms fail, the group will dissolve and be replaced by other groups with a more robust social structure.

The specific form of between-group selection appears to vary from culture to culture. Competition among lineages is relentless in feuding societies and has favored social structures that promote solidarity within lineages. The replacement of "weak lineages" by "strong lineages," either by warfare or sheer economic superiority, was commonly noted by the ethnographers in our survey. For example, on the Micronesian island of Truk

> there are also cases where a strong lineage took possession of a desirable plot held by a weak lineage which could not contest the action . . . In each of these cases the seizing party acquired full title through failure of the legal residual title holders to assert their rights. The Trukese seem to feel that acquisition of title by conquest is a legitimate form of transfer. Cases of seizure from weak groups by strong ones when there is otherwise no quarrel between them, however, do not meet with their approval. (Goodenough 1951, p. 52)

This passage reveals that interactions *among* lineages are not entirely without social constraints, but are governed more by "might is right" considerations than are interactions *within* lineages. Lineages compete against each other as corporate units, much as we expect individuals to compete when they are allowed to act as free agents. Individuals are not free agents, however, but are so regulated by social norms within each lineage that they often act more like organs than organisms.

It should be obvious from this example that multilevel selection theory does not lead to the fulfillment of a romantic vision of universal niceness. Conflict and competition are not eliminated but merely elevated in the biological hierarchy, where the problem of social dilemmas appears all over again at an even grander (and potentially more destructive) scale. Human population structures often have many tiers, with groups forming into metagroups that compete as corporate units. Space does not permit us to discuss multitiered hierarchies in detail, but suffice it to say that the fundamental ingredients of natural selection must be examined at each level.

Another kind of between-group replacement process occurs when individuals can freely change their membership in a group. If individuals or families do not like the group they are in, they often can simply leave to join another group or form their own, along the lines of the assortative interaction model that we presented in Chapter 4. In this fashion, individuals who attempt to exploit others within their group are simply deserted. This kind of replacement process is described in societies as seemingly different as the Mbuti of Africa (Turnbull 1965) and the Nootka of the Pacific Northwest (Drucker 1951). For the Mbuti, a legend tells of an elephant hunter who "loses his normalcy" by failing to accept food offered to him and failing to join in the conversation. He is promptly deserted as unsafe company. In real Mbuti life, quarrels and other ruptures of social organization often result in group fissioning. The Nootka were a hierarchically organized society in which the chiefs seemed to have great power, including "absolute ownership of all important economic and ceremonial rights." However, their power depended on the allegiance of their tenants, who were as free to abandon their chief as the Mbuti were to abandon the elephant hunter.

The possibility that existing social norms have resulted from a process of groups replacing other groups was even stated explicitly by one ethnographer in our survey: "A characteristic feature in the social life of the Chaco [Toba] Indians is the strong feeling of solidarity which unites all members of the community. This solidarity appears to be the natural outcome of the Darwinian law, that in a hard struggle for existence such societies have the best chance to survive in which the sympathetic feelings are most strongly developed" (Karsten 1923, p. 29).

These examples do not prove that social norms have evolved by between-group selection. Nevertheless, it is remarkable that the descriptions of social norms in our survey are so often accompanied by descriptions of a *selection process* by which the norms are supposed to have arisen. Later we will review a study that documents in more detail the process of one culture replacing another.

Design Features of Adaptive Groups

Adaptations must often be complex and sophisticated to perform their functions. An eye must be complex in just the right way to function as a visual organ, a heart must be complex in just the right way to function as a pump, and so on. Complexity and sophistication of design are unlikely to occur by chance and are therefore often cited as evidence for the process of adaptation by natural selection. In

Chapter 3, we encouraged thinking about functional design at both the group and individual levels. In this spirit, it is useful to think about social norms as a complex and sophisticated machine designed to forge groups into corporate units.

The idea of culture as a sophisticated machine is expressed by many ethnographers in our survey: "A healthy, integrated, comprehensively developed personality with an integrated religio-social world-outlook, creating a perfect harmony of the personal and social interests and the driving forces of life—such is the hidden, but mighty mainspring of the clan mechanism" (Gilyak; Shternberg 1933, p. 116); "Their ancestor cult, an elaborate body of beliefs, rites, and ceremonies, is minutely woven into their everyday existence, but it does not pursue them with phantom terrors. On the contrary, it is the principal means provided by Tale culture for keeping the individual and the group on an even keel" (Tallensi; Fortes 1945, p. 9); "The basic controls, however, do not reside in him, nor for that matter in *anyone* in the society. Rather, they reside in the organization of the society as a whole, or, more concretely, in the thinking of the interacting individuals, where the social organization exists as a symbolic system" (Toba; Henry 1951, p. 218).

In both biology and anthropology, functional explanations are often criticized as "just-so stories" when they are proposed and accepted without evidence. Our survey provides only a brief glimpse (5–10 pages) of social norms for each culture, which is insufficient to *prove* any functional explanation. Nevertheless, in addition to the general sentiments quoted above, our survey provides a number of intriguing hints about specific design features that are worth discussing so they can be studied more extensively in the future.

Groups are vulnerable to subversion from within in a number of different ways. Many forms of exploitation require a degree of privacy. Power imbalances make it easy for the strong to exploit the weak. A simple ethic of mutual aid is vulnerable to freeloaders who take and don't give. Adaptive social norms, if they are to benefit the group as a whole, must have specific design features that guard against these possibilities.

Managing privacy. Enforcement of norms often requires an ability to monitor the behavior of individuals. If members of a group are free to go off by themselves, then behaviors such as meat sharing can be difficult to enforce. Privacy is automatically limited in small face-to-face groups, but it can be limited even further by social norms. Eating alone is the ultimate bad behavior in many hunter-gatherer societies, something that only an insane person would do. Among the Senoi of the Malay Peninsula, a myth tells about a god who was the chief instrument in bringing people out of their presocial state by telling them that to eat alone was not proper human behavior (Howell 1984, p. 184). In Samoa, "those things done alone are at least suspect, if not downright wrong. The expectant mother, the young chief, the bride-to-be, must never be alone. Any one whose conduct is of importance to other people must be sheltered from solitude because no one would conceivably wish to be alone unless on evil bent" (Mead 1930, p. 81).

These and other examples from our survey suggest that social norms can go beyond enforcing specific behaviors to controlling the conditions that make specific behaviors possible. Not only can social norms limit privacy, but they can also force individuals to be sociable when they are together. For the Mbuti, "if anyone is silent and untalkative, this is considered unusual and unfortunate" (Turnbull 1965, p. 118). In the Mbuti legend about the elephant hunter cited before, failure to join in the conversation was grounds for desertion as unsafe company. If the social norm of a group is to be physically present and socially engaged, then the deck is stacked in favor of prosocial behaviors from the beginning. These norms may seem horrible and coercive to those who love their privacy, but the people who are governed by the norms may or may not see it that way. If all the rewards of social life require an individual to be present and socially engaged, then individuals may sincerely want to be present and socially engaged and may never even dream of taking a walk alone in the woods.

Managing the balance of power. Powerful individuals do not need privacy to exploit weaker members of their group. One of the most important design features of social norms might therefore be to maintain a balance of power among group members. Just as privacy is automatically limited somewhat in small face-to-face groups, so also is any single individual fairly weak in comparison with the entire group. However, social norms often appear designed to limit the amount of power that any individual can have. Especially in hunter-gatherer societies that approximate the ancestral human condition, social norms act to diminish rather than accentuate power differences, at least among members of the same sex and age class. Turnbull's (1965) classic ethnography of the Mbuti, which by chance was included in our survey, vividly illustrates how the most gifted and powerful individuals were expected to behave and how they were treated by members of their group:

> Njobo was an undisputed great hunter, knew the territory as well as anyone and had killed four elephants single-handed. He was a good enough Mbuti not to attempt to dominate any hunting discussion in the forest, merely to take a normal part. If he ever appeared to be overly aggressive or insistent he was shouted down and ridiculed, although highly popular. He was also the one chosen to represent the band to the villagers. Ekianga, on the other hand, was less generally popular and was the source of some friction, having three wives (one the sister of another prominent member of the band), but he was a fine hunter, endowed with exceptional physical stamina, and he too knew the territory well. Even at the height of his unpopularity he was one of the most effective "leaders" of the hunt. So was Nikiabo, a youth who had achieved some notoriety by killing a buffalo when barely out of childhood. Although a bachelor, he had a net of his own and took a prominent part in all hunting discussions. Makubasi, a young married hunter, was also accorded special respect because of his hunting prowess and his physical strength, combined with his knowledge of the territory. But while these four can be singled out as exceptional, they could either separately or together be outvoted by the rest of the hunters. On such occasions they were compelled either to give their assent to the popular decision or to refrain from joining the hunt that day. None of them had the slightest authority over the others. Nor was any moral pressure brought to bear in influencing a decision through personal considerations or respect. The only such moral consideration ever mentioned was that when the band arrived at a decision, it was considered "good," and that it would "please the forest." Anyone not associating himself with the decision was, then, likely to displease the forest, and this was considered "bad." Any individual intent on strengthening his own argument might appeal to the forest on grounds that his point of view was "good" and "pleasing"; only the ultimate general decision, however, would determine the validity of his claim. (p. 180)

This kind of egalitarianism appears to describe the majority of hunter-gatherer societies (Boehm 1993; Knauft 1991). Many of the cultures in our survey are more stratified, however, with "chiefs," "big men," and "elders" who seem to enjoy more privileges than others. Social stratification can be interpreted in two very different ways. On the one hand, some individuals might have escaped social control and succeeded in establishing their dominance over other individuals. In this case, social stratification is a product of competition within groups (step 2 of the procedure outlined in Chapter 3) and should not be interpreted as functional at the group level. On the other hand, it is common for functional systems of all kinds to become more hierarchical and differentiated as they increase in size (Simon 1962, 1981). If this rule applies to human societies, then stratification may actually benefit the group (step 1 of the procedure) and does not automatically signify a process of competition within groups.[9]

Our survey (and common experience) suggests that large stratified societies are indeed vulnerable to exploitation by powerful members and therefore cannot be explained entirely by step 1 of the procedure. Nevertheless, it is equally true that social stratification often contributes to group-level success and cannot be explained entirely by step 2 of the procedure. It is likely that human social groups must differentiate to remain functional as they increase in size. The big men, chiefs, and others who seem to be powerful are usually not despots who have totally escaped social control, but often behave in a capacity that in some significant degree benefits the group. For many of the cultures in our survey, leaders are controlled at least as much as they do the controlling. Among the Kpelle, for example, older brothers usually serve as family head, but they must consult with a council of elders to make important decisions. According to Fulton (1969, p. 70), "Much the same thing could be said of Kpelle leaders on all levels." Apache leaders were expected to exemplify the value of sharing by precept and example. Gifts received by leaders were largely redistributed or used to fulfill requirements of hospitality. Leaders could advise but not command. "The leader had no coercive force at his disposal which could be used to control others or maintain his position" (Basehart 1974, p. 145). The Nootkan chief was expected to be amiable, nonaggressive, and esteemed by his so-called subjects, who otherwise would not cooperate in economic and ceremonial affairs. It was considered disgraceful for a chief to engage in physical conflict within the group (Drucker 1951, p. 453).

Our survey suggests at least two specific design features of social norms that constrain the power of leaders. First, a person must be "above reproach" to become a leader. Leaders and other powerful individuals appear to be held to higher moral standards than the average person. This makes sense from a group-level adaptive standpoint, because a person with a long history of exemplary behavior is less likely to exploit the group when placed in a position of power. Given the importance of gossip as a mechanism of social control, it is interesting that gossip focuses on the moral conduct of *important* people (Barkow 1992).[10]

Second, no matter how exemplary their behavior, powerful individuals are often not allowed to decide the fate of other members of their group. A norm of personal autonomy, in which no one can be told what to do, exists side-by-side with norms for mutual aid and cooperation (Boehm 1993; Knauft 1991). It appears that decisions are seldom made by despotic individuals and simply imposed on other members of the group, at least until a society becomes very large.[11] Instead, important decisions are discussed in a group setting with an effort to achieve consensus (see also Boehm 1993, 1996). Individuals who deviate from a consensus decision lose their moral standing and can be punished, but decisions imposed without consensus are themselves immoral and strenuously resisted, especially when they appear unfair from the standpoint of within-group selection. Leaders and other seemingly powerful people have not entirely escaped from these social controls.

There is no guarantee that social norms will succeed in preventing exploitation within groups. Indeed, norms may become increasingly difficult to enforce as societies become larger, shifting the balance in favor of within-group selection. Since cultural evolution is a fast-moving process, we should expect to find many historical examples of well-organized societies that collapse as within-group selection intensifies and runs its course. These dysfunctional societies may in turn succumb to alternative social organizations that are less vulnerable to subversion from within. Our goal is not to show that everything is for the good of the group, but to see the process of multilevel selection in all its complexity. It is important to avoid extreme group-level functionalism, in which everything is interpreted as for the good of the group, and extreme individual-level functionalism, in which the welfare of the group is treated as irrelevant.

Managing the problem of freeloading. In addition to managing privacy and the distribution of power, adaptive social norms must also prevent simple freeloading. The freeloading problem is the classic argument against altruism in both human and nonhuman societies. Many of the social norms promoting generosity and cooperation that we have discussed seem highly vulnerable to freeloading. On closer examination, however, the simultaneous emphasis on group welfare and personal autonomy provides safeguards against unrestricted requests for aid and material goods. A good example comes from the Navaho:

> An obligation to cooperate is binding on all adult Navajos, regardless of kinship status or situation. Also, not only is there an expectation that an individual requested to perform a task will do so out of good will and generosity, but there is little or no effort to calculate the "debits" or "credits" resulting from the sum of exchanges between two individuals. On the other hand, an expectation exists that an individual is free to decide for himself whether to honor a request; conversely, the requester is obligated not to "push his request" and infringe on this autonomy. As a consequence of this paradox that all are generally obligated to cooperate but the individual is free to decide, the request situation is full of ambiguity. (Lamphere 1977, p. 57)

A result of the ambiguity is that individuals do not make requests lightly, lest they suffer the public humiliation of being refused. Our survey includes other examples of social norms that limit the benefits of freeloading.

We are aware that functional explanations are easy to invent and hard to prove, both in biology and the human sciences. Using hints provided by our survey as a guide, we have merely speculated about the design features of adaptive groups. Evolutionary biologists have developed a set of methods for evaluating whether a trait is an adaptation (Endler 1986), and these can also be applied to cultural traits. We believe that our hypotheses are testable and can be subject to more rigorous tests in the future.

Cultural Diversity and Adaptive Function

In Chapter 4 we showed that a single trait, such as wing length in fruit flies, can evolve by a diversity of proximate mechanisms. The natural world is full of species that have evolved to do the same kind of thing in different ways. Similarly, human cultures may have evolved to function as adaptive units via many proximate mechanisms.

For some cultures in our survey, prosocial behavior seems to require constant oversight and shaming. Amhara peasants are said to avoid immoral acts primarily to avoid the humiliation of being caught (Levine 1965, p. 82). In other cultures, prosocial behavior seems to be so internalized that it becomes second nature. The Cuna are said to be frank people whose neighborliness is sincere (McKim 1947, p. 44). In some cultures, the use of violence to maintain honor is one of the highest moral virtues. The Amhara peasant who avenged an insult is "following an ethic of cardinal importance" (Levine 1965, p. 83). Yet violence is prohibited as a means of conflict resolution in other cultures. Navahos are expected to accept compensation for wrongs and never to resort to physical violence against another Navaho (Shepardson and Hammond 1970, p. 129). Some cultures, such as the Lesu, are so dominated by the principle of reciprocity that "nothing, with the exception of food between kindred and friends, can be given or done for nothing" (Powdermaker 1933, p. 210). Other cultures, such as the Navaho, make "little or no effort to calculate the 'debits' or 'credits' resulting from the sum of exchanges between two individuals" (Lamphere 1977, p. 57).

Even varieties of a single religion such as Christianity can differ dramatically in the norms that encourage prosocial behavior. Our random sample of cultures includes the Amish, whose understanding of religious commitment is very different from other Christian sects. According to Hostetler (1980, p. 298), "Most fundamentalist churches and independent revivalistic movements stress individual liberation from sin more than submission to the corporate community of believers. They stress enjoyment rather than suffering, assurance of salvation rather than hope, a subjective rather than a submissive experience, a vocal rather than a nonverbal (silent) experience."[12]

Once again, these and other examples from our survey illustrate the cultural diversity that is the hallmark of our species. However, the proximate mechanisms that produce prosocial behaviors may be more diverse than the adaptive functions that the behaviors subserve.

Summarizing the Survey Results

Our survey provides only a glimpse of human societies around the world. It is a cultural collage based on less than 250 pages of ethnographic material. Nevertheless, it has the virtue of being a *random* collage. Barring a freak sampling accident, we could discard the entire survey, randomly select another twenty-five cultures from the HRAF, and much the same picture would emerge again. Furthermore, despite the small amount of material grounding our survey, the patterns that emerge appear to be very strong. In most human societies, individuals are not limitlessly free to employ any behavioral strategy that they please. Their behavior is strongly regulated by social norms that define what is acceptable for that culture. Social norms are powerful because rewards and punishments can be imposed at low cost. A wide range of behaviors can be rendered advantageous if they are sanctioned by social norms, and once they become established they present the observer with a form of cultural diversity that defies narrow functionalist explanation. Despite this diversity, virtually all cultures possess strong social norms that appear designed to promote the well-being of the group. Our survey provides some evidence for the actual process of between-group selection that is required for adaptive social systems to evolve. It also hints at some of the specific design features that allow human groups to function as adaptive units. All of these patterns support the basic assumptions and predictions of the theoretical models that we presented in earlier chapters. However, much more work is needed to delimit the process and products of group selection.

More Evidence for Groups as Adaptive Units

The patterns that emerge from our survey are sufficiently strong that we expect them to survive the more rigorous methods of analysis that we described at the beginning of this chapter. Furthermore, our results only appear strange against the background of individualism. Anthropology has been somewhat less influenced by individualism than other branches of the human sciences, and many anthropologists have reached conclusions similar to ours after much more extensive research. We did not discuss these studies at the beginning of the chapter because we wanted to avoid the bias of picking and choosing among cultures to prove our particular points. Now that we have supported our claims with a random sample of cultures, we will consider the work of two anthropologists who have access to far more information than our cultural collage can provide.

Egalitarianism in Small-Scale Societies

Christopher Boehm is uniquely qualified to comment on the evolution of human social behavior. He began his career as a cultural anthropologist and spent three years studying the feuding society of Montenegro (Boehm 1983, 1984). He then turned to the study of primates to explore the biological roots of human morality (Boehm 1978, 1981, 1992) and is currently director of the Jane Goodall Research Center at the University of Southern California. In addition, he has continued to study human morality by reviewing and synthesizing ethnographies of cultures from around the world (Boehm, 1993, 1996, 1997a,b). Boehm's grasp of the anthropological literature makes our own sampling of cultures appear paltry indeed, and it is therefore gratifying that he comes to the same conclusions.

According to Boehm (1993, 1996), most human groups uphold specific lists of behaviors that are desirable or undesirable. These may be called moral values (Kluckhohn 1952), and distinctive local themes amount to an "ethos" (see Kroeber 1948). Armed with such widely shared opinions, people use a wide range of sanctions to encourage proper behaviors and to discourage improper ones. As moral communities, most human groups are influenced as much by what is considered right and wrong as by their external environment. Boehm's view is obviously close to our own focus on social norms and low-cost rewards and punishments.

In a review of small-scale human societies, Boehm (1993) examined the ethnographic literature for "all locally autonomous small-scale communities that seemed to have a low level of ranking or stratification by class and an absence of authoritative leadership" (p. 228). From these he selected forty-eight societies that provided enough information to warrant further analysis. His survey is not a random sample and his analysis is based on qualitative interpretation rather than quantitative methods. As with our survey, however, the patterns that emerge are so strong that they will probably survive the application of more rigorous methods. In particular, the vast majority of these small-scale autonomous societies were egalitarian in character, at least in terms of interactions among adult males (see also Knauft 1991). These societies were egalitarian, not for lack of a desire to become dominant, but because dominance was high on the list of behaviors considered to be immoral.

Although Boehm excluded obviously stratified societies from his sample, one might have expected the societies that he included to have displayed a range of nonegalitarian social structures. For example, one might expect a certain amount of overt competition within groups, which would lead to a strong dominance hierarchy in which a few individuals monopolize the resources. This is the most common social organization of nonhuman primates. The egalitarian nature of small human groups documented by Boehm (1993) and Knauft (1991), therefore, is a radical departure from the social life exhibited by our own ancestors.

To enforce the ethic of egalitarianism, a society must emphasize two seemingly contradictory values. First, there must be an ethic of group welfare that causes individuals to work for the common good and avoid exploiting others (although members of other groups are often fair game). Second, there must be an ethic of personal autonomy that prevents people from being told what to do. The social norms must be powerful, but they must also be agreed upon by consensus and not forced upon any member of the moral community. Abandoning personal autonomy opens the door to exploitation within groups.

Boehm provides many examples of behaviors and social conventions that limit status (and therefore fitness) differences within human social groups. These include gossip, criticism and ridicule, disobedience, ostracism, abandonment, and execution. The Kalahari San "cut down braggarts." Among the Hazda, "when a would-be 'chief' tried to persuade other Hazda to work for him, people openly made it clear that his efforts amused them." Among the Iban "if a chief tries to command, no one listens." Among the Nambicuara, "if a chief could not keep food in supply or was too exacting or monopolized the females, the families under him went to another band." Similarly, the Mescalero "would join other bands if their chief was dishonest, unreliable or a liar." Australian aborigines "traditionally eliminated aggressive men who tried to dominate them." In New Guinea, "the execution of a prominent individual who has overstepped his prerogatives is secretly arranged by other members of the multiclan community, who persuade the target's own kinsmen to accomplish the task." These examples are so pervasive in the literature reviewed by Boehm that he summarized his findings with the following powerful statement (1993, p. 236):

> The data do leave us with some ambiguities, but I believe that as of 40,000 years ago, with the advent of anatomically modern humans who continued to live in small groups and had not yet domesticated plants and animals, it is very likely that all human societies practiced egalitarian behavior and that most of the time they did so very successfully.

As we have seen, egalitarian societies give way to more stratified societies as they increase in size, but social stratification needs to be interpreted in terms of multilevel selection. Bee hives, coral colonies, and single organisms benefit from the differentiation of their parts, and perhaps human societies do also. Boehm (1997a) agrees with our own conclusion that departures from egalitarianism often—although by no means always—contribute to group-level functioning. As Godelier (1986; quoted in Boehm 1993, p. 237) observed for the Baruya of New Guinea, "differences between individuals are only permitted . . . insofar as they work for the common good."

Cultural Group Selection in Action: A Smoking Gun

Our survey provided only hints about the actual process of between-group selection that is ultimately required for human groups to evolve as adaptive units. Perhaps the best-studied example of cultural replacement is the Nuer, an African pastoralist society that was rapidly expanding at the expense of the neighboring Dinka tribe when they were studied by the British anthropologist E. E. Evans-Pritchard. Evans-Pritchard's (1940) analysis of the Nuer became a classic ethnography that stimulated many subsequent studies. In a book entitled *The Nuer Conquest,* Raymond Kelly (1985) synthesized four decades of research on the Nuer-Dinka interaction and convincingly identified the factors that allowed one culture to outcompete another in between-group interactions.

At first sight, Kelly's book does not seem very hospitable to our evolutionary perspective. Written at the height of the acrimonious debate that followed the publication of E. O. Wilson's (1975) *Sociobiology,* Kelly's book was proclaimed as a victory for those who emphasized culture over biology. The eminent anthropologist Marshall Sahlins, a longtime critic of sociobiology, chortled with glee in his review, which is quoted on the back cover of the paperback edition: "This is in many respects a stunning work—brilliantly argued, carefully researched—of general (even critical) value to the field of cultural anthropology . . . Beside all this, it should evoke cries of pain from ecologists (natural and cultural), sociobiologists, materialists and others." Nevertheless, Kelly's analysis coheres quite well with our own evolutionary approach.

From historical and linguistic evidence, it is virtually certain that the Nuer and Dinka are more closely related to each other than to any other culture. The Nuer originated as an offshoot from the Dinkas and eventually became distinctive enough to be identified as a separate tribe. This makes the competitive superiority of the Nuer all the more intriguing, because it reflects an historically recent change in social organization.

Both the Nuer and the Dinka subsisted on a mixed economy of cattle and grain (millet). The primary impetus for territorial expansion was cattle raiding and ultimately the appropriation of grazing land. Of course, both tribes had an incentive to take cattle and land from each other, but the Nuer felt the impetus more strongly than the Dinka. This asymmetry was due to a difference in social organization rather than a difference in the physical environment inhabited by the two tribes. Both tribes had a bridewealth system that required the transfer of cattle between families during marriage, but the bridewealth payment was larger and included more male cattle for the Nuer than for the Dinka. As a result, the two tribes managed their herds differently and the Nuer herds overpopulated their grazing land more quickly than the Dinka.

This difference in social organization provided an extra incentive for the Nuer to invade Dinka territory, but it does not explain why they were successful. In fact, the transfer of many cattle for each marriage had the effect of making Nuer population density 36.5 to 45 percent *lower* than Dinka population density in adjacent districts. Nevertheless, despite their numerical disadvantage, the Nuer were consistently superior at invading Dinka territory and defending themselves against retaliatory raids. The reason is that the Nuer were able to field a larger fighting force, despite their lower population density. Large-scale Nuer raids were conducted by a force of fifteen hundred men organized into five columns that attacked a number of Dinka settlements simultaneously and came together to resist any counterattack that might be launched. Dinkas were never able to organize such a large force, either offensively or defensively, without the assistance of the colonial government.

The advantage of the Nuer was based on sheer numbers rather than novel military strategies or fighting techniques. The Dinka were able to repel smaller Nuer raiding parties, were familiar with Nuer tactics, and employed the same multicolumn organization of forces on their own infrequent raids. Nuer and Dinka oral traditions fail to mention a single battle that was won by a clever new strategy. The Nuer won their territory as a result of hundreds of raids by numerically superior forces taking place over a period of generations.

Despite the basic similarity in military tactics, there were some important differences in the social organization of Nuer and Dinka raiding parties. Consider the following account of one of the few successful Dinka raids (from Titherington 1927, p. 199, quoted in Kelly 1985, p. 53): "a large [Raik Dinka] raiding party surprised a Nuer village and, Dinka-like, sat down to a happy day of wrangling over the spoils. It was their last, for meanwhile the Nuer surrounded them, and in the ensuing panic slaughtered them to the last man."

In contrast, the division of cattle by Nuer raiding parties was postponed until the danger of counterattack was over. Then the cattle were divided among the households by a contest in which anyone who could seize an animal, tether it, and cut its ear had an absolute claim to it. As might be expected, fights broke out when two men seized a cow at the same time, but severe injuries were limited by a rule that only clubs and not spears could be used. A fair division of resources was not meticulously regulated in this fascinating example of organized chaos; however, competition was postponed until after the shared goal had been achieved, and at the same time the costs of contention were contained. As a result, the division of spoils was more equal than it might otherwise have been.

In general, the Nuer were known for their discipline, bravery, and their ability to withstand heavy casualties in battle. Kelly reports one instance in which a Nuer raiding party was intercepted by a government patrol and lost eighty-four men in the initial engagement. Nevertheless, they counterattacked the same night and again the following day before withdrawing to home territory. By contrast, not only were the Dinka disorganized in battle, but they tended to exploit each other's misfortune. Colonial records consistently report that Dinkas who were displaced from their land were relieved of their remaining cattle by other Dinkas. The Nuer also fought among themselves but somehow had the ability to resolve their differences and unite in the face of external aggression. Evans-Prichard (1940) described Nuer society as segmented in a way that allowed a hierarchy of levels to become functionally organized, from a single village to the entire tribe, depending on the scale of the threat they were facing.

To summarize, the Nuer social system replaced the Dinka social system because it was *better organized at the level of large groups*. It is important to stress that the Dinkas were unable to become organized at this level even after the need arose. Their very existence as a culture was threatened, and still they could not muster a large fighting force or suppress their internal squabbles. Something about the Dinka social system prevented functional organization at this level, in contrast to the Nuer social system. One of the most important differences involved the role of lineages in maintaining a bond between smaller social units in the Nuer but not the Dinka.[13] A minimal segment of the Dinka social system was conceived as a group of agricultural settlements that jointly herded their cattle in common wet-season pasture areas. The size of the minimal segment was environmentally limited by the extent and distribution of available high ground. In contrast, a minimal segment of the Nuer social system was conceived in terms of a lineage system and was not tied to the use of common pastures. As a result, the size of the minimal segment was not environmentally limited for the Nuer. Similar factors allowed the Nuer to add an extra level to their segmented social organization. In short, the Nuer lineage system allowed social groups to remain functionally integrated after they had become isolated with respect to cattle grazing.

The cultural differences between tribes persisted despite a massive influx of Dinka into Nuer society. According to Evans-Pritchard (1940; p. 221, quoted in Kelly 1985, pp. 64–65), Dinka captives, immigrants, and their descendents formed at least half of the Nuer population. Assimilation was so rapid that Evans-Pritchard found it difficult to distinguish second-generation descendants of Dinka from other Nuer. Nevertheless, despite the massive movement of *individuals* between tribes, the cultural and behavioral differences between *the tribes* remained stable. According to Kelly (1985, p. 65), "the key features of Nuer social and economic organization that were instrumental to territorial expansion remained unaltered by the massive assimilation of Dinka and Anuak tribesmen." This observation supports the general point we made in Chapter 4, that phenotypic differences between groups can be more heritable than phenotypic differences between individuals.

Kelly describes many other features that allowed the Nuer to surpass the Dinka as large-scale adaptive units. He also attempts to trace their interconnections and explain how the Nuer system evolved from the Dinka system. Kelly's approach to cultural systems is strikingly similar to that of evolutionary biologists who emphasize the importance of history, complexity, and development in biological systems (e.g., Gould and Lewontin 1978). In both cases there is skepticism about single-factor explanations and an appreciation that everything is connected to everything else. Cultural systems are not invariably adaptive, and the Dinka could have persisted indefinitely if the Nuer social system had not happened to originate. The Nuer social system was not perfectly adapted but merely more competitive than the system that it replaced. Phenotypic traits such as "raiding party size" cannot be conceptualized as single cultural "memes" (Dawkins 1976), much less as single genes; they arise from multiple lower-level traits that interact in a complex fashion and produce many side effects. All of these points are legitimate and we have tried to emphasize them ourselves. Despite the complexity of the cultural mechanisms that produce the group phenotype, however, the phenotype itself can be interpreted in functional terms, as Kelly also appreciates. The Nuer cultural system was in the process of replacing the Dinka cultural system because it caused human social groups to function better in between-group conflict.

Kelly's analysis illuminates many of the assumptions and predictions of our theoretical models, but we have discussed it in detail for one especially important reason. Evolutionary biologists are often forced to study the *products* of natural selection—adaptations—without having direct access to the actual *process* of natural selection, which occurred in the past. Methods exist to demonstrate that a trait is an adaptation, but the ultimate proof is to watch it evolve. Kettlewell's (1973) research on industrial melanism in moths earned its place in textbooks because it was among the first to document natural selection in action. More recent investigations—on Darwin's finches (Grant 1986) and other species—have monitored the process of natural selection in even more detail (reviewed by Endler 1986). Our survey and Boehm's more extensive research focus on the *products* of group selection—groups that seem to function as adaptive units. We claim that human social groups are so well designed at the group level that they must have evolved by group selection. Except for a few hints about between-group replacement processes that we described earlier, we cannot produce the smoking gun of group selection in action. Kelly's analysis of the Nuer conquest is one such smoking gun—a social system in the process of replacing another that is less adaptive at the level of large-scale groups. It is likely that the historical record contains many other examples that can be documented with equal or even more impressive detail. As Darwin (1871, p. 166) realized long ago, "at all times throughout the world tribes have supplanted other tribes."

Summing Up

The history of life on earth appears to have been marked by a number of major transitions in which previously independent lower-level units coalesced into functionally integrated higher-level units (Maynard Smith and Szathmary 1995). More recently, insects and even a few species of mammals, such as the naked mole rat (Sherman, Jarvis, and Alexander 1991), have made the same transition, coalescing into functionally integrated colonies. Whether human social groups can be added to this list may appear plausible or heretical, depending on one's intellectual background. The fact that opinions on the matter are so variable indicates a failure of scholarship (including science) to answer one of the most fundamental questions that can be asked about human nature.

By far the strongest theoretical case *against* human groups as adaptive units comes from the field of evolutionary biology. It is claimed that human groups do not have the same genetic structure as other ultrasocial species and therefore cannot evolve as adaptive units. We have examined the nature of this argument in great detail. Our first task was to resurrect the theory of group selection, which was prematurely rejected by evolutionary biologists during the 1960s. Our second task was to examine human groups from the standpoint of multilevel selection theory, which describes factors additional to genealogical relatedness. We concluded that the fundamental ingredients of natural selection—phenotypic variation, heritability, fitness consequences—can plausibly be said to exist for human groups, allowing them to evolve into adaptive units.

In this chapter we attempted to put empirical flesh on the theoretical bones by reviewing some of the vast literature on human societies around the world. Our survey of social norms supports the assumptions and predictions of our models. The factors that can make group selection a strong force in the absence of genealogical relatedness seem to be abundantly present, especially in the small face-to-face societies that existed for most of our evolutionary history. Human groups do not invariably function as adaptive units, and human nature may be as well-suited to maximizing an individual's relative fitness within groups as it is to forming a group-level organism. Nevertheless, most traditional human societies appear designed to suppress within-group processes that are dysfunctional for the group, and as a result natural selection has operated and adaptations have accumulated at the group level. Human social groups often function as adaptive units and are perceived as such by indigenous people and the ethnographers who study them.

Our survey is modest in scale and our interpretation needs to be checked with more rigorous methods. However, we did avoid the most important bias of picking and choosing among cultures to support our own view. Because we based our survey on a random sample of cultures, it is likely that our results apply to the hundreds of remaining cultures in the Human Relations Area File. Furthermore, at least some anthropologists with far more knowledge of the ethnographic literature have reached the same conclusions. The evidence includes not only the products of group selection—groups that function as adaptive units—but the smoking gun itself: the process of some groups replacing other groups. As strange as it may seem against the background of individualism, the concept of human groups as adaptive units may be supported not only by evolutionary theory but by the bulk of empirical information on human social groups in all cultures around the world. Perhaps our species can be added to the list of examples in which lower-level units (individuals) have significantly coalesced into functionally integrated higher-level units (groups).

It is important to stress that *everything* we have established so far has been without reference to the thoughts and feelings that guide human behavior. Evolutionary biologists often work from the outside in. They begin by trying to understand the phenotype, which determines survival and reproduction in the real world. Phenotypes can be studied without reference to the proximate mechanisms that produce them. As long as the proximate mechanisms result in heritable variation, adaptations will evolve by natural selection. There is a sense in which the specific proximate mechanism doesn't matter. If we select for long wings in fruit flies and get long wings, who cares about the specific developmental pathway? If the brainworm has evolved to sacrifice its life so that its group will end up in the liver of a cow, who cares how (or if) it thinks or feels as it burrows into the brain of the ant? Similarly, if humans have evolved to coalesce into functionally organized groups, who cares how they think or feel? The fact is that they do it, just as brainworms burrow into the ant's brain and fruit flies develop wings. It is difficult to adopt this mechanism-free attitude, but we must if we are to study humans as we do other species.

Of course, at some point we will want to understand the proximate mechanisms in order to gain a richer understanding of both the phenotype and the evolutionary process. Evolutionary biologists interested in human behavior made this transition about ten years ago. At first they leapfrogged over the subject of psychology and attempted to explain human behavior directly in terms of biological fitness. This approach yielded many insights but ultimately became limiting. To proceed further, it became necessary to study the evolution of proximate mechanisms and the field of evolutionary psychology was born (e.g., Barkow, Cosmides, and Tooby 1992; Buss 1994, 1995). Now it is time for us to delve below the level of behavior and consider the psychological mechanisms that cause individual humans to coalesce into functionally organized groups. Do people feel the direct concern for others that we typically associate with psychological altruism? Or can individual- and group-level adaptations be explained entirely in terms of mechanisms that we typically associate with psychological egoism? To answer these questions, we must review how psychological altruism and egoism have been studied by philosophers and psychologists. Then we will develop an evolutionary theory of psychological mechanisms that goes beyond the traditional approaches. Even before we begin Part II, however, we can conclude Part I with an important statement about human nature: At the behavioral level, it is likely that much of what people have evolved to do is *for the benefit of the group*.

Reprinted by permission of Prentice-Hall Inc.
from *The Best of Rube Goldberg* (1940), pp. 98–99.

II

Psychological Altruism

➤ 6 ➤

Motives as Proximate Mechanisms

In this second part of the book we shift focus—from altruism as an evolutionary problem to altruism as an issue in psychology. This reorientation involves more than simply looking at the same question from a different angle; rather, the phenomenon under study undergoes a transformation as well.

An organism behaves altruistically—in the evolutionary sense of the term—if it reduces its own fitness and augments the fitness of others. In contrast, the concept of psychological altruism applies, in the first instance, to motivational states, and only derivatively to the behaviors those motives may cause. This shift from altruism as a property of behavior to altruism as a property of motives may seem trivial, but in fact it introduces a new set of issues. In part, this is because psychologists and philosophers have discussed the motivational question without connecting it with the theory of evolution. We will need to assess how altruism has been addressed in psychology and philosophy; evolutionary ideas will come into play, but it would be a disservice to the problem to ignore the way it has developed outside of evolutionary biology.

Proximate and Ultimate Mechanisms

When a behavior evolves, a proximate mechanism also must evolve that allows the organism to produce the target behavior. Ivy plants grow toward the light. This is a behavior, broadly construed. For phototropism to evolve, there must be some mechanism inside of ivy plants that causes them to grow in one direction rather than in another. We have here a three-step causal chain—from evolutionary process to internal mechanism to behavior:

Selection for phototropism → Internal mechanism → Growing toward the light

The behavior evolved in an ancestral lineage because it was favored by natural selection; within the lifetime of an organism, the behavior now occurs because there is an internal mechanism inside the organism that causes it.

This causal chain entails that questions about the explanation of behavior can be answered in two ways (Mayr 1961). Why do present-day ivy plants grow toward the light? One answer traces this behavior back to earlier facts concerning the organisms' ontogeny; they grow in a particular direction because they contain internal mechanisms that make them do so. A second answer traces the behavior back to still earlier facts about the organisms' phylogeny; ivy plants now grow toward the light because natural selection favored this behavior in their ancestors. These two answers are not in conflict; they merely advert to relatively proximal and relatively ultimate links in the causal chain.

Ivy plants presumably do not have minds, but the same point applies to behaviors that occur in creatures that do. Human beings obtain nutrition from their environment, just as ivy plants obtain energy from the sun. If behaviors that enable us to obtain nourishment are products of evolution, then there must be a device inside of individual human beings that causes us to eat some things rather than others. This is a job that our minds do for us. People eat what they do because of the beliefs and desires they have. An evolutionary perspective on human behavior requires us to regard the human mind as a proximate mechanism for causing organisms to produce adaptive behaviors.[1]

Having explained what evolutionists mean by the distinction between proximate and ultimate causes, we now need to explain the psychological distinction between the ultimate desires and the instrumental desires that people have. This is not a technical and esoteric idea but a concept that is familiar from everyday life. Some things we want for their own sakes; other things we want only because we think they are means to some more ultimate end. For example, consider the question of why people want to avoid pain. It would be a mistake to think that people seek to avoid pain only because they think this will contribute to some *other* goals they have in mind. On the contrary, the fact is that people simply don't like pain; avoiding painful sensations is something we want to do for its own sake.

Although the desire to avoid pain is an ultimate goal in terms of how people think about pain, it is not ultimate in the evolutionary sense of that term. There is an evolutionary reason that we have the desire to avoid pain. Pain is typically associated with bodily injury. Present-day organisms seek to avoid pain because this strategy was favored by natural selection. Wanting to avoid pain is psychologically ultimate, but evolutionarily instrumental. Just to nail down this difference between the evolutionary and the psychological meanings of the term *ultimate,* let's consider the following diagram:

Natural selection for avoiding bodily injury ⟶ Desire to avoid pain ⟶ Avoidance of bodily injury

In one sense, the desire to avoid pain is *not* ultimate; it traces back to an earlier evolutionary process. In another sense, the desire to avoid pain *is* ultimate; its existence does *not* trace back to any more fundamental *desire* that the organism has. There is no conflict between these two ideas. The desire to avoid pain is a proximate mechanism that evolution has brought into being, but it also is an ultimate desire in terms of the organism's psychological economy; its existence does not depend on the organism's thinking that avoiding pain contributes to the attainment of some other goal.

We will further clarify the psychological distinction between ultimate and instrumental desires at the end of the present chapter. But for now, we can use it to formulate the main question about psychological egoism and altruism that the remainder of this book seeks to answer. Do people ever have altruistic desires that are psychologically ultimate? Or do people want others to do well only because they think that this will provide a benefit to self? *Psychological egoism* is the theory that all our ultimate desires are self-directed; the motivational theory called *psychological altruism* maintains that we sometimes care about others for their own sakes. The theories agree that people sometimes want others to do well; the debate concerns whether such desires are always instrumental or are sometimes ultimate.

How the Psychological and Evolutionary Concepts Are Related

What is the relationship between psychological egoism and psychological altruism on the one hand and the concepts of evolutionary altruism and selfishness that we discussed in the first half of this book? How is the motivational distinction related to the distinction that is drawn in terms of fitness consequences?

We may begin by reminding ourselves of an obvious difference between the evolutionary and the psychological concepts. An organism need not have a mind for it to be an evolutionary altruist. The examples discussed in Chapter 1 illustrate this point. The brain worm burrows into the brain of the ant in which it lives, sacrificing its own life and thereby helping the other parasites in the ant to survive and reproduce. A *Myxoma* virus that is less virulent than the other viruses that live in the same host has a lower reproductive rate, thereby helping the group of viruses to spread to new hosts. And if two female wasps found a nest, the one that produces more daughters than sons will be less fit than the one that produces an even sex ratio, though the one that has more daughters will enhance the group's productivity. You don't have to have beliefs and desires to be an evolutionary altruist.

Even if we restrict our attention to organisms that *do* have minds, we need to see that there is no one-to-one connection between egoistic and altruistic psychological motives on the one hand and selfish and altruistic fitness consequences on the other. Just to make this point crystal clear, we will describe four examples; they show that every combination of psychological egoism/altruism with evolutionary selfishness/altruism is possible. In the first two, the evolutionary and psychological categories coincide; in the last two, they part ways.

Example 1: Suppose an organism wants to gather as much food for himself as he can. This desire is self-directed; if the organism has this goal as an end in itself, then the organism, as so far described, is a psychological egoist. If some organisms in a population have this desire, while others do not, what are the fitness consequences of this difference? Organisms that gather as much as possible for themselves are evolutionarily selfish, whereas organisms that restrain their gathering of food and so leave more for others are evolutionary altruists. Thus, the desire to gather as much food as you can is an example of a motive that is psychologically egoistic and evolutionarily selfish.

Example 2: Consider a person who cares about the welfare of the people in his group as an end in itself. This desire is an example of psychological altruism. If we compare this trait with the trait of not caring about other people, except when caring for others increases one's relative fitness within the group, then group selection will be required for the trait of psychological altruism to evolve. If so, caring for others as an ultimate end is an example of evolutionary altruism.

Example 3: A person who helps everybody in his group, but does so only because helping makes him feel good, would be an evolutionary altruist and a psychological egoist. We want to describe a more subtle example, however, one that provides more insight into how the evolutionarily and psychological concepts diverge.

Consider an individual who wants to help build a stockade. This fort, let us suppose, will provide everyone in the group with defense against predators or marauders, but that is not why the individual in question wants to help. He wants to help build the fort solely because the fort will thereby be stronger, and his sole ultimate concern is for his own safety. This individual is a psychological egoist.

Now let's consider the fitness consequences of helping to build the fort and the consequences of not helping to build it. Since everyone in the group—builder and nonbuilder alike—gets to enjoy the benefits of the fort (it is what economists term a *public good*), builders are evolutionary altruists and nonbuilders are free riders—they are evolutionarily selfish. Within any group, builders are less fit than nonbuilders, because everyone benefits from the stockade though only the builders incur the energetic cost of constructing it. Wanting to help build the fort just because you think it will make you safer is an example of a trait that is psychologically egoistic but evolutionary altruistic.

When psychological egoists deliberate about whether to help, they compare how well off they'd be if they help and how well off they'd be if they did not. This involves assessing two hypothetical situations, only one of which will be made actual. The stockade example illustrates how psychological egoists who aim to maximize their *absolute* fitness sometimes will decide to perform actions that lower their *relative* fitness within the group (Wilson 1991; Sober 1998b). The egoist in our example cares about his own absolute well-being, not whether he does better or worse than others. This is an important difference between what psychological egoists care about and what the process of within-group selection "cares" about.

Although egoistic deliberation involves comparing *hypothetical* situations, the question of what trait will evolve is resolutely grounded in the realm of the *actual*. When the question is whether helping or not helping will evolve, the relevant comparison is how well off helpers and nonhelpers actually are; one compares the fitnesses of the two traits in the actual population at hand. Biologists sometimes use psychological egoism as a heuristic for predicting what trait will evolve; it is easy to see that faster running speed will evolve to replace slower running speed in a population of zebra, since an egoist aiming to maximize fitness would prefer to run fast. However, the stockade example shows that this rule of thumb can lead to the wrong answer. If evolution is governed by within-group selection, *not helping* is the trait that will evolve; if group selection plays a determining role, then *helping* will evolve. If both group and individual selection occur, then the question of what will evolve can be answered only by assessing the magnitude of each force. The choice made by a psychological egoist does not always correspond to the trait that is favored by within-group selection.[2]

Example 4: Think of a parent who cares irreducibly about the welfare of her children. This is a concern for someone other than the agent herself, so it is an altruistic psychological motive. Now consider a population in which some parents care about their children while others do not. Suppose the first set of parents is more reproductively successful than the second. This means that caring about one's children is an example of evolutionary selfishness, because the trait makes its bearer fitter than the alternative trait does.[3]

These four examples are summarized in the following table:

		Evolutionary	
		Selfishness	Altruism
Psychological	Egoism	1. *S* wants to gather as much food as possible.	3. *S* wants to help build the fort.
	Altruism	4. *S* cares about the well-being of her children.	2. *S* cares about the well-being of others in the group.

The take-home message is that every motive can be assessed from two quite different angles. The fact that a motive produces a behavior that is evolutionarily selfish or altruistic does not settle whether the motive is psychologically egoistic or altruistic. And the fact that a behavior is produced by a given psychological motivation leaves open the question of why the behavior evolved.

In Chapter 2, we discussed Frank's (1988) remark that "group selection is the favored turf of biologists and others who feel that people are genuinely altruistic." The automatic assumption that individualism in evolutionary biology and egoism in the social sciences must reinforce each other is as common as it is mistaken. More care is needed to connect the behaviors that evolved because of multi-level selection with the psychological mechanisms that evolved to motivate those behaviors. The multiple connections that may link the evolution of behavior and the psychology of motivation mean that we could have developed this second half of our book in two ways. One possibility would be to address the full architecture of the human mind as a product of multi-level selection; this architecture may or may not include psychological altruism, but it almost certainly includes much more. The alternative, which is the one we have adopted, is to evaluate the case for and against psychological altruism *per se*. This narrower focus on psychological altruism—not on the full range of motives that might produce helping behavior or adaptive behavior in general—has the virtue of being more manageable, and it is an important subject in its own right. Many regard genuine psychological altruism as a figment of the romantic imagination; we will be well satisfied if we can dismantle this position and establish the plausibility of psychological altruism as part of the architecture of the human mind.

The Problem of Functional Equivalents

The idea that an evolved behavior requires an evolved proximate mechanism can be supplemented with a second idea. If evolution causes an organism to produce a given range of behaviors in a given range of environmental circumstances, in principle there are many proximate mechanisms that may allow the organism to produce this pairing of behaviors with environments.

Consider, for example, a marine bacterium that avoids oxygen—or dies if it cannot. Evolution apparently has furnished this organism with a behavioral control device that allows it to swim away from higher concentrations of oxygen. It is not hard to imagine many pieces of machinery that would have done the trick. First and most obviously, evolution might have endowed the bacterium with an oxygen detector. However, there are many indirect mechanisms that also might have worked. For example, oxygen concentrations tend to be higher near the surface of bodies of water. This means that any device for detecting the difference between up and down could allow the bacterium to avoid oxygen. Many anaerobic species tell up from down by detecting the earth's magnetic field (Blakemore and Frankel 1981).[4]

The point may be generalized. Suppose an organism has evolved the strategy of producing behaviors B_1, B_2, . . . , B_n in environments E_1, E_2, . . . , E_n, respectively. In principle, any proximate mechanism that effects this pairing of behaviors with environments might be present. One such mechanism allows the organism to detect E_1, E_2, . . . , E_n directly. However, if E_1, E_2, . . . , E_n are correlated with other properties in the environment—with I_1, I_2, . . . , I_n—then the organism may pursue the indirect strategy of detecting these I properties. Since E_1, E_2, . . . , E_n might be correlated with *many* other properties, there are *many* possible indirect proximate mechanisms that in principle might have evolved. These alternative mechanisms are said to be more or less *functionally equivalent*—they all deliver roughly the same behaviors in the same circumstances. It is a standard biological problem to try to figure out which of these proximate mechanisms the organism uses.

This issue arises in the motivational problem of psychological egoism and altruism. For example, if natural selection has led parents to take care of their children, it could have achieved this result by any number of proximate psychological mechanisms. Many organisms manage to do this without having anything like the mental capacities that human beings possess. But in the case of human parental care, a sophisticated cognitive and affective system is at work. People take care of their children because of the beliefs and desires they have. What sort of beliefs and desires might furnish the proximate mechanism for human parental care? The first and most obvious device would be to have parents be altruistically inclined toward their children—to want their children to do well as an end in itself. However, an egoistic motivational mechanism can be imagined that generates roughly the same behaviors. Suppose the organism feels pain when its children do poorly and feels pleasure when they do well. If the parent's only ultimate desire is to feel pleasure and avoid pain, it will take care of its children. The psychologist's problem is to discover which of many conceivable motivational mechanisms the organism in fact deploys.

In the case of the marine bacterium, the problem of determining which proximate mechanism evolved can be solved in two ways. First, we can run an experiment in which we break the usual relationship of magnetic field to oxygen concentration and see how the organism behaves. If we put the bacterium in a pool of water that has the same amount of oxygen at the top and at the bottom, does the bacterium still swim down? If we reverse the magnetic field, does the bacterium swim up? A second strategy is to dissect the organism and determine by observation what devices it contains. If we know enough about bacterial anatomy, we may be able to tell by observing the organism's internal makeup whether it has an oxygen detector or a magnetosome.

Can either of these procedures be followed in the case of human motivation? The first has been used repeatedly. Psychologists have run experiments intended to provide evidence about whether people have altruistic or egoistic motives; philosophers have constructed thought experiments that seek to address this question as well. We will examine these efforts in Chapters 8 and 9. In contrast, the second procedure—opening up the organism and gazing within—is currently not a viable one. Even if we cut into the human brain and examined what is there (or, less invasively, used a PET scan or a functional MRI device), we would not know what to look for. What brain structures mark the difference between altruistic and egoistic motivation? Even if neurobiology will answer this question one day, it offers little guidance now.

We will return to the parallel between the magnetotaxis of marine bacteria and the question of psychological motivation in Chapter 10. For now, we mention it simply to point out that the problem that psychologists and philosophers have addressed concerning psychological altruism and egoism is not unique to the domain of psychological motivation. The problem of *functional equivalents* always looms on the horizon when one tries to determine what proximate mechanism underwrites a given range of behaviors.

Beliefs and Desires

We have emphasized the importance of viewing psychological motives as proximate mechanisms for regulating behavior; in this respect, they resemble the magnetosomes in marine bacteria and the devices in ivy plants that make them grow toward the light. In spite of this generic similarity, however, the fact remains that psychological motivation constitutes a distinctive type of proximate mechanism. For the remainder of this chapter, we will discuss what is special about the mind as a proximate mechanism for controlling behavior. The points we will make in this connection describe a shared framework within which the theories of psychological egoism and altruism are each constructed.

The egoism-altruism debate assumes that *beliefs and desires are items in the mind that produce behavior*. A debate about what our ultimate motives are presupposes that we have motives and that they are causes of the way we act. Although this assumption has not gone unchallenged, we (unsurprisingly) accept it here as a reasonable working hypothesis. If science someday establishes that beliefs and desires do not exist, just as it earlier decided that phlogiston and ghosts do not exist, then we expect the debate about psychological egoism and altruism to be tossed on the rubbish heap of history.[5]

In accepting that beliefs and desires cause behavior, we do not mean that every behavior traces back to beliefs and desires. As we have already seen, this will not be true if behavior is defined broadly enough. When the doctor hits your knee with a hammer, your knee jerks, but this is not because of the beliefs and desires you happen to have. Likewise, when an ivy plant grows toward the light, this is something the plant *does,* but not because it has thoughts and wants. If not all behaviors are produced by beliefs and desires, what makes beliefs and desires distinctive? The feature we wish to describe here is that *beliefs and desires have propositional content*. Believing and desiring are *propositional attitudes*.[6]

Consider what it means to say that Jack believes that there is water in the glass. If we take this statement at face value, we will say that it means that Jack bears a particular relation—the relation of believing—to a proposition (the one expressed by the phrase "there is water in the glass"). Desire also is a propositional attitude. When we say that Jill wants there to be water in the glass, we are saying that Jill bears a certain relation—the relation of wanting true—to a proposition (the one expressed by the phrase "there is water in the glass"). Jack and Jill have different attitudes to the same proposition.

Although the sentence "there is water in the glass" is part of a particular human language (English), the proposition expressed by that sentence is no more a part of English than of any other spoken language. Jack's Parisian cousin Jacques may also believe that there is water in the glass, even though he speaks no English. In saying that he has this belief, we are saying that he is related, not to a particular English sentence, but to the proposition that this English sentence happens to express. The declarative sentences in a language are true or false in virtue of the truth or falsity of the propositions they express. A proposition and a sentence are just as different as you and your name.[7]

When we attribute beliefs or desires to people, we thereby say that they are related to propositions in a certain way. What type of relation are we talking about here? The relation is *representational*—believing and desiring involve forming mental representations of the proposition in question.[8] This helps pinpoint what distinguishes belief and desire from other causes of behavior. The ivy grows toward the light, but without forming a representation of what it wants to achieve and another representation of how it will set about attaining that end. Hammer taps likewise manage to produce knee jerks without the mediation of mental representations.

Organisms are able to formulate a given belief or desire only if they have the capacity to represent its constituent concepts. To form a belief or a desire whose propositional content is "there is water in the glass," you must have available the concept of *water* and the concept of a *glass* (as well as the concept of *physical containment* and the concept of *existence*). Individuals who don't have these concepts can't have these thoughts and wants. This has interesting implications for the attribution of beliefs and desires to nonhuman organisms. When we say that an individual believes (or wants true) some proposition, we inevitably describe the proposition in a human natural language.

If a dog does not have the concepts expressed by the English terms *water* and *glass,* then it will be false, strictly speaking, to say that the dog forms a belief whose propositional content is that there is water in the glass. However, it doesn't follow from this that nonhuman organisms don't have beliefs and desires. Rather, what follows is that human natural languages may not be well-equipped to describe the representations these organisms use to formulate their thoughts and wants. Perhaps cognitive ethology will one day develop methods for characterizing the thought worlds of nonhuman organisms more accurately; the concepts needed to do this will have to be invented by science, if the language of everyday life does not provide such resources ready-made.

How Beliefs and Desires Work Together

There is a set of concepts that an organism uses to formulate its beliefs; there also is a set of concepts that an organism exploits in formulating its desires. One might view these sets of concepts as comprising a language out of which various representations that express propositions can be constructed. There is a vital reason why these two sets of concepts should be the same. A common language should be accessible to the organism in formulating both its beliefs and its desires so that the organism can engage in means-end *deliberation.* Aristotle discusses the example of a man who wants a covering and believes that a cloak would make a good covering. This desire and belief together lead him to form the intention of obtaining a cloak. In this example, the concepts out of which the desire is constructed and the concepts that figure in the belief *overlap*. If desires and beliefs were built out of disjoint vocabularies, it is hard to see how they could work together in deliberation. If Jill wants *X*, but none of her beliefs tell her how to get *X*, how is she to figure out what she should do?

Psychologists in different subfields have defended *modularity theses* (Fodor 1983; Barkow et al. 1992). Rather than postulate an all-purpose mechanism that governs all mental activity, many theorists have defended the idea that there are separate modules for language acquisition, for face recognition, and for other tasks. What we are now suggesting, however, is that belief and desire formation are not rigidly segregated from each other; a common vocabulary must provide the same (or roughly the same) conceptual resources to both the devices that construct beliefs and the devices that form desires. Although believing and desiring are separate activities, they do not exist in thoroughly isolated modules.

When we talk in what follows about "deliberation" and "means-end reasoning," we do not restrict these terms to lengthy episodes of self-conscious reflection. A reasoned decision about what to do can be experienced as virtually instantaneous. Suppose a lifeguard sees a drowning child and jumps in the water to help. Even though the action occurs quickly, it still is plausible to think that beliefs and desires are consulted and processed in accordance with some decision rule. The fact that the agent's behavior is so well suited to the accomplishment of a goal strongly suggests that a rational process is occurring. Why did the lifeguard swim to the child's aid? Why did the lifeguard grab a life preserver as she jumped from the dock? These questions suggest that the lifeguard acted on the basis of what she believed and what she wanted to accomplish.[9]

When people are interviewed after they intervened heroically in emergency situations, they often report that they "didn't think" but just helped spontaneously. This comment, if taken literally, would be a problem for both the hypothesis of psychological egoism and the contrary hypothesis that says we sometimes have altruistic ultimate motives. Both theories assume that people act on the basis of their desires; they act as they do *because* of what they want to achieve and *because* they have certain beliefs about how to achieve their goals. We suggest that most emergency interventions are best explained by viewing the helper as thinking about means and ends. Thinking can occur in a flash, but it is thinking nonetheless.

Desire and the Concept of Satisfaction

The concept of desire we have been describing does not mention any feeling or sensation. To be sure, feelings sometimes accompany the desires we have. Hunger sometimes accompanies the desire to eat, but a moment's reflection shows that the one can occur without the other. Desires need not be accompanied by disagreeable sensations that disappear once the desire is satisfied.[10]

The term *satisfaction* gets used in two ways. One can say that people *feel* satisfied and also that their desires *are* satisfied. If Nancy wants it to rain tomorrow, her desire is satisfied if and only if it rains (Stampe 1994). Of course, even if it rains, Nancy may not learn that it has, so she may not get to feel good. Her desire is satisfied, though she does not obtain a feeling of satisfaction. To see whether Nancy feels satisfied, you must know what is going on in her mind. To know whether her desire for rain is satisfied, you must look at the weather.

If you lose sight of this point about the concept of satisfaction, you run the risk of being taken in by a fallacious argument about human motivation. From the premise that people want their desires to be satisfied, nothing follows concerning whether their ultimate desires are egoistic or altruistic. In particular, the fact that people want their desires to be satisfied does not entail that all they really want is to obtain a feeling ("the feeling of satisfaction"). A defender of psychological egoism must do better than this.

Is Desire Always a Propositional Attitude?

Although we have claimed that desire is a propositional attitude, the way people talk about desire sometimes makes this point less than obvious. For example, when we see a dog begging for the bone his master holds, we may say "Fido wants the bone." This seems to mean that Fido is related to a particular physical object, not to a proposition. Saying that Fido *wants* the bone seems to imply no more conceptual sophistication on Fido's part than the claim that Fido *chews* the bone.

In reply, we suggest that statements like "Fido wants the bone" are elliptical; some of the semantic constituents of the statement fail to receive overt expression. When we say that Jane wants the glass of water, this is short for the claim that Jane *wants to have the glass of water*. The word *have* leaves open what Jane wants to do with the glass once it is hers. When we say that Jane *wants to have the glass of water,* it is but a short step to recognize that Jane wants true the proposition "Jane has the glass of water."[11] Parallel remarks apply to Fido. He wants the bone, but that is short for saying that he wants to have the bone. And from there we are obliged to recognize that he wants true a proposition. It may be difficult to formulate the propositional content of Fido's desire with any precision; however, if what Fido has is a desire, and not just a "tendency" to pick up bones, then he must have formed a representation that has propositional content.

The ellipsis just noted is available in English when we describe agents who want things for themselves. The sentence "Jane wants to have the glass of water" shortens to "Jane wants the glass of water," but no such saving of ink is available for the statement "Eve wants Adam to eat the apple." Presumably, Eve's desire and Jane's desire are states of the same type; just as Eve wants a proposition to come true when she wants something for Adam, so Jane wants a proposition to come true when she wants something for herself.[12]

We conclude that the appearance that there are nonpropositional desires is just that—an *appearance*. When people want things (either for themselves or for others), they want a proposition to be true. Furthermore, for an individual to be able to want a proposition to be true, the individual must be able to use the concepts that figure in a representation of the proposition. Although we talk about organisms wanting objects, without thereby pinpointing what concepts the organisms are using, this should not lead us to think that there is such a thing as "nonconceptual desire." Desiring involves the formation of representations that have propositional content.[13]

Desire and the Concept of "I"

John Perry (1979) tells the story of a visit he made to a supermarket. As he walked up and down the aisles pushing a shopping cart, he noticed that there were tracks of sugar on the floor. He came to realize that someone in the store was carrying a bag of sugar with a hole in it; this person was leaving a thin trail of sugar behind him as he walked. As Perry moved around the store, he came to believe that the sugar-trailer had visited a number of aisles, that this person was unaware of the mess he was making, and so on. In this way, Perry formed a more and more detailed picture of what the sugar-trailer was doing. Then Perry looked at his own shopping cart and suddenly realized—"*I* am the person who is trailing sugar." The point of this story is that this last belief—the one that features the concept of "I"—provides a *new* piece of information; it differs from all the beliefs that Perry had before. Perry's thesis is that a first-person belief is not equivalent to any conjunction of third-person beliefs, no matter how elaborate. More generally, his claim is that a belief formulated by using "indexical" expressions ("I," "this," "here," "now," etc.) says something different from what a nonindexical belief is able to convey.

When Jane wants a drink of water, how does she represent to herself the proposition that she wants to come true? The obvious guess is that she uses the first-person representation "I have a drink of water." Jane probably does not represent what she wants by using her own name. And even if she does think of herself in this peculiar way, perhaps verbalizing her intention by saying "Jane will have a drink of water," this can't express the full content of what is on her mind. After all, who is Jane? Jane must realize that *she* is Jane, if she formulates her desire by wanting "Jane" to have a drink of water. Using the concept of "I" is as crucial for Jane to represent what she wants as it is for the sugar-trailer to represent what he discovers.

When people want things for themselves, we expect them to want propositions to come true that they represent to themselves by using the concept of "I." However, it would go too far to say that an organism can have desires only if it has this concept. Oldenquist (1980) describes the interesting example of a hypothetical organism whose desires are strictly general and impersonal. For example, suppose this organism wants green things to receive food. If the organism itself is green, then this desire will lead the organism to provide food to itself. And if the organism is the *only* thing in its environment that is green, then it will provide food *only* for itself. Its *behavior* will be self-directed even though it lacks the concept of "I." Let's imagine that this organism is equipped with other such desires, each of them general and impersonal, which are so arranged that the organism manages to deal successfully with the problems that its environment presents. Although this organism has desires, it cannot be said to be egoistically motivated (Oldenquist 1980). The organism's desires give no special status to self-interest over the interests of others. It isn't that the organism recognizes that I and Thou are different but accords them equal importance. Rather, the organism thinks about all green things in the same way; it fails to conceptualize the distinction between self and other.

If an organism can have desires that lead it to take care of itself without possessing the concept of "I," why did the capacity to formulate such indexical desires evolve? One reason may be that desires that use the concept of "I" automatically apply uniquely to the organism itself, whereas nonindexical desires do so only if they happen to be rich enough. If the organism thinks "I want food," it will try to provide food to itself, rather than to any of the other objects in its environment. However, if the organism's desire is that green things receive food, it will achieve the same end only if the organism happens to believe that it is the only thing in the environment that is green. This can easily fail to be the case. If other things are green, how can the organism have a nonindexical desire that applies uniquely to itself? The obvious strategy is for the organism to formulate desires that are conceptually rich. Perhaps the organism will succeed in giving food to itself alone if it wants all green, left-handed, grey-haired individuals who wear pinky rings to have food. This arrangement is precarious, however. Even if this desire manages to single out the organism uniquely, what will happen if the organism changes its characteristics? And what desires will motivate its children to take care of themselves? With sexual reproduction, offspring do not exactly resemble their parents. If the desire for food is inherited, how are parents and offspring each to represent themselves as deserving recipients of nutrition? The concept of "I" makes all this quite straightforward.[14]

The other reason to expect indexical desires to appear early in the evolution of mind derives from the fact, noted earlier, that an organism's beliefs and desires are constructed from a common vocabulary. Indexical concepts are important constituents of perceptual beliefs. Suppose Fido is looking at a tree and forms the belief that there is a bone behind the tree. How does Fido conceptualize this tree that his belief is about? A plausible guess is that he represents this tree by relating it to *himself;* he thinks of the tree as the one in front of *him.* Fido has the concept of "I," or something like it, as a device for orienting his perceptual beliefs. If the concept of "I" (or something like it) is used to form perceptual beliefs, and if perceptual beliefs emerged relatively early in the evolution of mental abilities, then the concept of "I" was available early on for the construction of desires. With this concept available, psychological egoism became possible.

What does it mean for an organism to have the concept of "I"? An important distinction needs to be drawn if one is to understand this issue. In their book *How Monkeys See the World,* Cheney and Seyfarth (1990, pp. 240–242) distinguish *self-awareness* from *self-recognition.* Self-awareness requires individuals to form beliefs and desires about the contents of their own minds. Individuals who are self-aware are "psychologists." They not only have beliefs and desires; in addition, they think of themselves as having beliefs and desires.

Self-recognition is a different achievement, one that does not require that the individual be a psychologist. Cheney and Seyfarth illustrate this idea by describing experiments with chimps that involve mirrors and television sets. If a sleeping chimp has a large spot of bright paint smeared on his forehead, the chimp, when he wakes up, will be able to use a mirror or a television image to figure out what has happened. It seems evident from the chimp's behavior that he does not simply think "*someone* has paint on his forehead" when he looks in the mirror or at the television, but rather thinks to himself something like "*I* have paint on my forehead." The chimps in this experiment make much the same discovery as the man pushing the shopping cart in Perry's story. Cheney and Seyfarth would say that the chimps have recognized themselves, but this does not mean that the chimps view themselves as possessing minds that contain beliefs and desires. Self-recognition does not require that the individual be a psychologist. This is how self-recognition and self-awareness differ.

Although passing the mirror test is evidence that the organism has the concept of "I," we do not think that this behavior is a necessary condition for having that conceptual competence. Even if dogs look blankly at their images in a mirror or on a television screen, they still may have the ability to represent themselves indexically in their perceptual beliefs. To be sure, it remains unclear what the scientific ground rules are for attributing thoughts to nonhuman organisms. But when Fido wants the bone that his master holds before him, it seems natural to attribute to him a desire whose content is "I have the bone." Similarly, when Fido barks at dogs that wander too close to where he lives, it is hard to resist attributing to him the belief that "this is my territory." These attributions may not exactly capture what Fido has on his mind, but attributing the concept of "I" to Fido is no more of an idealization than attributing to him the concept of "bone" or "territory."

In the debate between egoism and altruism, everyone agrees that some of the desires that people have are egoistic in character. We have suggested that individuals are able to have these uncontroversial desires only if they have the concept of "I." Being able to use this concept does not require self-awareness, but merely self-recognition. Organisms that form perceptual beliefs that represent objects in their environment in terms of how those objects are related to self may plausibly be thought of as employing the concept of "I."

Ultimate versus Instrumental Desires

One of the most important conceptual distinctions that figures in the debate between egoism and altruism is the one noted earlier between ultimate and instrumental desires. The egoist does not deny that people sometimes want others to do well. For example, if Susan and Otto are business partners, Susan may want Otto to meet with success. The egoist will insist, however, that Susan's desire that Otto do well is only instrumental; the only reason Susan wants Otto to do well is that she thinks this ultimately will benefit Susan.

Defenders of the egoism hypothesis affirm claims of the form "*S* wants *O* to do well only as a means to obtaining a benefit for *S*." Proponents of the altruism hypothesis deny some statements that have this form; they sometimes advance claims of the form "*S* wants *O* to do well as an end in itself, not just because this will provide *S* with a benefit." Thus, both theories require that we understand the difference between ultimate and instrumental desires. Here is how we understand the idea:

> (1) *S* wants *M* solely as a means to satisfying *S*'s desire for *E* if and only if (a) *S* wants *M,* (b) *S* wants *E,* and (c) *S* wants *M* only because *S* believes that obtaining *M* will promote obtaining *E*.

We say "promotes" because the agent need not think that getting *M* will *suffice* for the attainment of *E*. For example, suppose that Alice pays attention to traffic while driving only because she wants to avoid automobile accidents. Paying attention isn't sufficient for avoiding an accident, but it causally contributes to that outcome.

The means-end relation links desires together in a chain; a desire that is relatively instrumental traces back to a desire that is more ultimate, and this desire may trace back to a third desire that is more ultimate still. Arnold wants to drive the car because he wants to go to the bakery. He wants to go to the bakery because he wants to buy some bread. He wants to buy some bread because he wants to make a sandwich. We will say that a desire is "ultimate" or "irreducible" or "an end in itself" if it is not held for purely instrumental reasons.[15] We mentioned earlier that people want to avoid physical pain, not just because this is a means to something else, but because they simply don't like pain; avoiding pain is an end in itself. Of course, even if this is one ultimate goal that people have, it remains to be seen whether there are others. Perhaps some organisms, ourselves included, have several means-end chains of desires, each with its own first member.[16]

Proposition (1) defines what it means for someone to want *M* solely as a means to obtaining *E*. This definition can easily be modified to define a slightly different concept. Suppose someone desires *M* for *two* reasons—because *M* contributes to E_1 and because *M* contributes to E_2. If so, it will be false that *M* is desired *solely* as a means for attaining E_1; however, it will be true that *M* is desired as a means for attaining E_1. This contrast will be important in what follows. Egoism maintains that we want to help others *solely* because doing so provides some benefit to self. An opponent of egoism may concede that the desire for self-benefit is *one* reason we help, but will deny that it is the *only* one.

Proposition (1) describes the means-end relation in terms of a static "snapshot"; the organism desires *M* as a means to obtaining *E* at a particular time if and only if its wanting *M* and wanting *E* are related in a certain way *at that time*. We now will supplement this picture with a dynamic description. The means-end relation can be characterized in terms of how the organism is inclined to *change:*

> (2) If *S* desires *M* solely as a means to attaining *E*, then, if *S* comes to believe that *M* won't provide *E*, *S* will stop desiring *M*, continue to desire *E*, and try to find new means for attaining *E;* what *S* won't do is stop desiring *E*, continue to desire *M*, and try to find new means for attaining *M* (Batson 1991).

Definition (1) includes the idea that the individual *believes* that attaining *M* will promote the achievement of *E*. The new perspective in (2) is afforded by asking what happens if the organism comes to *disbelieve* that this relation obtains.

There is an additional dynamic implication of the means-end relation. It concerns what happens when one desire is satisfied but the other is not:

> (3) If *S* desires *M* solely as a means to attaining *E*, then, if *S* attains *E* but does not attain *M* (or more precisely, if *S* *believes* that this has happened), *S* will stop desiring *M*; however, if *S* attains *M* but does not attain *E*, *S* will not stop desiring *E*.

Whereas (2) describes a situation in which the agent has not yet attained either *M* or *E*, (3) describes what the agent will do if one is attained but the other is not.[17] If Arnold wants to visit the bakery only because he wants a loaf of bread, proposition (3) tells us we can make two predictions. He will stop wanting to make the trip if someone delivers a loaf to his door; if he goes to the bakery and they are sold out, he will not stop wanting bread.

Propositions (2) and (3) describe an important difference between instrumental and ultimate desires: *when someone desires M solely as a means to attaining E, he or she is less disposed to give up E than to give up M*. *E* is more deeply entrenched than *M* is—it is less subject to displacement when the subject acquires new information. Anything that removes *E* from the agent's set of desires is apt to displace *M*; in contrast, removing *M* from the set of desires probably won't extinguish the desire for *E*.

This does not mean that an individual's ultimate goals are totally unalterable. A blow to the head, a psychotropic drug, or a normal developmental transition may change an individual's ultimate desires. Indeed, the mere passage of time can make an ultimate desire wither away. Consider Wanda. At age 20, she wants to be a harpsichordist and also wants to be a painter; both of these she regards as ends in themselves, not as means to attaining anything else. Eventually, Wanda comes to see that she must decide which of these goals to pursue; there isn't time enough for both. Suppose she decides to be a harpsichordist. What will happen to Wanda's desire to paint? It is possible that she will retain this desire and that it will remain unsatisfied. However, it also is possible that as Wanda immerses herself more and more in music, her desire to paint will gradually diminish and finally disappear.

In the *Treatise of Human Nature* (1739), Hume claims that "reason is, and ought only to be the slave of the passions." A natural gloss of this famous pronouncement is that reason cannot lead one to adopt, modify, or abandon an ultimate desire. Reason helps us decide what means we should pursue, given the ends we have; our ultimate ends, however, are beyond the reach of reason (Enç 1996). Hume's use of the terms *is* and *ought* suggests that he was, in fact, making two claims, not just one. There is a *descriptive* claim about the effects that psychological processes of reasoning in fact have; there is also a *normative* claim about what reasoning *ought* to achieve. We think the descriptive claim may well be wrong.

For consider Silas, who wants wealth for its own sake, not as a means to any more ultimate end. Suppose a philosopher suggests to Silas that he ought to change how he feels about money because the pursuit of wealth for its own sake will not make him happy. Perhaps this argument should strike Silas as irrelevant. After all, if Silas is as described, then the argument should fail to impress him, since he does not regard the pursuit of wealth as a mere means for attaining happiness; the fact that money doesn't buy happiness should leave him unmoved. However, this point about how Silas *should* react to the argument does not guarantee how he will react *in fact*. The argument may trigger psychological processes that cause Silas to change how he feels about money.

We take no stand on whether Hume's normative thesis is true. If reasoning could change a person's ultimate desires, what would be wrong with its doing so? Is it really so clear that a change in ultimate desires induced by a reasoning process must be due to a malfunction? In any event, we do think the descriptive claim is probably false. As the example of Silas illustrates, it seems possible for reasoning to alter a person's ultimate aims; they are not *completely* impervious to rational reflection. The important point, reflected in propositions (2) and (3), is that ultimate aims are usually *less* changeable than the means we adopt to achieve them.

Propositions (2) and (3) describe how the means-end relation is *predictive*. If we know that Sally desires *M* solely as a means to obtaining *E*, this allows us to predict how she *will* behave.[18] In this respect, saying that a subject's desires are arrayed in this way resembles attributing a dispositional property—a tendency to behave in a certain way—to an object. When we say that a sugar cube is *now* water-soluble, we are saying how it *will* behave if it is immersed. It is interesting that dispositional claims generally, and attributions of means-end relations to a person's desires in particular, are not *retrodictive.* To say that an object is water-soluble is to say how it *will* behave if immersed; the claim says nothing, *per se,* about how the object got to be that way. The attributions, so to speak, are future-directed, not directed toward the past. An object may now be water-soluble because it was always thus. Alternatively, the object may be water soluble now because it underwent a process that changed its internal constitution.

This point has special relevance to the issue of egoism and altruism. If Fred now wants Ned to do well, and this is an ultimate, not an instrumental, desire on Fred's part, then Fred is now an altruist. On the other hand, if Fred now wants Ned to do well purely as a means to producing some selfish benefit for Fred, then Fred, as so far described, is an egoist. We might try to test which of these propositions is true by availing ourselves of the machinery provided by propositions (2) and (3). However, if we decide, after performing the experiment, that Fred is now an altruist, this says nothing about what caused him to become one. Fred may have been altruistically inclined from birth (like the object that was water-soluble throughout its lifetime), or Fred may have acquired this altruistic inclination at some point in growing up. It is even conceivable that Fred initially acquired his concern for Ned because he was rewarded for doing so, and that this concern then took on a life of its own; what begins as a purely instrumental desire may get transformed into a desire that is *functionally autonomous* (Slote 1964; Kavka 1986).

That a desire may change in status is not an esoteric theoretical possibility. We often start to pursue an activity for one reason, but continue to do so for another. It is a familiar fact about parenthood, and about other relations of affection, concern, and love, that people often begin caring about others for one reason, but continue caring about them for another. A mother and father start caring for their baby, rather than for the other babies in the maternity ward, because they believe that *this is their child.* However, after living for some years with the child, how many parents would stop caring if they suddenly learned that a mistake had been made at the hospital—that they had been sent home with the wrong baby? Some parents might stop. The point, however, is that many would not. Parents begin caring for children because the child bears a particular biological relation to them; later on, they care for their children for reasons having nothing much to do with that.

This scenario suggests another. Perhaps people care about others for hedonistic reasons early in life, but then their desires change—they come to care about others as ends in themselves. Even if children are hedonists or egoists, this does not mean that the adults those children become cannot be altruists.[19] Of course, the conceivability of this transformation does not show that it actually occurs. Nonetheless, we must recognize that a desire that is acquired for hedonistic reasons may or may not retain that status later on.[20]

7

Three Theories of Motivation

The present chapter will map out three psychological theories of motivation—hedonism, egoism, and altruism. We intend to proceed carefully, since evaluation of these theories is often short-circuited by biased definitions and spurious arguments. Just as it is easy to solve the evolutionary problem of altruism by defining "selfishness" as *whatever evolves,* so it is easy to solve the psychological problem of motivation by defining "self-interest" as *whatever people want.* More subtle biases can and do enter discussion of the psychological issue. The care we take here in formulating our hypotheses will pay dividends in subsequent chapters when we examine different attempts to resolve the motivational question.

In the course of delimiting the three motivational theories, we will have to address a variety of related issues. How is altruism related to morality? How is it related to the emotions of empathy and sympathy? Does egoism assume that agents always rationally calculate what is in their best interests? These details are important to our project because they help specify exactly what the traits are that need to be investigated. Before we can evaluate whether people ever have altruistic ultimate motives, or ask what evolutionary theory has to say about this motivational question, we must have a clear view of the phenotypes that require analysis.

Defining Hedonism

The discussion at the end of the last chapter of what it means for a desire to be ultimate or instrumental makes it easy to define hedonism. Hedonism says that the only ultimate desires that people have are the desires to obtain pleasure and avoid pain. All other desires are purely instrumental with respect to these two ends. Construed in this way, hedonism is a descriptive theory, not a normative one; it does not recommend actions or say whether it is good or bad that people are as they are. The theory merely attempts to describe how the mind is structured.[1]

When referring to our aversion to pain, hedonism uses the term *pain* quite inclusively. In ordinary parlance, there are many aversive sensations besides pain—nausea, dizziness, anxiety, depression, and a host of others. The hedonist has no trouble with the idea that avoiding nausea can be an end in itself. Yet, it may sound a little odd to say that nausea is a type of pain. The solution is to understand the term *pain* as encompassing all aversive sensations. Any sensation that a person dislikes experiencing the hedonist will dub an instance of "pain."

Similar remarks pertain to the word *pleasure.* If pleasure is a sensation, it is very hard to say which particular sensation it is. People "take pleasure" in a variety of experiences. One can enjoy the taste of a peach, but also be pleased to learn that others are doing well (or ill). In what sense do these two experiences "feel the same"? Hedonism need not insist that they involve a single type of sensation. Both can count as instances of pleasure, if "pleasure" names any experience that a person enjoys having (Sidgwick 1907).

The distinctive feature of hedonism is that it says that ultimate desires are always *solipsistic.* What we ultimately care about is limited to states of our own consciousness; what goes on in the world outside the mind is of instrumental value only.

Defining Egoism

Egoism maintains that the only ultimate goals an individual has are *self-directed;* people desire their own well-being, and nothing else, as an end in itself. If you care about the well-being of others, this is only because you think the well-being of others is instrumentally related to benefiting yourself. Strictly speaking, the ultimate desires postulated by egoism say nothing about the situation of others. Malevolence is as alien to the egoist's basic outlook as benevolence is (Butler 1726).

Egoism's use of the term *self-directed* requires clarification, but two properties of the egoistic theory should be clear from the outset. First, an irreducible concern for the welfare of others is incompatible with egoism. Second, an irreducible concern for obtaining pleasure and avoiding pain is quite consistent with egoism. In other words, the altruism hypothesis is incompatible with psychological egoism, whereas hedonism is a species of egoism. Although all hedonists are egoists, not all egoists are hedonists. An egoist may have the desire to accumulate money as an end in itself, but a hedonist may not. The same would be true of the ultimate goal of climbing Mount Everest. Egoists need not have states of their own consciousness as the only things they care about as ends in themselves.

To determine whether a desire is "self-directed," we must attend to the propositional content that the desire has. If Sam wants to eat an apple, this is a self-directed desire, because the proposition *Sam eats an apple* mentions Sam but no one else. Similarly, if Sam wants Aaron to eat an apple, this is an other-directed desire, because the proposition *Aaron eats an apple* mentions Aaron but not Sam himself. An egoistic ultimate desire is self-directed; an altruistic ultimate desire is other-directed.

This construal of egoism encounters problems when we consider desires that mention both self and other. Suppose that Aaron wants to be famous, not as a means to anything else but as an end in itself. A little reflection on what "famous" means indicates that the content of Aaron's desire involves a relation between self and other; Aaron wants others to know who he is. Even though Aaron's desire isn't purely self-directed, it may sound odd to conclude that Aaron is not an egoist.[2] A parallel difficulty arises if we define altruism as the claim that some of our ultimate desires are purely other-directed. Suppose Sam wants the apples to be distributed equally between himself and Aaron, not as a means to some further end but as an end in itself. Even though Sam's desire isn't purely other-directed, it may sound odd to conclude that this desire is not altruistic.

These problems might disappear if the desires just mentioned were merely instrumental. If Aaron wants to be famous only because he thinks this will provide him with pleasurable experiences, then he is an egoist. Similarly, if Sam wants the apples to be distributed equally in part because he has the ultimate goal that Aaron's situation be improved, then he is an altruist. Unfortunately, these suggestions evade the problem posed. How should we categorize desires that have relational facts as their propositional contents *when these desires are ultimate?*

To cram this variety of ultimate desire into egoism, or into altruism, would be difficult to justify, and the attempt to do so might appear to bias the case in favor of one position or the other. In consequence, we propose to add a third category to egoism and altruism; *relationism* is the view that people sometimes have ultimate desires that certain relational propositions (connecting self and specific others) be true. If the reader thinks that some cases of relationism are properly viewed as subspecies of altruism or of egoism, we invite the reader to adjust the conceptual taxonomy we are suggesting. Our assessment of these theories will not be affected by such amendments.[3]

We have defined egoism so that the distinction between conscious and unconscious desires plays no role. An individual whose ultimate desires are *conscious* and an individual whose ultimate desires are *unconscious* will both be egoists if those ultimate desires are all directed solely toward self-benefit. The fact that individuals sincerely want to help others, and do not consciously experience this desire as involving a sacrifice or a conflict with their authentic selves, does not tell us what their ultimate motives really are. There is nothing in the egoism hypothesis that prohibits other-directed desires from being fully integrated into the agent's personality. Those desires must be instrumental, but people need not experience them as alien intrusions.

We hope it is obvious that the egoism hypothesis we have described is not the same as the view that might be called "vulgar egoism." Vulgar egoism maintains that people are moved solely by the goal of securing *material* benefits. We think it is obvious that this version of egoism is too narrow; the desire for material benefit is *one* motive that people have, and so it helps explain *some* aspects of human behavior. But there is much that it cannot explain. Egoism, unlike vulgar egoism, deploys a wider notion of self-benefit, one that includes internal (psychological) payoffs as well as external (material) ones.

Egoism is sometimes criticized for viewing happiness as a one-dimensional state (LaFollette 1988), but this criticism does not apply to the version of the theory we have described. If someone wants to discover a cure for cancer, to climb Mount Everest, and to experience the euphoria of romantic infatuation, the egoist need not pretend that these three goals somehow boil down to the same thing. Each of these desires is self-directed; if they exhaust what the agent's ultimate desires are, then the agent is an egoist, regardless of whether these desires reduce to a deeper unity. Symmetrically, the altruism hypothesis is not committed to the idea that people think of the welfare of others as a simple and one-dimensional matter.

We have one last comment on our definition of egoism. It is better to describe egoism as holding that all ultimate ends are self-directed than to say that it views all ultimate ends as "selfish" (Henson 1988). When Jim wants his tooth to stop aching, it is misleading to say that Jim is being selfish in feeling this way. The term *selfish* carries with it overtones of disapproval. This is not, strictly speaking, what egoism maintains; egoism is a descriptive, not a normative, theory.

Short-term and Long-term Egoism

Consider Ronald. He is now deliberating about whether he will continue to smoke cigarettes. He realizes that any single cigarette will have only a negligible impact on health; but over the long run, smoking many cigarettes may well have a devastating effect. Ronald is all too cognizant of the pleasure he gains from each and every cigarette. As he thinks about whether to stop smoking altogether, his mind drifts to a more immediate question. He holds an unlit cigarette before him and considers whether he should light it and smoke. According to Ronald's way of seeing things, smoking produces a short-term benefit. But he also realizes that over the long term smoking is apt to impose a substantial cost. If Ronald cares only about the here and now, he will light up. If he cares enough about the long-term quality of his life, he will stop smoking, starting today.

Our interest here is not to predict what Ronald will do but to ask what egoism says about this problem. Notice that the short-term and the long-term considerations that weigh with Ronald are both self-directed. Wanting the pleasure produced by nicotine is a self-directed desire, but so is the desire to be healthy and long-lived. Although egoism says that our ultimate desires are always self-directed, it does not say how much importance people assign to present versus future benefits. Some may see this lack of specificity as a defect in the egoistic theory; this is a criticism that we will analyze in Chapter 9. For now, we merely register our opinion that this flexibility in egoism is unobjectionable. Whatever Ronald does in the situation just described will be consistent with egoism. Egoism does not say which specific desires people have as their ultimate ends; it says merely that they strive for a certain *type* of ultimate goal.[4]

Defining Altruism

The altruism hypothesis maintains that people sometimes care about the welfare[5] of others as an end in itself. Altruists have irreducible other-directed ends.

The word *sometimes* marks a logical difference between the altruism hypothesis and the hypotheses of hedonism and egoism. Hedonism and egoism are claims about *all* the ultimate desires an individual has, whereas altruism makes no such universal claim. Egoism says that all ultimate desires are self-directed, but the theory we are calling altruism does not say that all ultimate desires are other-directed. Of course, it is possible to construct a monolithic theory of this type, but no one would believe for a moment that it is true. Rather, we should construe altruism as part of a *pluralistic* theory of motivation that maintains that people have ultimate desires about others as well as about themselves. Egoism and hedonism, on the other hand, are rightly understood as (relatively) *monistic* doctrines.[6]

The thesis of altruism, as we understand it, says that some people at least some of the time have the welfare of others as ends in themselves. This does not entail that *most* people are altruistic *all* the time, or that *some* people are altruistic *most* of the time, or that the altruism that people sometimes experience is especially *strong*. A person who is prepared to make a very small sacrifice in order to provide a huge benefit for someone else may be altruistic, but this person will be less altruistic than someone who is prepared to make a larger sacrifice. Our version of altruism is quite compatible with the existence of widespread selfishness (a point to which we will return).

It might be suggested that the altruism hypothesis, so construed, is too modest a theory to be of much interest. If the theory claims merely that people sometimes have irreducibly altruistic motives, but says nothing about how strong or pervasive those motives are, why is it worth discussing? To be sure, there is more to the psychology of altruism than the altruism hypothesis we have just identified. However, we believe that this hypothesis is fundamental because more ambitious claims about the importance of altruism are committed to this modest thesis. In addition, this apparently modest claim is precisely what psychological egoism denies—it is the very nub of the issue.

Who are the others about whom altruistic individuals have ultimate concerns? The most obvious examples involve desires that focus on the welfare of another *person.* But consider people who care irreducibly about "the environment," meaning the well-being of the entire earth (both living and nonliving). Is this altruism? And what about people who care about a nation, a religion, an ethnic group, or a cultural tradition, not just as means but as ends in themselves?[7] True, such concerns sometimes count as "selfless." But are they altruistic? Although we will concentrate on the altruistic regard that human beings may have for other human beings, there is no reason to rule out these other candidates. The principal point about altruism, as we understand it, is that it attributes to people ultimate desires concerning the welfare of individuals other than themselves. We are prepared to be quite liberal concerning what might count as an "individual." This decision won't affect the main arguments we will advance in what follows; readers are invited to construe altruism more narrowly if they so wish.

The altruism hypothesis says that we have other-directed ultimate desires, whereas psychological egoism says that all of our ultimate desires are self-directed. But egoism and altruism are usually understood to involve more than this. For example, if Iago views the destruction of Othello as an end in itself, then one of Iago's ultimate aims is other-directed. Nonetheless, it would be odd to call Iago an altruist, since his other-directed desire is *malevolent.* Similarly, people whose only ultimate aim is to harm or destroy themselves would not normally be counted as egoists, since what they want for themselves is not their own good. This is why our definitions of egoism and altruism go beyond the distinction between self-directed and other-directed ultimate desires. Egoists ultimately desire only what they think will be good for themselves; altruists have ultimate desires concerning what they think will be good for others.

The benevolent intentions associated with altruism can take two forms. An altruist may want others to have what they actually want for themselves; alternatively, an altruist may want for others something they have never thought of, or have considered and rejected. If Sheila buys Oscar a particular book simply because Oscar has been wanting that book, we may have a case of the first type. If Stanley wants Olivia to take her medicine even though she does not want to do so, we may have a case of the second.

The concept of altruism is sometimes restricted to cases in which an individual helps others without any expectation of receiving an *external* benefit, such as money or power (Macaulay and Berkowitz 1970); this definition entails that individuals are altruistic when they help just because they think that helping will make them feel good. We disagree with this way of defining the concept. To see why, consider a heroin addict whose every action is ultimately aimed at securing the pleasant states of consciousness that the drug produces. The addict, as so far described, is a hedonist. But now let us perform a thought experiment. Let us place this person in an environment in which the only way to get the drug is by helping people. This is very different from the real world that addicts usually inhabit, but the hypothetical situation is worth considering for the light it sheds on the conceptual issue. An addict who helps others only because the effect of helping is a drug-induced euphoria is not thereby an altruist. The same point applies to people in the real world who do not take heroin; if they are "hooked" on helping because of the pleasure that helping affords and the pain it allows them to avoid, their actions do not make them altruists. We must not muddy the waters by treating altruism as a form of hedonism.

Another definition of altruism also merits comment. Altruism is sometimes defined by saying that individuals have the ultimate desire that other people's desires be satisfied. This resembles a formulation used in the social sciences according to which an altruist is someone whose utility function "reflects" the utility functions of others. These definitions entail that altruists must have representations of the mental states of others. We believe that this formulation of the theory is stronger than it should be. Suppose Stanley wants Olivia to take the medicine simply because Stanley believes that it would be good for her. Stanley is not thinking about Olivia's *desires* but about her *health*. Perhaps Olivia doesn't want to take the medicine; she is so despondent that she doesn't even want to get well. Nonetheless, Stanley is an altruist because he has Olivia's well-being as an end in itself. Maybe altruists are psychologists, as discussed in Chapter 6; however, we don't think this is true as a matter of definition.

Now that hedonism, egoism, and the pluralistic theory of motivation that embeds the altruism hypothesis have been formulated, let us reflect on how they are logically related. First we note an asymmetry. Hedonism entails egoism, but egoism does not entail pluralism; indeed, egoism and pluralism (which includes altruism) are incompatible. On the other hand, there is a symmetry here, which can be identified by considering what each hypothesis entails about the ultimate motives that might be involved in explaining why an individual performs a particular action. These views are *nested.* Suppose Lois helps someone. Hedonism says that Lois did this because she cares ultimately about the state of her own consciousness, and about nothing else. Egoism grants that this may be *part* of the explanation, but says that it need not be the whole story. According to egoism, Lois helped because she cares ultimately about her own situation, not about the welfare of others. Psychological pluralism grants that this may be *part* of the explanation, but denies that it must be the whole truth. The transition from hedonism to egoism, and from egoism to pluralism, involves canceling restrictions on the set of ultimate desires that might explain an agent's behavior.

Our construal of egoism and altruism entails that these two types of motivation are not exhaustive. Besides ultimate desires for one's own well-being, and ultimate desires concerning the well-being of this or that other individual, there are additional possibilities to consider. We have mentioned *relationism,* which maintains that people ultimately desire that certain relations between self and specific others obtain. Later in this chapter, we will argue that the ultimate desire to uphold a general moral principle should be regarded as neither altruistic nor egoistic.

Empathy, Sympathy, and Personal Distress

Empathy and sympathy are emotions. When they occur, do they trigger altruistic desires? Common sense suggests that they do; empathy and sympathy sometimes elicit helping behavior, and it makes sense to see this behavior as tracing back to the desire to improve the other person's situation. The causal chain seems to be this:

Emotions of empathy and sympathy ⟶ Desire to help ⟶ Helping

Psychologists have reached the same conclusion; see Batson (1991, pp. 93–96) for a review. However, even if empathy and sympathy have these effects, the question remains of whether the resulting desire to help is ultimate or instrumental. Perhaps empathy and sympathy are able to evoke altruistic desires because people don't like experiencing these emotions and therefore wish to do what they can to extinguish them. Thus, the existence of empathy and sympathy does not resolve the debate between psychological egoism and altruism. Nonetheless, it is worth getting clear on what empathy and sympathy are and why they differ from the altruistic desires they sometimes cause.

The term *empathy* entered English in 1909 as E. B. Titchener's translation of *Einfühlung* (Wispé 1987); since then, its meaning has gone through several metamorphoses in different branches of psychology and it has been absorbed into everyday English as well. The term *sympathy* has an older provenance, but it too has been used in different ways and has been expropriated as a term of art in various psychological theories. Although the definitions we will suggest coincide with ordinary and scientific usage in some respects, they depart from that usage in others. Given the fact that both terms have been put to multiple uses, it would be quixotic to expect a single pair of definitions to agree with what each and every person using the terms has meant. Rather, our response to the present Babel of meanings is to try to single out what is fundamental. In any event, the categories we will describe are more important than the labels we will use to name them.

Empathy is sometimes contrasted with sympathy by saying that empathy involves identifying with others, whereas sympathy involves a more detached variety of emotional connection. What does "identification" mean here? The idea is sometimes explained by saying that empathy makes the boundary between self and other disappear. We believe that this is almost always a poetic overstatement (so does Batson 1991). When Barbara learns that Bob's father has just died, she may empathize with Bob without losing sight of the fact that they are two different people, not one and the same person. As much as Barbara's heart goes out to Bob, Barbara understands perfectly well who it is who has just lost a parent. When people confuse the real misfortunes of others with their own more fortunate situation, we do not praise them for their ability to empathize; empathy is not the inability to keep track of who is who.

An everyday example of what it means to "identify with" another individual is provided when people talk of identifying with a sports team. This doesn't involve the delusion of believing that one is identical with the New York Yankees. Rather, it means regarding one's self as part of a whole to which the team also belongs and caring about the fate of that whole. Good deeds performed by Yankees make one proud, foul deeds make one ashamed. "What they do reflects on me," or so Yankees fans seem to feel. This same pattern of thinking is present in deeper and more pervasive types of identification—with family, clan, ethnicity, nationality, and religion. The "I" is defined by relating it to a "we." Human beings don't simply *belong* to groups; they *identify* with them. This is an important fact about human experience.

Whether or not empathy entails identification, we suggest that empathy involves sharing the emotion of another. Barbara's empathy involves her feeling sad because Bob is sad. Of course, it is perfectly possible that Barbara may empathically connect with one of Bob's emotions, but not with another. Suppose that Bob feels both sad and guilty about his father's death and that Barbara empathizes with the sadness, but not with the guilt, of which she is unaware. No precise degree of similarity between the two individuals' overall emotional states is built into the concept of empathy (Eisenberg and Miller 1987; Eisenberg and Strayer 1987).

Empathy is sometimes said to require "perspective-taking." We agree that when people feel empathy, they typically have some understanding of why others see the world as they do. However, we don't want to build this in as a requirement. If *O* feels terror or sadness, *S* may see that this is so without knowing what it is in *O*'s situation that elicited these emotions. *S* may respond empathically; *S* may "feel for" *O* with the end result that their emotions match. Empathy requires one to understand *that* the other person is experiencing an emotion; it need not involve a deeper grasp of *why* this is so.

Although empathy involves emotion matching, more is required. Suppose that O feels so depressed and anxious that he is unable to think of anyone but himself; *S* learns about this, and this information somehow causes *S* to go into precisely the same state. *S* is now matching O's emotion, but *S* is not empathizing with O. The reason is that *S* is not even thinking about O. It is one thing for *S* to feel sad, quite another for *S* to feel sad *for* O. The same distinction can be drawn with respect to other emotions; for example, when *S* feels frightened, this doesn't necessarily involve *S*'s feeling frightened *for* O. This point about empathy, as well as the other observations we have made, are consolidated in the following definition:

> *S* empathizes with O's experience of emotion *E* if and only if O feels *E*, *S* believes that O feels *E*, and this causes *S* to feel *E* for O.[8]

What does it mean for one individual to feel a certain emotion "for" another? Here we may exploit the idea that belief involves the formation of representations that have propositional content (Chapter 6). If *S* feels sad for O, then *S* forms some belief about O's situation and feels sad that this proposition is true. When Barbara empathizes with Bob, he is the focus of her emotion; she doesn't just feel the same emotion that Bob experiences. Rather, Barbara feels sad *that Bob's father has just died;* she feels sad about what has made Bob sad.

Our definition of empathy does not require that the shared emotion be negative. When O feels sorrow, *S* can empathize, but the same is true when O feels joy.[9] Herein we find one difference between empathy and sympathy. It sounds odd to talk about someone's sympathizing with another by feeling happy. If *S* sympathizes with O, then *S* must feel bad. There is, however, a more important difference. Consider the fact that your heart can go out to someone without your experiencing anything like a similar emotion. This is clearest when people react to the situations of individuals who are not experiencing emotions at all. Suppose Walter discovers that Wendy is being deceived by her sexually promiscuous husband. Walter may sympathize with Wendy, but this is not because Wendy feels hurt and betrayed. Wendy feels nothing of the kind, because she is not aware of her husband's behavior. It might be replied that Walter's sympathy is based on his imaginative rehearsal of how Wendy would feel if she were to discover her husband's infidelity. Perhaps so—but the fact remains that Walter and Wendy do not feel the same (or similar) emotions. Walter sympathizes; he does not empathize.[10]

Even though sympathy does not require emotion matching, it still is not the same as a dispassionate grasp of someone else's misfortune. We therefore propose the following definition:

> *S* sympathizes with *O* precisely when *S* believes that something bad has happened to *O* and this causes *S* to feel bad for *O*.

This definition, like the one proposed for empathy, makes use of the idea that one person feels a certain emotion "for someone else." Feeling bad for someone requires that one feel bad. That is, one must experience an "aversive" emotion, such as sadness or anger. Aversive emotions are feelings that people dislike having, but this is not to deny that people often think they should be experiencing such emotions. When bad things happen to those we care about, aversive emotions are the ones we think it is appropriate for us to feel. We'd rather not experience them in the sense that we'd rather the world not contain situations that prompt them.

Thus defined, sympathy and empathy both differ from *personal distress,* a point first made in the psychology literature by Daniel Batson (see, e.g., Batson 1991). When the perceived misfortune of another causes one to feel bad and one quite forgets the other person, the self-focused emotion that results is neither empathy nor sympathy. Personal distress involves feeling bad without feeling bad *for someone else.* There is evidence that this difference between empathy and sympathy on the one hand and personal distress on the other is associated with various physiological differences. Sympathetic and empathic concern for others is associated with *lowered* heart rate, whereas personal distress (even when triggered by the situation of another person) is accompanied by *increased* heart rate; different facial expressions and degrees of skin conductance are associated with the emotional difference as well (Eisenberg and Fabes 1991). These physiological correlates of empathy and sympathy are consistent with the more general pattern of *somatic quieting,* which often accompanies an individual's focusing attention on the external environment (Lacey 1967; Obrist et al. 1970).[11]

Our definitions entail that empathy and sympathy are emotions that involve a cognitive component; each requires the formation of a belief. How, then, should we describe infants a few days old, who often cry when they hear other infants crying (Simner 1971; Hoffman 1981b)? Perhaps the causal chain in such cases is something like the following:

$$O \text{ is unhappy} \longrightarrow O \text{ cries} \longrightarrow S \text{ is unhappy} \longrightarrow S \text{ cries}$$

S is unhappy *because* O is unhappy. Even if *S* forms a belief in this instance, it isn't so clear that *S* forms the belief that O is unhappy or that *S* believes that something bad has happened to O. We take no stand on the empirical question this raises about child development.[12] Right now, we merely point out a consequence of our definitions. Maybe reactive crying is a precursor of empathy and sympathy, rather than the genuine article (Hoffman 1981a; Eisenberg and Miller 1987; Eisenberg and Strayer 1987; Thompson 1987).

Although empathy and sympathy both require the formation of a belief, the requisite types of belief are different. Empathy entails a belief about the emotions experienced by another person. Empathic individuals are "psychologists" (Chapter 6); they have beliefs about the mental states of others. Sympathy does not require this. You can sympathize with someone just by being moved by their objective situation; you need not consider their subjective state. Sympathetic individuals have minds, of course; but it is not part of our definition that sympathetic individuals must be psychologists.[13]

Empathy and sympathy do not automatically entail the existence of altruistic desires. Nancy Eisenberg (personal communication) has suggested a simple way to see why. It is possible to enter these emotional states by thinking about problems that have already been solved. Suppose Wendy discovers her husband's infidelity, divorces him, and then creates a good life for herself. If she then recounts this sequence of events to Walter, he may find himself empathizing with the Wendy of a few years past. What does this empathy motivate Walter to do? Apparently, Walter empathizes without forming the desire that Wendy's situation be improved.

Even if empathy and sympathy are causes of altruism, other causes may be possible. Perhaps one can want the situation of another person to improve without feeling anything. Something like this more detached form of altruism may occur when people learn about disasters in distant places. People often feel empathy or sympathy when they meet suffering face-to-face; reading about suffering in the newspaper can fail to elicit this emotional reaction. Perfectly decent people are able to go about their daily activities after they learn of the horrible misfortunes that beset strangers. It isn't that the information fails to elicit desires concerning the welfare of others; what may be true is that the bad news fails to make them feel bad. The emotions of empathy and sympathy most commonly arise when people directly perceive individuals in trouble or have a personal connection with them that allows a third-person report to register powerfully. However, perhaps we have the ability to care about suffering that we neither see nor hear, and that afflicts individuals who are not our near and dear. It is possible that other-directed desires come into existence without the mediation of an empathic pathway (Karniol 1982).

Altruism and Morality

Morality and altruism are sometimes equated, both at the level of action and at the level of motivation. The first equation says that morality always requires us to sacrifice self-interest for the sake of others. The second says that to be motivated by an altruistic desire is the same thing as being motivated by a moral principle. Both these equations are mistaken; there *is* a relationship between morality and altruism, but we must consider the issues more carefully.

What is a moral principle?[14] Moral principles, like all principles, properly so called, are *general*. They specify general criteria or relevant considerations for deciding what one ought to do. Consider, for example, a distributive principle that is central to Rawls's (1971) theory of justice: *the difference principle* says that the resources in a society may be allocated unequally only if this benefits those who are worst off. Notice that this principle does not mention specific individuals. It does not say that Earl should receive a government subsidy or that Sarah should have her taxes increased, though the principle, in conjunction with specific facts about Earl and Sarah, may have precisely this implication. In this respect, moral principles formally resemble scientific laws of nature. Newton's law of gravitation is a general principle because it covers all objects that have a certain property (mass); the principle does not mention the Earth and the Sun, although the principle, in conjunction with specific facts about the Earth and the Sun, entails that they generate a particular gravitational force.[15]

Moral principles, if they are general in the way just described, conform to an abstract universalizability criterion. They entail that if it is right for one individual to perform an action in a given circumstance, then it is right for anyone else who is relevantly similar to perform that action in the same circumstance.[16] Of course, moralities differ in what they take the relevant similarity to be. Different moral principles specify different criteria; what is good according to one may be abhorrent according to another. For example, a tribal morality may lay down obligations that one has to group members but not to outsiders. Other moralities may claim that one has certain obligations to all human beings, or to all sentient organisms. Despite such substantive differences, however, these moral systems have in common the fact that they set forth principles of the form "anyone with such-and-such characteristics is to be treated thusly." Universalizability is an invariant feature.[17]

If moral principles must be general, then it is clear that an individual can have altruistic desires without being motivated by moral principles. This is because altruistic desires are often directed at specific individuals, whereas moral principles, in virtue of their generality, are about no one in particular. Suppose two parents want their child to do well, not for egoistic reasons but because they take the well-being of their child to be an end in itself. It is possible that the parents have this altruistic desire[18] without embedding it in any moral system at all. They may never formulate the thought that all parents should care about their children; nor need they think that if some other child were theirs, they would have an obligation to take care of that child as well. Perhaps this is especially clear when the parents in question are *nonhuman animals*. Specific desires need not be accompanied by endorsements of general principles.

This difference between one's general moral principles and one's altruistic concerns is easy to discern with a little reflection. Consider two women, Alma and Beth, who know each other only slightly. Each has a child; unfortunately, each child dies. Alma will almost certainly grieve more acutely for her own child than she will for Beth's. And if Alma is honest, she will admit that she wanted her own child to survive more than she wanted Beth's to do so. But in spite of these feelings and desires, Alma may recognize that, from a moral point of view, what happened to her and her child is no worse than what happened to Beth and to Beth's child. Moral principles involve a kind of *im*personal assessment that differs from the personal perspective that frequently accompanies our emotions and desires.

Just as an individual can be an altruist without being moved by moral principles, the converse is also possible. People sometimes believe that moral principles are binding for reasons that have nothing to do with how obeying those principles will affect the well-being of others. Some may find this deontological position wrong-headed, but the fact remains that many people (including influential philosophers, such as Kant), rightly or wrongly, have been deontologists. Examples may be found in moral beliefs that are grounded in theistic convictions. Many people believe that certain actions are required simply because God commands them. You are supposed to act in certain ways, not because God will punish you if you do not, and not because your conformity will benefit other people (or God), but simply because of God's say-so. People who accept this idea act on principles, but this does not entail that they have altruistic ultimate motives.

Another gap between altruism and morality is worth noting. Altruistically motivated actions can be morally wrong. It is easiest to see this when helping someone involves harming a third party. Suppose Alan cheats Betty at cards because Alan wants to use the money to buy something for Carl. Alan may be motivated by an altruistic concern for Carl, but that may not be enough to morally justify the way he treats Betty. A macabre illustration of this point is provided by accounts of the training that Nazi concentration camp guards and physicians received. They were taught that they had to overcome their feelings of revulsion because the atrocities they were committing were for the good of the German people (Lifton 1986). If these individuals helped implement the Final Solution in part because they had the ultimate goal of helping the *Volk,* they provide a striking example of how psychological altruism can help underwrite moral evil.

Just as altruistically motivated actions can be immoral, it also is possible for selfishly motivated actions to be the ones that morality requires. For example, consider the utilitarian maxim that goods should be distributed so as to maximize the collective happiness. Suppose a single dose of a medicine is available and that it will go to either Boris or Morris. Utilitarianism says that the drug should go to the person who will receive the greater benefit. If it is up to Boris to decide who will get the drug, and if the medicine would benefit him more, then utilitarian principles require him to take it. However, let's imagine that Boris doesn't think about utilitarian principles or about any other morality; he is merely a selfish person and so he keeps the drug for himself. According to utilitarianism, he has done the right thing (though not for the right reason). Moralities rarely require complete self-abnegation. Utilitarianism is an example; it says that self-interest counts no more *and no less* than the interests of others.[19] Moralities of this type have some implications that coincide with the dictates of egoistic desires, while other implications coincide with the dictates of altruism. Once again, it is the *im*personal character of moral principles that distinguishes them from the personal character of altruistic and self-interested desires.

We have argued that morality sometimes conflicts with self-interest and that at other times it conflicts with self-sacrifice. Although both clashes are possible, it is interesting that people often react differently to these two types of conflict. In our story about Boris and Morris, we said that utilitarianism requires Boris to take the medicine. If Boris fails to do this, and he selflessly gives the medicine to Morris, we might not feel great moral outrage. Our reactions would probably be different if morality required Boris to give the medicine to Morris, and Boris selfishly kept the drug for himself. Commonsense morality seems to set minimum standards concerning how much self-sacrifice is required, but it allows individuals to sacrifice *more* if they wish. We suspect that this is not a parochial feature of contemporary society but a fairly pervasive characteristic of people and societies generally. What might have led morality to take this form? Why doesn't morality place a lower bound on how much *selfishness* we are required to exhibit, but allow people to be more selfish if they wish? Surely the social function of morality is central to explaining this asymmetry—a point that connects with our discussion in the first part of this book of human beings as a group-selected species.

Satisficing and Irrationality

Hedonism, egoism, and the altruism hypothesis are all claims about the ultimate desires that people have. As such, none of them says anything directly about what people do. For any of these theories to generate predictions about an individual's behavior, it must be supplemented with assumptions concerning what the individual believes and the processes whereby beliefs and desires generate behavior.

One hypothesis that describes how beliefs and desires lead to action is that individuals are *rational maximizers*. This is the idea that people choose the action that their beliefs indicate will get them the most of what they want.[20] This assumption, much used in the social sciences, has come in for heavy criticism.

The idea that people are rational maximizers presupposes a kind of computational omniscience that mortal creatures do not possess. If the options under consideration are sufficiently complex, people may fail to figure out which option will provide the most of what they want. This problem led Herbert Simon (1981) to suggest *satisficing* as a more realistic principle. Individuals satisfice when they accept the first option that comes to mind that is *good enough*. A satisficer need not survey and analyze the entire field of alternatives. The savings in search time and computation can be considerable. Additional reasons to doubt that agents are rational maximizers come from the growing body of evidence that people systematically deviate from rational modes of inference. It isn't just the occasional lapse in attention that makes people draw an invalid conclusion from a set of assumptions. Rather, people often seem to reason by exploiting heuristics that work well in some contexts but lead to systematic error in others (Kahnemann, Slovic, and Tversky 1982).

Although psychological egoism is sometimes criticized for holding that people are rational maximizers, we feel that this objection is off the mark. *All* the motivational theories we are considering require a view about how beliefs and desires lead to action. Egoism is no more wedded to an unrealistic conception of this process than is the pluralistic theory in which the altruism hypothesis is embedded. In what follows, we usually will portray individuals as rational maximizers; however, we adopt this assumption strictly as a matter of convenience. When this idealization becomes problematic, adjustments can be made. The point of importance is that the inadequacy of the idealization is a problem for *all* the theories we need to consider. What is a problem for everyone is no one's problem in particular.

How Desires Interact

Individuals often have more than one desire that is relevant in an episode of deliberation. How should we understand the idea that desires "interact" in the production of behavior?

Desires are said to "push" or "incline" agents in different "directions." When two desires conflict, it is the stronger one that determines the behavior that ensues. This commonsense description of how desires work together may not entirely capture the rich phenomenology of our inner lives. Yet, it is a highly serviceable idealization, one whose implications we want to explore. The idea is that desires are related to action in the way that component forces impinging on an object are related to its resulting motion in Newtonian mechanics. If you push a billiard ball due north and someone else pushes it due south, the direction of motion is determined by which component cause is stronger.

The idea of conflict between desires is especially relevant to understanding the concept of altruism. As noted earlier, the altruism hypothesis is best thought of as part of a pluralistic theory of motivation. We need to be able to conceptualize the conflicts that can arise between a concern for self-interest and a concern for the welfare of others. This will help us get clearer on what the altruism hypothesis does and does not entail.

Consider a hypothetical example. Suppose you are thumbing through a magazine one day and see an advertisement. The ad asks you to send a check for $25 to a charity that helps starving children. You feel that you could afford this donation. Of course, you could find other things to do with $25. But the picture in the ad is pathetic. You believe that a $25 contribution will make a real difference for the children (it won't solve the whole problem, of course). You think for a moment and then send a $25 check to the charity.

There are at least two motives that may have moved you to action. Perhaps your motive was altruistic; maybe you cared about the welfare of the children, not as a means to some benefit for yourself, but as an end in itself. On the other hand, perhaps your motive was selfish.

Maybe you wrote the check in order to obtain a glow of satisfaction—a nice feeling about yourself—and to avoid feeling guilty. Indeed, it is conceivable that your action was produced by both motives acting at once. We now will describe three possible relationships that might obtain among these two possible desires—an individual might have the altruistic desire but not the selfish desire, an individual might have the selfish desire but not the altruistic desire, and an individual might have both. This last category is the one of greatest interest, since it involves the kind of pluralism that allows the interaction of desires to be examined.

We will represent desires as *preferences*. If you want the children to be better off, this means that you prefer their being better off over their being worse off. If you want to feel the glow of satisfaction, this means that you prefer feeling the glow over not feeling it. Note that the first of these preferences is other-directed, while the second is self-directed.

The first preference structure we want to describe characterizes people who care nothing about the welfare of others; the only thing that matters to them is their own situation. This purely egoistic preference structure is depicted in the following 2-by-2 table. The table answers two questions about such individuals. What preference do they have as to whether they will receive some putative benefit (feeling good)? What preference do they have as to whether the children do better rather than worse? The numbers in the table represent the *ordering* of the agent's preferences; a situation with a higher number is preferred over a situation with a lower number. The absolute values have no meaning; we could have used the numbers "8" and "6" instead of "4" and "1." Think of the four cells in this table as four states that the world might occupy; the egoist's ranking of these four possibilities is as follows:

The Egoist

		OTHER +	OTHER –
SELF	+	4	4
SELF	–	1	1

Egoists care only that they receive more (+) rather than less (−) of the benefit that is at issue. It is a matter of indifference to such people whether the children do better (+) or worse (−). These individuals are not benevolent; they also are not malevolent. Their attitude to others is one of indifference.

If you were an egoist, would you give the $25 to charity? Donating the money would have two effects. You would receive a glow of satisfaction and the children would be better off. In similar fashion, not donating the money also would have two effects. You would feel bad and the children would be worse off. (For simplicity, we are ignoring whether you prefer to retain the $25.) In this situation, there are two possible actions, giving or not giving, whose consequences are represented in the table by the upper-left entry (+ to self and + to other) and the entry in the lower-right (− to self and − to other). Given the options available, egoists will choose the first action; they will donate the $25 to charity. They therefore choose an action that benefits others. However, this benefit to others is not the goal of their action; it is a mere side-effect. If you were an egoist, you would help the starving, but your ultimate motive would be to make yourself feel good.

The second preference structure is the mirror image of the egoist's. Pure Altruists care nothing about their own situation; their only desire is that other people be better off:

The Pure Altruist

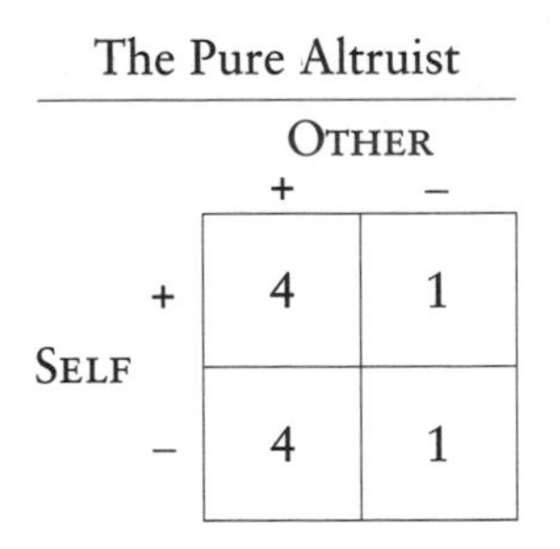

		OTHER +	OTHER −
SELF	+	4	1
SELF	−	4	1

What would Pure Altruists do if they had to choose between the actions represented by the entries in the upper-left and lower-right cells of the table? One action will benefit both self and other; the other will benefit neither. Pure Altruists choose the former action. A consequence of this choice is that Pure Altruists will feel good about themselves, but this benefit to self is a side-effect of the act, not the act's real motive.[21]

Egoists have only one ultimate preference; the same is true of Pure Altruists. Now let's consider preference structures in which an irreducible concern for self coexists with an irreducible concern for others. There are two cases to consider; the first we call "E-over-A Pluralism":

The E-over-A Pluralist

		Other +	Other −
Self	+	4	3
	−	2	1

E-over-A pluralists prefer that they be better off rather than worse off (since 4 > 2 and 3 > 1 in the preference ranking). They also prefer that other people do better rather than worse (since 4 > 3 and 2 > 1). These individuals are pluralists because they have both self-directed and other-directed preferences.[22]

We call this preference structure "E-over-A" to describe what these individuals will do if their own welfare *conflicts* with the welfare of others. Suppose the agent faces a choice between two actions. The first action provides a benefit to self but prevents other people from benefiting; this is the outcome represented in the upper-right cell. The second action confers a benefit on others but deprives self of a benefit; this is the lower-left outcome. When self-interest and the welfare of others conflict, E-over-A Pluralists give priority to themselves (since 3 > 2). Their egoistic preference is stronger than their altruistic preference. Notice how E-over-A Pluralists and Egoists differ. Egoists do not care *at all* about the situation of others. E-over-A Pluralists *do* prefer that others do better rather than worse. However, both say "me first" when self-interest conflicts with the welfare of others.

The last preference structure we want to mention also is pluralistic, but the weight it gives to self and other is the reverse of the pluralism just described:

The A-over-E Pluralist

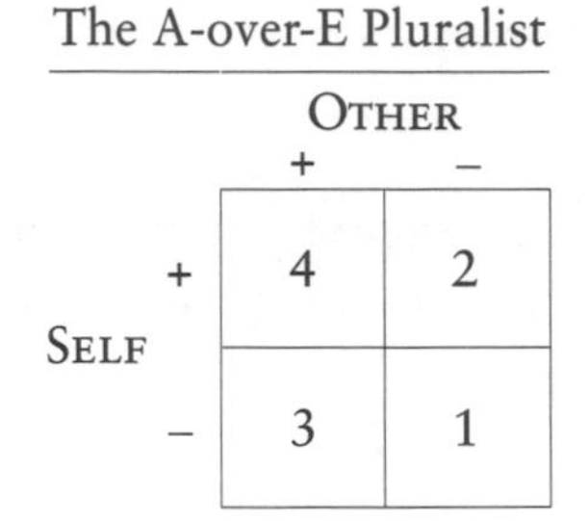

A-over-E Pluralists care about others and about themselves as well. When self-interest and the interests of others conflict, however, they sacrifice their well-being to advance the interests of others. When A-over-E Pluralists have to choose between upper-right and lower-left, they choose lower-left (since 3 > 2).[23]

The four preference structures just described—Egoism, Pure Altruism, E-over-A Pluralism, and A-over-E Pluralism—all produce the same behavior when the choice is between upper-left and lower-right; all agents choose an action that benefits both self and other over an action that benefits neither. This choice situation is one in which self-interest *coincides* with the welfare of others. However, when self-interest and the welfare of others *conflict,* Egoists and E-over-A Pluralists perform one action, while Pure Altruists and A-over-E Pluralists perform the other.

Of these four preference structures, only one is consistent with the egoism hypothesis, whereas three are consistent with the altruism hypothesis. This uneven split is due to the fact that egoism is a (relatively) monistic theory, whereas the altruism hypothesis is compatible with pluralism. The altruism hypothesis says that people sometimes have an irreducible regard for the welfare of others; whether they have other irreducible preferences is left open. It therefore is a mistake to interpret the altruism hypothesis as saying that people sometimes help for *purely* other-directed reasons; the hypothesis does not rule out the possibility that instances of helping are sometimes or even always accompanied by ultimate motives that are self-directed.

You can see from these four preference structures why it is inaccurate to describe the altruism hypothesis as saying that people are sometimes "disposed" to sacrifice self-interest to benefit others. This is not true of E-over-A Pluralism. In that preference structure, there is an irreducible desire that others do better rather than worse; however, this preference is so *weak,* compared with the desire for self-benefit, that the individual will never produce self-sacrificial behavior. E-over-A Pluralists are *not* disposed to sacrifice self-interest; yet, they have irreducibly altruistic motives.[24] The altruism hypothesis leaves unspecified whether altruistic ultimate desires are stronger or weaker than ultimate desires that are self-interested.

The two pluralistic preference structures show why the terms "because" and "only because" must be used carefully in describing how motives are related to behavior. When self-interest and the welfare of others coincide (that is, when the choice is between upper-left and lower-right), it is true that pluralists help *because* doing so benefits self. However, it will not be true that they help *only because* helping benefits self. The latter, exclusive, claim—that self-interest is the *only* motive—is true of Egoists and of them alone.

In setting out this typology, we are not suggesting that people fall into the same category in every situation they encounter. Even if people are sometimes A-over-E Pluralists, this does not mean that they always are. A person may be willing to sacrifice self-interest for the sake of others in some situations, but not in others. What the debate over egoism and altruism requires us to ask is whether there are *any* circumstances in which concern for others is anything more than instrumental.

Applying these different preference structures to the simple example of donating money to charity helps clarify why it is so difficult to infer what someone's ultimate motives are from what the person does. When self-interest and the welfare of others *coincide,* all four preference structures predict the same behavior. This means that the observed behavior—person *X* helps person *Y*—is thoroughly uninformative about whether the egoism or the altruism hypothesis is true. When self-interest and the welfare of others *conflict* (that is, the individual has to choose between upper-right and lower-left), Egoism and E-over-A Pluralism make one prediction while Pure Altruism and A-over-E Pluralism make another. But even here, if the agent avoids self-sacrifice, this result fails to distinguish between monistic Egoism and E-over-A Pluralism.[25] Perhaps it is possible to discern what preferences people have by observing their behavior, but it remains to be seen how this can be done. Those who think that human behavior makes the egoism hypothesis obvious should think again.

Interacting Desires as Interacting Causes

When desires interact to produce a behavior, this is a special instance of causes interacting to produce an effect. If it is hard to tell what people's motives are by observing their behavior, this difficulty may trace to generic problems that pertain to inferring causes from effects.

Consider a farmer who grows two fields of corn. In the first, the corn plants are of identical genotype (G1) and they receive one unit of fertilizer (F1). In the second field, the corn plants also are genetically identical, but they have genotype G2; in this second field, the plants receive two units of fertilizer (F2). At the end of the growing season, the farmer sees that the plants in the first field are one unit tall on average, while those in the second field average four units of height. These observations may be summarized in two cells of a 2-by-2 table:

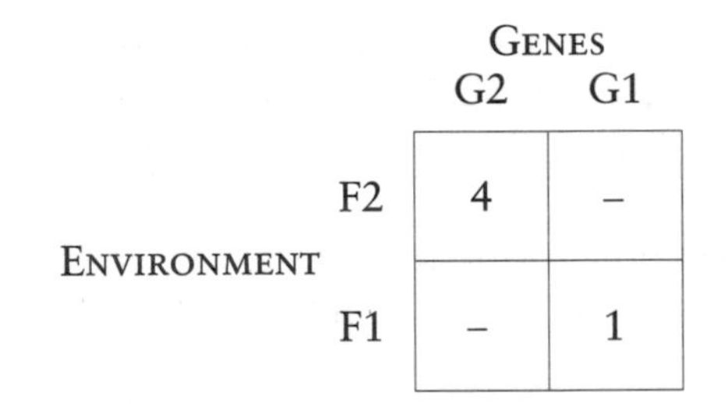

		Genes	
		G2	G1
Environment	F2	4	–
	F1	–	1

Suppose the farmer wants to answer a question about the importance of *nature* and *nurture:* Do the corn plants in the two fields differ in height because they are genetically different, because they grew in different environments, or for both these reasons? With the data described so far, the farmer has no way to tell. The reason is that the genetic and the environmental factors are perfectly *correlated;* G1 individuals always inhabit F1 environments and G2 individuals always live in F2 environments.

The way for the farmer to make headway on this problem is to break the correlation. The farmer should plant a third field in which G1 plants receive two units of fertilizer and a fourth field in which G2 plants get one unit of fertilizer. The results can be entered in the other two cells of the 2-by-2 table just displayed. With observations about what happens in all four treatment cells, the farmer can make an inference concerning how genetic differences and differences in fertilizer treatment contribute to variation in plant height. Here are four outcomes that this experiment might produce:[26]

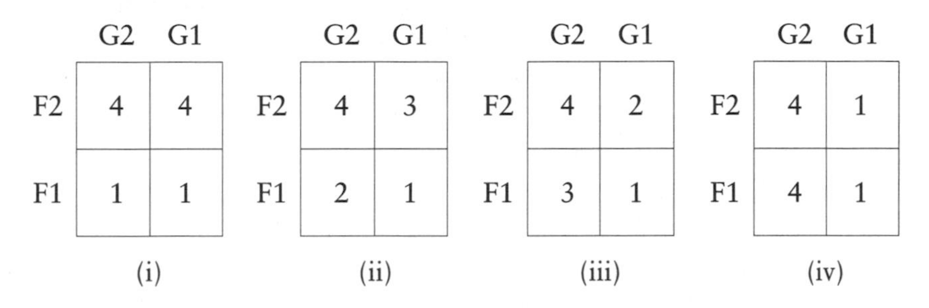

(i)	G2	G1
F2	4	4
F1	1	1

(ii)	G2	G1
F2	4	3
F1	2	1

(iii)	G2	G1
F2	4	2
F1	3	1

(iv)	G2	G1
F2	4	1
F1	4	1

In outcome (i), the genetic factor makes no difference. Whether the plants have genotype G1 or G2 does not affect their height; it is the environmental factor—the amount of fertilizer the plants receive—that explains all the observed variation. Outcome (iv) is the reverse of (i). In (iv), the fertilizer treatment makes no difference; genetic variation explains all the variation in height. Outcomes (i) and (iv) support *monistic* explanations of the variation in plant height; each suggests that only one of the factors considered made a difference in the observed outcome.

Outcomes (ii) and (iii), on the other hand, support *pluralistic* conclusions. Both suggest that genetic and environmental factors made a difference. However, they disagree about which factor mattered more. In outcome (ii), changing the fertilizer treatment yields two units of change in height, whereas changing from one genotype to the other produces only a single unit of change. In this case, the environmental factor makes more of a difference than the genetic factor. By the same reasoning, we can see that outcome (iii) suggests that genetic variation was more important than the environmental factor considered.[27]

We hope the analogy between the egoism-altruism problem and the puzzle faced by our farmer is clear. When self-interest and the welfare of others *coincide,* it will be impossible to say whether the resulting behavior was produced by egoistic motives, by altruistic motives, or by both. This impasse resembles the farmer's initial situation. When genetic and environmental factors are perfectly *correlated,* it will be impossible to say whether the resulting variation in height was caused by genetic differences, by environmental differences, or by both. In the problem concerning egoism and altruism, the obvious experiment to perform is to put individuals in situations in which self-interest and the welfare of others *conflict.* The parallel procedure for the farmer is to plant two more fields of corn so that the upper-right and lower-left cells in the 2-by-2 table can be filled in; in this way, the initial, confounding correlation is *broken.*

Although this way of understanding the egoism-altruism debate is quite fundamental, it would be a mistake to exaggerate its resemblance with the farmer's problem. One disanalogy between the farmer's problem and the debate over motivation is that the farmer is trying to explain the variation that exists within a population of individuals, whereas arguments about egoism concern the different motives that exist within individuals themselves. A second difference between the two problems is that the egoism and altruism hypotheses are more abstract than the one that the farmer is considering. As noted earlier, egoism and altruism do not say which *specific* desires people have; rather, they specify the *types* of desires that people have as their ultimate aims. This makes the hypotheses of egoism and altruism harder to test.

But perhaps the most fundamental difference between the farmer's problem and the egoism-altruism debate is this: In the farmer's problem, the candidate causes can be identified in advance of knowing their effect on plant height. The farmer measures out the fertilizer; the local seed distributor passes along information about the genotypes of the seeds planted. No such independent access is readily available in the problem of discerning people's motives. We infer people's motives from their behavior; aside from this, we have little or no access to what their motives really are. This does not mean that the question of altruism versus egoism is insoluble; it does mean that we must tread carefully, since the inference problem is a difficult one.

8

Psychological Evidence

In this chapter and the next, we will review a number of scientific and philosophical arguments that have aimed to resolve the controversy concerning psychological egoism and altruism. These arguments are a motley assemblage. Some come from empirical findings in experimental social psychology. Others involve science fiction thought experiments. Still others appeal to methodological principles that describe how one should evaluate rival hypotheses when observation is indecisive. Although there is much to be learned from these arguments, we will conclude that none of them settles whether human beings ever have altruistic ultimate motives. This verdict of "not proven" is where this chapter and the next one end, but it will not be the conclusion of the book as a whole. In Chapter 10, we will bring evolutionary considerations to bear on the question of motivation.

The present chapter explores three approaches to the egoism-altruism debate that empirical psychology requires us to consider.[1] The first concerns introspection. Can we tell what our ultimate motives are simply by gazing within our own minds? The second involves the law of effect. Does this psychological principle show that hedonism must be the motivational structure of an organism that is capable of learning from experience? The third approach will be to review the experimental literature in social psychology. What do these experiments teach us about psychological egoism and motivational pluralism?

Is Introspection the Answer?

Before we begin examining the details of psychological experiments and philosophical arguments that seek to determine what our ultimate motives are, we want to address a reaction that some readers may have to this problem. Can't people tell by introspection what they want as ends in themselves? If so, the debate concerning egoism and altruism is easy to resolve; we can simply gaze within our own minds and see whether we are egoists or pluralists.

We need to be clear about what problem introspection is being asked to solve here. The problem is not just to determine what people want, but to tell what their *ultimate* as opposed to *instrumental* desires are. Social psychologists have asked people who gave money to charity and who did volunteer work for philanthropic causes why they did so. Helpers often reply that they "wanted to do something useful" or to "do good deeds for others" (Reddy 1980). Even if these introspective reports were true, they would not tell us whether the reported desires are ultimate or instrumental. Psychological egoism can grant that people want to help others; it claims that these desires are merely instrumental. Asking "Why did you help?" is the wrong question, if the point is to assess this theory. However, even the direct question "What are the ultimate motives behind your helping?" may fail to produce the information we seek, if people lack introspective access to their ultimate motives.

Introspection has had a bad reputation in psychology for a long time. When psychology broke away from philosophy at the end of the nineteenth century and became an autonomous discipline, one of the ways in which it developed its credentials as an objective science was by rejecting introspective methods. Like other objective sciences, psychology was expected to attend exclusively to data that are publicly accessible. Behaviorism took this flight from introspection to its logical extreme. Not only were introspective reports thought to provide no evidence about the inner workings of the mind; in addition, behaviorism rejected the very goal of elucidating mental states. Instead, behaviorism sought to explain behavior solely on the basis of environmental stimuli. Many nonbehavioristic approaches to psychology shared behaviorism's rejection of introspective methods, even if they retained the goal of understanding inner mental states and processes.

For example, Freud and his school maintained that the unconscious conceals and systematically distorts mental contents; one of Freud's deepest influences on psychology was to cast doubt on the reliability of introspection.

Quite apart from the tradition of shunning introspection that has developed in psychology, it is important to realize that the reliability of introspection is, in the end, a contingent matter. A well-grounded opinion about this issue should be based on evidence; a kneejerk faith in the unreliability of introspection is no more defensible than a complacent confidence in its infallibility. Furthermore, we have to realize that introspection may be more reliable in some domains than in others. For example, even if introspection is unable to reveal why people commit verbal slips, it is a separate question whether introspection can answer the question posed by egoism and altruism.

If the reliability of introspective reports about our ultimate motives must be decided by evidence, how should we proceed? The most direct way to evaluate the reliability of a report requires that one have independent access to the state of affairs that the report is supposed to describe. For example, to determine directly whether a thermometer accurately reports temperature, you must know what an object's true temperature is. Similarly, to assess directly the reliability of introspective reports about motives, one must know independently what people's true motives are. Quite obviously, the reliability of introspective reports about ultimate motives cannot be decided directly, if we don't already know how to resolve the debate between egoism and altruism.

Is a more indirect strategy available? The reliability of a thermometer can be checked even when you don't know what the temperature of any object is. Suppose you know that a certain manipulation leaves an object's temperature unchanged. You don't know what the temperatures are of the objects on your desk, but you are prepared to say that whether an object is on the right or the left side of your desk does not influence its temperature. If so, you can randomly assign objects to the left and right sides of your desk, use the thermometer on each, and see if there is a significant difference between the two sets of measurements you obtain. If the thermometer is reliable, and if your assumption that location does not affect temperature is correct, then there should be no difference.

This type of experiment could be performed to help decide whether people have reliable introspective access to their own ultimate motives. For example, suppose we have the subjects in our experiment fill out a questionnaire, which we tell them measures how empathic they are. We then throw away these questionnaires without looking at them and randomly divide the subjects into two groups. We clearly explain to each what the hypotheses of psychological egoism and motivational pluralism assert. We then tell the subjects in group 1 that they scored high on the empathy measure; we ask them to determine introspectively whether they are egoists or pluralists. We tell the people in group 2 that they scored low on the test and ask them to figure out by introspection whether they are egoists or pluralists. On the assumption that one's ultimate motives are not affected by being told whether one scored high or low on an empathy test, there should be no difference between these groups in terms of their introspective reports, if introspection is highly reliable. On the other hand, if the groups differ in their reports, this suggests that people are *suggestible*. When people think they are introspecting, they in fact are applying to themselves a theory obtained from the outside.

This experimental design needs to be fine-tuned in several ways. For example, we need to control for the possibility that subjects do not tell the truth to others about what they introspect; perhaps they know their own minds by introspection but tailor their verbal reports to fit what they think the experimenter wants to hear. One way to address this problem might be to have subjects not put their names on the reports about introspection that they write, thus assuring their anonymity. Another strategy might be to further subdivide the two groups, telling half the people in group 1 and half the people in group 2 that the experimenter is trying to prove that psychological egoism is true and telling the other subjects that the experimenter is trying to prove that psychological pluralism is correct. Although some wrinkles need to be ironed out here, it seems reasonably clear that the reliability of introspective reports about ultimate motives is amenable to empirical study.[2]

To our knowledge, the type of experiment we have just described has not been performed. Even so, it is important to realize that skepticism about introspection is not an undefended prejudice on the part of psychologists; there is considerable evidence for thinking that people often have erroneous conceptions of what is going on in their own minds. We'll describe one such finding, drawn from the useful article by Nisbett and Wilson (1977), "Telling More Than We Can Know—Verbal Reports on Mental Processes." One of the most striking results of research on situational factors that influence helping behavior is the so-called bystander effect (reviewed by Latané, Nida, and Wilson 1981). A bystander's probability of helping another person who is perceived as being in need declines as the number of other bystanders increases. Psychologists often suggest that this lowering of the probability of helping is due to a diffusion of perceived responsibility; when there are more bystanders, an agent is more likely to think that someone else should do the helping. Whether or not this is the right explanation, the bystander effect has been confirmed when the needy other is a stranger, but also when the relationship is quite intimate. For example, people who need kidney transplants are more likely to find a sibling who agrees to donate if they have *fewer* siblings (Simmons, Klein, and Simmons 1977, p. 220).

In their work on the bystander effect, Latané and Darley (1970) asked experimental subjects whether their inclination to help was influenced by how many bystanders were present. Subjects consistently denied that this was so, and also denied that other people are influenced by this consideration. If behavior is determined by the agent's beliefs and desires, then the fact that people behave differently in two situations must mean that they have different beliefs or different desires in those situations. If subjects are not aware that their behavior when bystanders are present would differ from their behavior when bystanders are absent, then they presumably are not aware that they would have different beliefs or different desires in those two circumstances. It doesn't follow from this that people are unaware of what their *desires* are, or that they are not aware of what their *ultimate desires* are. Still, this last possibility cannot be dismissed. If the mind is not an open book, then why think that one chapter in that book—the one in which one's ultimate desires are inscribed—can be read infallibly by introspection?

In the culture we inhabit, some people are sincerely convinced of psychological egoism, while others are convinced that they and others have altruistic ultimate motives. Defenders of egoism often think that pluralists are trapped by a comforting illusion. Pluralists sometimes entertain a reciprocal hypothesis—that egoists embrace a darker view of human motivation because they enjoy thinking that they have the fortitude to do without comforting illusions. Of course, neither of these suggestions tells us what our ultimate motives in fact are. However, both indicate that sincere introspection is not enough.

Proponents of both positions believe that their pet theories apply to others and to themselves as well. It is conceivable that defenders of egoism are right about themselves *and* that defenders of pluralism are right about themselves. It is even conceivable that people who change their minds about whether they are egoists or pluralists have true views about themselves both before *and* after. Yet the possibility remains that one side or the other is mistaken in the claims they make about their own motives. Sincere avowals, by both parties, must be set to one side. Introspective claims should be regarded as just that—as claims, whose accuracy must be judged on other grounds.

The Law of Effect

Hedonism is sometimes defended by saying that the theory describes what an organism must be like if it is to be capable of learning from experience. The idea is that learning requires organisms to experience positive and negative sensations; experiencing the former and avoiding the latter must constitute its ultimate goals in behavior. According to this proposal, learning takes the form of a conditioning process. If a behavior is followed by a positive sensation, this raises the probability that the organism will repeat the behavior. If the behavior is followed by a negative sensation, this lowers that probability. This is the law of effect, first proposed by E. L. Thorndike (see Dennett 1975 for discussion). Without this feedback loop through the experiential consequences of behavior, there is no way for the organism to change the way it acts.

It is a curious historical fact that the law of effect, which adverts to the positive and negative experiences that accompany behavior, was embraced as a central principle by behaviorists, who also demanded that psychology stop trying to talk about inner mental states. Be that as it may, the law of effect is of interest to our present inquiry because of its connection with hedonism; the claim we need to assess is that hedonism must be the true theory about human motivation if organisms that learn must obey the law of effect.

The law of effect does not say that every behavior occurs because the organism was conditioned earlier; that would mean that no behavior ever occurs for the first time. This is not only absurd on its own terms; it also conflicts with the very idea of conditioning. A conditioning process requires that the organism perform the target behavior at least once *before* it receives the conditioning rewards and punishments. Rather, what the law of effect says is suggested by a simple example of operant conditioning. Consider a pigeon in the controlled environment that has come to be called "a Skinner box." At first, the pigeon's pecking is unrelated to whether a light in the box is on. If the pigeon is rewarded for pecking when the light is on, however, the pigeon's behavioral pattern will change. As the pigeon is repeatedly rewarded, the probability increases of its pecking when the light is on. Eventually, it pecks precisely when the light is on. Before the conditioning process, the pigeon's probability of pecking is *independent* of whether the light is on; after the conditioning process is over, the probability of pecking if the light is on is close to 1, and the probability of pecking if the light is off is close to 0. Understood in this way, the law of effect does not rule out the occurrence of behaviors that were never conditioned; rather, what it rules out is the existence of probabilistic dependencies between behavior and environment that were not caused by a conditioning process. It also rules out the possibility that a conditioning process could fail to induce such dependencies.

Both these implications are problematic in the context of behaviors that are strongly influenced by "innate" or "instinctual" factors.[3] Consider Konrad Lorenz's famous example of imprinting in greylag geese. A gosling will follow the first adult goose or human it sees that gives calls in response to the gosling's calls. However, goslings will not treat a rock as "Mom," nor will they imprint on a model chicken that emits prerecorded calls, if the calls are not produced in response to the goslings' calls (Lorenz 1965). That is, the probability of the imprinting behavior differs according to the environmental cue, but this is not because a conditioning process occurred.

Just as a behavior can depend on an environmental stimulus without its having been rewarded earlier, so it can fail to occur even though it was rewarded before. Garcia and Koelling (1966) exposed rats to a complex stimulus consisting of flashing lights, noise, and saccharin-flavored water. Afterwards, the rats were exposed to X-rays, which made them sick. The rats thereby acquired an aversion to foods flavored with saccharin, but not to food that was accompanied by noise or by flashing lights. In another experiment, the same compound stimulus was used, but this time the rats received an electric shock to their feet. As a result, they developed an aversion to the noise and the flashing lights, but not to saccharin. The same pattern has been recorded for a variety of species, our own included (Breland and Breland 1961; Hineline and Rachlin 1969; Sevenster 1973; Gallistel 1980). For example, it is easier to condition human infants to be afraid of snakes, caterpillars, and dogs than of opera glasses or cloth curtains (Rachman 1990, pp. 157–158), and it is easier to condition a physiological response to angry faces than to happy ones (Ohmman and Dimberg 1978).

These results cast doubt on the law of effect's commitment to what learning theorists call "the equipotentiality thesis"—the idea that conditioning can successfully pair any stimulus with any behavior. This is not to deny that *some* behaviors can be explained by the law of effect. For example, Moss and Page (1972) ran an experiment in which people on a busy street were asked for directions to a well-known local store. Most complied. Of those who provided directions, some were thanked with a smile while others were abruptly interrupted and told that their directions were incomprehensible. A short time later, the people who provided directions encountered a person who had just dropped a small bag. More than 93 percent of those who had been graciously thanked when they provided directions helped the person who had dropped the bag, whereas only 40 percent of those who had been rebuffed and scolded offered help. Moss and Page also found that helping in a control group—people who had not been positively or negatively reinforced—occurred at a rate of 85 percent. These are patterns one would expect, given the law of effect.

It is important to remember that the law of effect is a *general* principle; the question is whether it is true of *all* behavior, not just of *some*. We suggest that it is not—being rewarded does not *always* raise the probability of a behavior's being repeated, and probabilistic dependencies between behavior and environment do not *always* stem from this type of conditioning process. In addition, even circumstances that conform to the law—such as the experiment that Moss and Page performed—provide no evidence for hedonism. Moss and Page's observations were consistent with the hypothesis that pleasure and pain are motivators. Their experiment does not show that people care *only* about pleasure and pain.

The law of effect describes one possible mechanism whereby an organism can modify its behavior in the light of experience. However, it is not the only one that is conceivable. Consider the fact that means-end deliberation can generate actions without conforming to the dictates of hedonism. Deliberation leads us to revise an instrumental desire by using more ultimate desires as leverage. Any more ultimate desire will do the trick. Consider Arnold, who suddenly acquires the desire to get into his car and drive to the bakery. He acquires this new desire because he wants to buy bread and he believes that the bakery is the best place for him to do this. Deliberation is able to produce new instrumental desires because the agent regards old desires as ends for which means must be sought. The same is true when an individual abandons a desire already held. If Arnold is about to drive to the bakery when his friend brings him a loaf of bread, Arnold may lose his desire to drive to the bakery. When deliberation leads us to acquire a new instrumental desire or to abandon an old one, this is something that our *other* desires accomplish for us. There is no requirement that our ultimate goals must include the desire to attain pleasure and avoid pain; still less is it required that attaining pleasure and avoiding pain must be the only ultimate desires we have.

This point about learning is an important one, for it allows us to identify an evolutionary question that otherwise might escape our notice. Pleasurable and aversive sensations constitute one mechanism that allows an organism to learn from experience. In principle, there are others. Why did evolution assign to pleasure and pain the roles they now play in learning? This question deserves a substantive answer; we render the question invisible if we reply that learning, by definition, is a conditioning process mediated solely by pleasure and pain.

We conclude that psychological egoism cannot be defended by appealing to the law of effect. That "law" is sometimes false. And even when behaviors recur *because* they were rewarded earlier, it does not follow that they do so *only because* they were rewarded. Even if people are motivated by the prospect of attaining pleasure and avoiding pain, it does not follow that this goal is the only thing that people ultimately care about.

Experiments in Social Psychology

Egoism is a (relatively) monistic theory, whereas the altruism hypothesis, as we understand it, is part of a more pluralistic view of human motivation. This logical difference between the two theories has implications for how each of them may be tested. An experiment demonstrating that people in fact possess a particular egoistic ultimate motive *does not* disconfirm the altruism hypothesis, but a hypothesis demonstrating that people possess a particular altruistic ultimate motive *does* disconfirm the egoism hypothesis. It does no good in this debate to show that people are motivated by egoistic concerns. Ideally, an experiment should put people in a situation in which they will behave one way if they have altruistic ultimate motives and behave another way if they do not. But this is not so easy to do, since, as we have noted, the egoism hypothesis is a flexible instrument that comes in a variety of forms.

The research in experimental social psychology that does the best job of coming to grips with the problem of testing egoism and altruism is that of Daniel Batson and his associates. This work was synthesized by Batson (1991) in his important book *The Altruism Question;* Batson and Shaw (1991) provides a useful summary. Batson is admirably alert to the risk of confusing real altruism with pseudo-altruism. His experiments are designed to track down and test different varieties of egoism; Batson is well aware that refuting one or two forms of egoism is not the same as refuting egoism *per se.* In discussing Batson's work, we also will examine the provocative work of Robert Cialdini and his associates, as well as some other experimental research.

Batson wishes to test a conjecture that he terms the *empathy-altruism hypothesis.* As noted in the previous chapter, Batson was the first person in experimental social psychology to delineate the distinction between empathy and personal distress. Whereas personal distress typically leads people to want to improve their own situations, the empathy-altruism hypothesis asserts that empathy causes people to have altruistic desires. Since this claim is supposed to be incompatible with egoism, the altruistic desires that it says are triggered by empathy must be *ultimate*. Batson's methodology is to test different versions of egoism against the empathy-altruism hypothesis. In each case, he argues that the version of egoism considered is disconfirmed by the data, but that the data support the empathy-altruism hypothesis.

The first version of egoism that Batson considers is the *aversive-arousal reduction* hypothesis. This is the idea that bystanders who see a needy other have unpleasant experiences that they wish to expunge. They help for the same reason you turn down the thermostat when the room is too hot. When you lower the thermostat, this is not because you care about the room. Helping others has precisely the same type of motivation—it is merely a means of achieving a better level of personal comfort.

The philosopher C. D. Broad (1952, pp. 218–231) argued against this version of egoism by describing a physician who travels to Asia to open a clinic for people suffering from leprosy. The physician knows that this line of work will bombard him with distressing experiences. Broad thought that this example straightforwardly refutes the egoistic hypothesis that helping is motivated just by the desire to escape from the unpleasant experiences occasioned by exposure to those in need. If one's only ultimate desire is to avoid or reduce aversive feelings, one would *avoid* any situation like the one the physician chose.

We suspect that many readers, and most philosophers, will find Broad's argument sufficient to refute the aversive-arousal reduction hypothesis. Broad's claim that people sometimes act the way the physician does in his example rings true; and the logic of his argument—that it refutes this egoistic hypothesis—sounds right as well. If this is correct, then there seems to be no need for further experiment or natural observation. Readers who feel this way may be surprised that Batson and his associates ran several experiments to test the aversive-arousal reduction hypothesis against the empathy-altruism hypothesis. Why did they do so? Psychologists often run experiments in which commonsense tells one what to expect; since common sense expectations are not always borne out, there is merit in testing propositions that people think are intuitively obvious. Let's examine how Batson and his colleagues proceeded.

In one of these experiments, subjects were told that they would be part of a study in which they would watch over closed-circuit television while a student receives ten electric shocks. Actually, each subject watched a videotape of an actress ("Elaine") who pretends to find the first two shocks quite distressing. The videotape then shows one of the experimenters tell Elaine that he is concerned about her discomfort; he suggests to her that she stop taking the shocks if the subject agrees to substitute for her. Elaine gratefully consents to this arrangement, the television screen goes blank, and a confederate then enters the room where the subject is and asks whether the subject would be willing to take Elaine's place.

Subjects in the "easy-escape" treatment of the experiment had been told before they started to monitor Elaine that they would be required to watch just two shocks; the confederate reminds them of this and offers them the opportunity of exiting the experiment if they do not want to substitute for Elaine. Subjects in the "difficult-escape" treatment had agreed at the outset to watch the entire sequence of ten shocks; the confederate reminds them of this promise and says that they will have to watch Elaine experience eight more shocks if they do not take her place.

Not only did subjects vary in terms of whether escape was easier or more difficult. They also were manipulated in ways designed to influence how much empathy they would feel for Elaine. This was achieved by a variety of means. One technique exploited the fact that people tend to empathize more with individuals whom they take to be similar to themselves (Stotland 1969; Krebs 1975). Subjects in the high-empathy treatment received a description of Elaine that closely matched what they had reported about themselves in an inventory of personal values and interests that they filled out before they saw Elaine. So-called low-empathy subjects were given a description of Elaine that failed to match. The idea behind this manipulation was not that all people in the high-empathy treatment would have high empathy for Elaine and that all in the low-empathy treatment would have little empathy, but that the average level of empathy would be higher in the former group than in the latter.[4]

Each subject in the experiment was given either the easy-escape or the difficult-escape treatment, and each was given the high-empathy or the low-empathy treatment. This means that each subject was placed in one of four circumstances. The experiment discovered how often people in each of these four treatment cells volunteered to help Elaine. Let's represent the frequencies of offers to help in the four circumstances as w, x, y, and z:

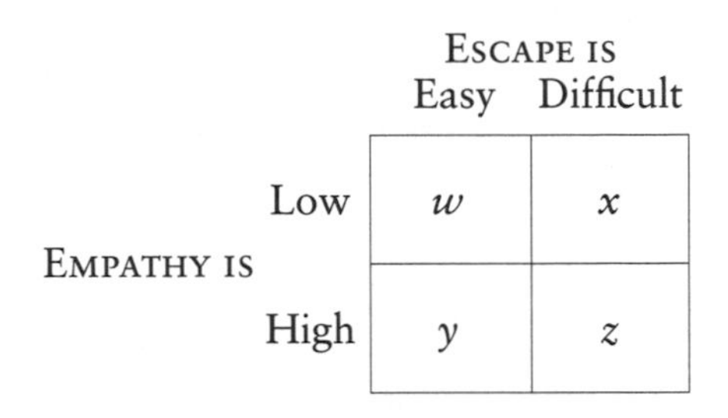

		ESCAPE IS Easy	Difficult
EMPATHY IS	Low	w	x
	High	y	z

Before describing what the results of the experiment were, we need to consider what the two hypotheses predicted. To do this, we must consider carefully what each hypothesis says.

Suppose we interpret the empathy-altruism hypothesis as saying just that a higher level of empathy makes people more inclined to help, at least sometimes. Construed in this way, the hypothesis predicts merely that $y > w$ or that $z > x$ or both. In parallel, suppose we interpret the aversive-arousal reduction hypothesis as saying just that people are sometimes more likely to help if it is difficult for them to escape from the needy other. Under this construal, the hypothesis predicts merely that $x > w$ or $z > y$ or both. Notice that if we interpret the two hypotheses in this modest way, their predictions do not conflict. The empathy-altruism hypothesis makes "vertical" predictions about the 2-by-2 table, whereas the aversive-arousal reduction hypothesis makes "horizontal" predictions. If this is all the hypotheses say, then they do not disagree; in fact, it would be wrong to view the aversive-arousal reduction hypothesis as a version of egoism, since it does not rule out the possibility that the empathy-altruism hypothesis is true. Batson avoids this problem by interpreting both hypotheses as making both "horizontal" and "vertical" predictions about the frequencies of helping. Construed in *this* way, the hypotheses come into conflict with each other.

Batson interprets the two hypotheses as disagreeing over whether empathy level will make a difference among subjects who are in the easy-escape treatment (i.e., over how w and y are related).[5] He reads the empathy-altruism hypothesis as predicting that empathy will augment helping when escape is easy; he interprets the aversive-arousal hypothesis as predicting that empathy will make no difference in this circumstance. The result of the experiment was that high-empathy subjects offer to help more than low-empathy subjects when escape is easy.[6] This disconfirms the aversive-arousal reduction hypothesis as an *exclusive* explanation of helping. Even if this motive sometimes plays a role, it cannot be the only one at work.

It does not follow, as Batson realizes, that the empathy-altruism hypothesis is correct; there may be *other* egoistic motives besides aversive-arousal reduction that can explain the experimental results. For example, perhaps high-empathy subjects in the easy-escape treatment realized that they would retain painful memories of the needy other if they declined to help, whereas low-empathy subjects in the easy escape treatment were less plagued by this worry. If so, we have an egoistic explanation for why high-empathy subjects more often offered to help. This point was made separately by three commentators on the Batson and Shaw (1991) article (Hoffman 1991; Hornstein 1991; Wallach and Wallach 1991).

Batson and Shaw (1991, pp. 167–168) formulate a reply to this suggestion. They argue that the egoistic hypothesis just described—that people offer to help because they know that refusing to do so will leave them with painful memories of the needy other—makes a prediction about what will happen in another experiment. The prediction is that high-empathy individuals will choose to receive news in the future about the situation of needy others only to the extent that they expect the news to be good. A third experiment, which we'll describe shortly, provides evidence that this isn't so. We agree that this argument supports Batson and Shaw's claim that individuals who choose to help rather than to escape aren't motivated solely by the desire to avoid having the belief that the needy other is doing poorly. However, this point about information seeking does not undercut a different egoistic explanation that appeals to guilt feelings. Just as exiting without helping can produce guilt feelings, so can refusing information about the situations of needy others on the grounds that the news might be bad. More on this soon.

Batson's second experiment focuses on a different egoistic hypothesis, which he calls the hypothesis of *empathy-specific punishment.* This conjecture comes in two forms; it says that empathically aroused individuals help because they want to avoid the censure of others or because they want to avoid self-censure. We'll focus on the second formulation. Batson reasons that if this hypothesis were correct, people would be less inclined to help when they are provided with a strong justification for not helping. In the experiment, subjects were asked whether they would help a needy other. Some were told that many others had declined to help when placed in the same situation; these subjects thus received a *high* justification for not helping. Other subjects were told that few people had refused to help, thus receiving a *low* justification for not helping. Subjects also differed in whether they had low or high-empathy for the individual who needed help. The 2-by-2 experiment thus placed subjects in four treatment cells, and the frequency of volunteering to help among subjects in each cell was tabulated:

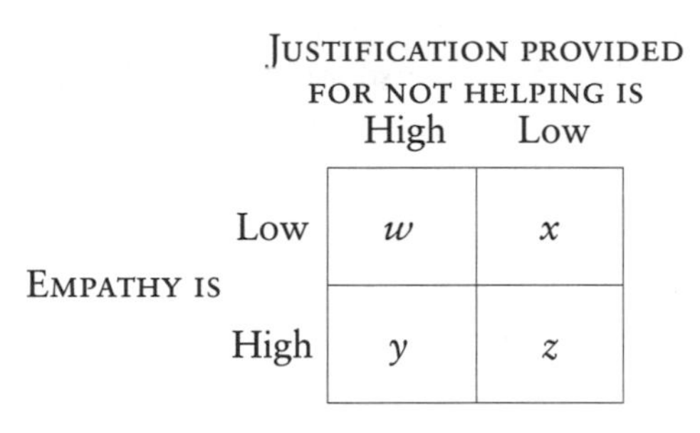

		JUSTIFICATION PROVIDED FOR NOT HELPING IS	
		High	Low
EMPATHY IS	Low	w	x
	High	y	z

Batson (1991, p. 136) interprets the empathy-specific punishment hypothesis and the empathy-altruism hypothesis as agreeing about a lot: both predict that empathy level makes a difference, regardless of how much justification the subject receives for not helping; both predict that $y > w$ and $z > x$. He also says that the two hypotheses predict that low-empathy subjects will help less when they have a strong justification for not helping ($w < x$). Where, then, do the hypotheses disagree? The nub of the matter, as Batson sees it, concerns how the level of justification for not helping affects high-empathy subjects (Batson and Shaw 1991, p. 116). The empathy-specific punishment hypothesis predicts that high-empathy subjects will help more when they have little justification for not helping than they will when they have a lot of justification for not helping ($z > y$). The empathy-altruism hypothesis, in contrast, is said to predict that high-empathy subjects will not be influenced by this; they should provide the same amount of help regardless of whether they receive high or low justification for not helping ($z = y$).[7]

Thus construed, the empathy-altruism hypothesis scores a victory; it turns out that the frequency of offers to help made by high-empathy subjects is not affected by whether they receive high or low justification for not helping. Yet, a central interpretive question remains: Does empathy promote helping by causing subjects to have an altruistic ultimate motive? Even if the wish to avoid disapproval cannot explain the outcome of this experiment, this question remains open.

The third egoistic hypothesis that Batson examines is the hypothesis of *empathy-specific reward,* which comes in two forms. The first says that we help only in order to receive a mood-enhancing reward, either from self or from others; a special case of this hypothesis asserts that we help only to secure the mood-enhancing news that the needy other is better off. The second version of the empathy-specific reward hypothesis says that empathy causes sadness, which we want relieved, so we seek some mood-enhancing experience that will do the trick.

Batson, Dyck, Brandt, Batson, Powell, McMaster, and Griffitt (1988) tested the first of these hypotheses by seeing how the mood (as determined from self-reports) of high and low-empathy individuals was affected by depriving them of the opportunity to help. As before, we can understand the study in terms of the predictions that the hypotheses make about what will happen in different treatment cells. However, now there are six treatments and the effects recorded in the cells are not frequencies of offers to help but of (self-reported) levels of mood:

		Help administered by		
		No one	Subject	Third party
Empathy is	Low	*a*	*b*	*c*
	High	*d*	*e*	*f*

The empathy-altruism hypothesis predicts that the mood of a high-empathy subject will depend on whether the needy other receives help, not on whether that help comes from the subject or from some third party; the prediction is that $e = f > d$.[8] The empathy-specific reward hypothesis, on the other hand, predicts that high-empathy subjects have their moods boosted only if they themselves provide the help—i.e., that $e > f = d$. The experiment came out as the empathy-altruism hypothesis predicted (Batson 1991, p. 150; Batson and Shaw 1991, p. 117). High-empathy subjects have their moods improve when they learn that the needy other has received help; it doesn't matter to them who the helper was.

One way to try to explain this result within the framework of egoism is provided by the *empathic joy hypothesis* (proposed by Smith, Keating, and Stotland 1989). This hypothesis says that people help, not to receive the rewards that come from helping, but to gain the good feelings that derive from sharing vicariously in the needy person's relief. Good news about the needy person's improved situation provides a boost in mood; it doesn't matter why the needy person now is better off. Batson, Batson, Slingsby, Harrell, Peekna, and Todd (1991) tested this hypothesis by seeing if the proportion of subjects choosing to have a second interview with a needy other would be influenced by whether they were told that the person has a 20 percent, a 50 percent, or an 80 percent chance of improvement before that second conversation. The authors reasoned that the empathic joy hypothesis and the empathy-altruism hypothesis make different predictions about how often individuals in six different treatment cells will elect to have a second interview:

		Probability that needy other will show improvement		
		20%	50%	80%
Empathy is	Low	*a*	*b*	*c*
	High	*d*	*e*	*f*

Both hypotheses predict that high-empathy subjects should request the second interview more often than low-empathy subjects, and the experiment bore this out.

The experimenters reasoned that the hypotheses make different predictions about how high-empathy subjects differ among themselves. They interpret the empathic joy hypothesis as predicting that high-empathy subjects are more likely to request a second interview when the probability is higher that an interview will provide good news about the needy other; the prediction is that $d < e < f$ (Batson 1991, pp. 161–162). In contrast, the empathy-altruism hypothesis is said to predict that high-empathy individuals should either not be influenced by how probable it is that they will receive good news, or that they should be most interested in receiving news when they are maximally uncertain. That is, the empathy-altruism hypothesis predicts that either $d = e = f$ or $d < e > f$. The result of the experiment matched this prediction of the empathy-altruism hypothesis—the frequency of requests for a second interview among high-empathy subjects is not an increasing function of the probability that the news will turn out to be good.

Even so, it is not difficult to invent an egoistic explanation of this outcome. Uncertainty can be a torment; this is a familiar experience when the question mark concerns our own welfare, and also when the uncertainty involves the well-being of those we care about. Of course, we'd rather receive good news than bad, but people also prefer receiving information over remaining in the dark. We may apply this idea to Batson's experiment by hypothesizing that high-empathy subjects choose to receive news because they want to reduce the disagreeable feelings that accompany uncertainty. In addition, declining the offer of information might make high-empathy subjects feel guilty. Apparently, the results of this experiment can be accommodated within the framework of egoism.

The second version of the empathy-specific reward hypothesis that Batson studied is due to the work of Cialdini and colleagues (Cialdini, Schaller, Houlihan, Arps, Fultz, and Beaman 1987; Schaller and Cialdini 1988). This is the *negative-state relief hypothesis,* which says that empathic individuals become sad when they witness a needy other; they then help in order to lift themselves out of their sadness. One of Cialdini's experiments is striking and unexpected in its results. The experiment begins with all subjects taking a "drug" (actually, a placebo). Some subjects are given perspective-taking instructions designed to produce high-empathy with a fellow student; others are placed in a low-empathy treatment. They then are told that the student needs help in going over her class notes. Before giving subjects the opportunity to volunteer to help, some of the subjects are told that the drug they had taken earlier will have the effect of freezing their mood for a half-hour or so. These students are thus in the "fixed-mood treatment." The other students are given no such story and thus are said to occupy the "labile-mood treatment." The experiment tabulated how often subjects volunteered to help in four circumstances:

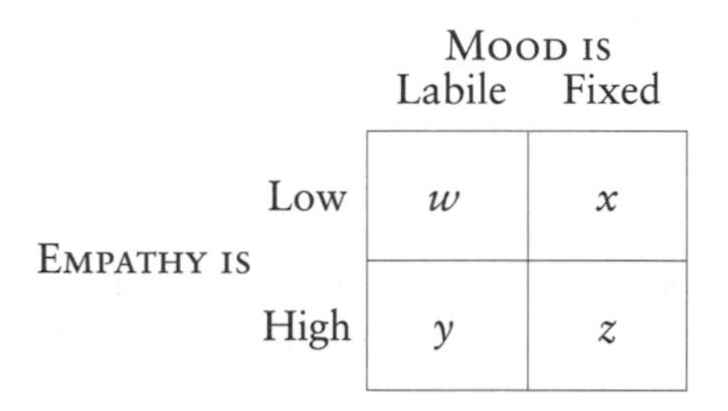

		MOOD IS	
		Labile	Fixed
EMPATHY IS	Low	w	x
	High	y	z

Cialdini et al. (1987) reasoned that the negative-state relief model predicts that high-empathy subjects should help more than low-empathy subjects when their mood is labile, but that empathy should make no difference in helping when subjects believe that their mood is fixed; that is, the negative-state relief model predicts that $y > w$ and $z = x$. In contrast, Cialdini and his coauthors suggest that the empathy-altruism hypothesis predicts that high-empathy should produce more helping, regardless of whether subjects think their mood is fixed or labile ($y > w$ and $z > x$). The point of difference between the two hypotheses, therefore, concerns whether subjects in the fixed-mood treatment volunteer to help more when they feel a high degree of empathy.

Cialdini et al. (1987) tabulated both the *proportion* of individuals who volunteered to help in each treatment cell and the *amount of time* they volunteered to spend. The data on amount of time supported the negative-state relief model, whereas data on proportions volunteering to help did not. Although Cialdini et al. viewed this and another experiment as favoring the egoistic hypothesis under test, Batson (1991, p. 166; Batson and Shaw 1991, p. 118) regards the results as more equivocal. Despite this difference in interpretation, both parties note that the results might have been due to *distraction*—after empathy was induced, the subjects were told for the first time that the drug they had taken earlier would freeze their moods. Perhaps this jarring information diminished the amount of empathy they experienced.

This possibility was confirmed by an experiment performed by Schroeder, Dovidio, Sibicky, Matthews, and Allen (1988), which closely resembled the Cialdini experiment, except that subjects were told about the mood-fixing effect of the drug when it was given to them and then were simply *reminded* of this fact just before they were asked whether they would help. Schroeder et al. found that high-empathy individuals volunteered to provide more time helping than did low-empathy individuals, both when their moods were labile *and* when they were fixed (though these observed differences were not statistically significant). The pattern of data on how often individuals volunteered to help was not terribly strong and was not well explained by either hypothesis. Schroeder et al. and Batson (1991, p. 168) drew the conclusion that these experiments do not favor Cialdini's negative-state relief model.

Batson, Batson, Griffitt, Barrientos, Brandt, Sprengelmeyer, and Bayly (1989) tested Cialdini's hypothesis in another way. As usual, subjects were assigned to a high-empathy or a low-empathy treatment. In addition, some individuals were told that they would receive a mood-enhancing experience (such as listening to music) regardless of whether they chose to help a needy other; other individuals, who also were provided with the opportunity to help, were not offered the chance to receive the mood-enhancing experience. The experiment produced data about the proportion of individuals choosing to help in these four treatments:

		Mood-enhancing experience was	
		Promised	Not promised
Empathy is	Low	w	x
	High	y	z

The negative-state relief hypothesis predicts that empathy will not affect helping when a mood-enhancing experience has been promised ($y = w$), while the empathy-altruism hypothesis predicts that empathy will make a difference ($y > w$). The data favored the empathy-altruism hypothesis (Batson 1991, p. 172; Batson and Shaw 1991, p. 119).

Here again, an egoistic explanation is not far to seek. If empathizing with a needy other makes a subject sad, why expect the subject to think that listening to music will be a completely satisfactory mood corrective? When we are sad, we usually are sad about something in particular. It is not surprising that the pain we experience in empathizing with the suffering of others is not completely assuaged by any old pleasant experience; however, this presents no difficulty for the egoism hypothesis.

The strategy behind Batson's research program is to show that each of the versions of egoism he has formulated encounters observations that it is unable to explain. How do these findings bear on the question of whether there is a set of observations that no version of egoism will be able to explain? Of course, there is no deductive entailment here; from the fact that everyone has a birthday, it does not follow that there is a single day on which everyone was born. The failures of simple forms of egoism don't *prove* that more complex formulations also must fail. Even so, it might be suggested that Batson's experiments *raise the probability* that no version of egoism can be observationally adequate. This may be so. Nonetheless, when we survey the ingenious experiments that social psychologists have constructed, we feel compelled to conclude that this experimental work has not resolved the question of what our ultimate motives are. The psychological literature has performed the valuable service of organizing the problem and demonstrating that certain simple egoistic explanations are inadequate. However, there is more to egoism than the hypotheses tested so far. What we find here is a standoff.

One reason the methodology that Batson uses may not be able to resolve the egoism-altruism debate becomes apparent when we compare his experimental design with the experiment we discussed at the end of the previous chapter. There we described a farmer who wants to know whether the difference in height observed between two fields of corn is due to a genetic difference, an environmental difference, or both. The farmer plants two additional fields of corn and thereby obtains data that allow this question to be answered. If the farmer has so little difficulty in disentangling nature and nurture in this experiment, why are Batson's experiments more equivocal?

The farmer experimentally manipulates genes and environment, just as Batson manipulates empathy level and some other factor (e.g., ease of escape). What can be discovered from this method, in the first instance, is that empathy makes a difference in the effect variable measured. However, this finding leaves unanswered the crucial question of whether empathy has its effect by triggering an altruistic ultimate motive. The same problem could arise for our farmer. Suppose the farmer isn't interested just in whether genes make a difference in plant height, but wants to test a more detailed hypothesis. This hypothesis has two parts; it says that there are genes that cause corn plants to grow tall *and* that these genes bring about this effect by producing enzyme *X*. If the farmer ran the experiment we described before and discovered that genes make a difference in plant height, one-half of the gene-enzyme hypothesis would be confirmed. But what about the other half? The farmer can tell from the observations that genes make a difference, but he can't tell whether they do so by activating enzyme *X*.

Of course, even if one type of experiment is incapable of disentangling psychological egoism and motivational pluralism, another design might be able to do the job. Nonetheless, it is tempting to claim that *any* behavior elicited in a psychological experiment will be explicable both by the egoism hypothesis and by the pluralism in which the altruism hypothesis is embedded (Wallach and Wallach 1991). We take no stand on this stronger thesis; what experimental psychology has been unable to do so far, new methods may yet be able to achieve. For now, however, the conclusion we draw is a discouraging one. Observation and experiment to date have not decided the question, nor is it easy to see how new experiments of the type already deployed will be able to break through the impasse.

It is essential to understand that just as the psychological literature has not established that the egoism hypothesis is false, it also has not established that pluralism is false. Why, then, has egoism struck so many as a theory one should believe unless one is forced to change one's mind? Perhaps there are compelling arguments, not based on observed behavior, that tip the scale in egoism's favor. Alternatively, the possibility needs to be faced that there is no good reason to regard egoism as a theory that is innocent until proven guilty. Historically, the egoism and altruism hypotheses have competed on an uneven playing field. In the next chapter, we will inquire further into the question of whether the privileged status assigned to egoistic explanations can be justified.

The Significance of the Psychological Question

The reason it is difficult to obtain experimental evidence that discriminates between egoism and motivational pluralism is that we have allowed egoism to appeal to *internal* rewards. If the view were more restrictive—if egoism claimed that we care only about *external* rewards such as money—the problem would be easier. We have already explained our reasons for choosing a definitional framework in which warm feelings are just as much a self-benefit as cold cash. The question we now want to address is, *why does it matter, given this definitional framework, what our ultimate motives really are?* The thought behind this question is that what matters, in human conduct, is how people treat each other; the motives behind helping really aren't important. If an altruist would be nice to you and an egoist would be nasty, you would want to know which type of person you were about to meet. But if altruists and egoists will treat you the same way, why should you be interested?

We have two replies. First, we think the question of whether people ever care about others as ends in themselves is theoretically significant. Of course, some people may find this question uninteresting, just as some may not be enchanted by theoretical issues in astronomy. However, if psychology is in the business of trying to explain behavior by elucidating mental mechanisms, it is hard to see how the structure of motivation can fail to be an important psychological problem.

Our second reply to people who think that what matters is whether people help, not why they do so, is practical. Consider the conjecture that what people believe about their ultimate motives influences how much they help. The hypothesis is that when people believe that egoism is true, they are inclined to be less helpful (Batson, Fultz, Schoenrade, and Paduano 1987). There is evidence in favor of this hypothesis. Frank, Gilovich, and Regan (1993) discuss several studies that compared economists with people in other disciplines in terms of a variety of measures of cooperative behavior. The pattern is that economists tend to be less helpful. It might be thought that this is because people who are less inclined to be helpful self-select for careers in economics. To rule this out, Frank et al. did a before-and-after study on students enrolled in two introductory economics courses and also on students in an introductory astronomy course.

Students were asked at the beginning of the semester, and also at the end, whether they would return to its owner an envelope they found that contained $100. The students also were asked whether they would inform a store about a billing mistake if they had been sent ten computers but had been billed only for nine. At the *beginning* of the semester, the economics students and the astronomy students said they'd perform the honest action about equally often. The economics and the astronomy students differed in how they *changed* during the semester. The willingness to act dishonestly increased among students in the economics classes more than it did among those in the astronomy class. This is evidence that studying economics inhibits cooperation. Of course, it is a further question whether economics has this effect by encouraging people to believe that psychological egoism is true. We think that this is a plausible guess, since this motivational theory plays a more prominent role in economics than in any other discipline.

Even if believing psychological egoism makes people less helpful, it does not follow that this theory is false. Maybe egoism is true, and the perception of its truth causes people to become a little more self-centered. The point we are making here is that if psychological egoism is false, then that may be a fact worth knowing, not just because it is theoretically important but because recognizing the falsehood of egoism may influence conduct (Batson 1991).

9

Philosophical Arguments

Philosophers sometimes tell people what they already know. For example, ordinary people typically believe that the physical world does not cease to exist when they close their eyes. Most philosophers feel the same way. Why, then, do they discuss the "problem of the external world"? The reason is that philosophers want to ascertain whether this piece of common sense can be rationally justified. The *conclusion* that physical objects exist independently of our perceiving them is not at issue; rather, philosophers are interested in the *arguments* that can be constructed for this quotidian thesis. Here philosophers attempt to justify what everybody already believes.

Some philosophy is like this, but discussion of egoism and altruism is different. In this case, philosophers have not simply affirmed what everyone else already believes; rather, they often have sought to undermine a widely held position about what our ultimate motives are. Many social scientists—and many of the rest of us—think that psychological egoism must be true. In contrast, there is and has been a strong tradition in philosophy for regarding this position as mistaken. Here is a context in which many philosophers reject what many people believe.

The present chapter does not provide a history of what different philosophers have said on this subject. Rather, we have assembled a set of distinctively philosophical arguments; most of them aim to establish that egoism is mistaken. Whether these philosophical arguments are any more successful than the psychological arguments reviewed in the previous chapter, we now must determine.

Butler's Stone

Joseph Butler (1692–1752) produced an argument that many philosophers think refutes hedonism. C. D. Broad's view of what Butler accomplished is not unusual: "he killed the theory so thoroughly that he sometimes seems to the modern reader to be flogging dead horses. Still, all good fallacies go to America when they die, and rise again as the latest discoveries of the local professors. So it will be useful always to have Butler's refutation at hand" (Broad 1965, p. 55).

As often happens with famous arguments in the history of philosophy, there is some disagreement concerning exactly what the argument says. So let us begin with Butler's own words:

> That all particular appetites and passions are towards *external things themselves,* distinct from the *pleasure arising from them,* is manifested from hence; that there could not be this pleasure, were it not for that prior suitableness between the object and the passion: there could be no enjoyment or delight from one thing more than another, from eating food more than from swallowing a stone, if there were not an affection or appetite to one thing more than another. (Butler 1726, p. 227)

Butler's point is that what we want is the *food,* not the pleasure that comes from eating. We will call this argument against hedonism *Butler's stone.*

Butler's reasoning echoes through much of the philosophy that came after him, and his influence on recent philosophy has been strong. For example, Broad claims that misers and politicians are living refutations of hedonism, since they desire money and power even when these items conflict with the attainment of happiness. Broad then remarks:

> It is no answer to this to say that a person who desires power or property enjoys the experiences of getting and exercising power or of

> amassing and owning property, and then to argue that therefore his ultimate desire is to give himself those pleasant experiences. The premise here is true, but the argument is self-stultifying. The experiences in question are pleasant to a person only in so far as he desires power or property. This kind of pleasant experience presupposes desires for something other than pleasant experiences, and therefore the latter desires cannot be derived from desire for that kind of pleasant experience. (Broad 1952, p. 92)

Broad goes on to assert that Butler's argument also refutes the hedonistic explanation of why people value self-respect and self-display.

Although Butler's stone concerns the relationship between the desire for pleasure and the desire for external things (like food), both of which are self-directed, the form of argument it embodies has enjoyed a wider application. In a much reprinted article, Joel Feinberg uses Butler's line of reasoning to argue against universal selfishness:

> Not only is the presence of pleasure (satisfaction) as a by-product of an action no proof that the action was selfish; in some special cases it provides rather conclusive proof that the action was *unselfish*. For in those special cases the fact that we get pleasure from a particular action *presupposes that we desired something else*—something other than our own pleasure—as an end in itself and not merely as a means to our own pleasant state of mind. (Feinberg 1984, p. 29)

In similar fashion, Thomas Nagel deploys a Butlerian argument in his discussion of altruism:

> There is one common account which can perhaps be disposed of here; the view that other-regarding behavior is motivated by a desire to avoid the guilt feelings which would result from selfish behavior. Guilt cannot provide the basic reason, because guilt is precisely the pained recognition that one is acting or has acted contrary to a reason which the claims, rights, or interests of others provide—a reason which must therefore be antecedently acknowledged. (Nagel 1970, p. 80)

Here it is the relationship between a self-directed desire (the desire to avoid feeling guilty) and an other-directed desire (the desire that another person's "claims, rights, or interests" be respected) that is addressed.

This is how we interpret Butler's stone:

1. People sometimes experience pleasure.
2. When people experience pleasure, this is because they had a desire for some external thing, and that desire was satisfied.

Hedonism is false.

We, of course, have no problem with the first premise. However, we will argue that the conclusion does not follow from the premises and that the second premise is false.

Consider the causal chain from a *desire* (the desire for food, say), to an *action* (eating), to a *resultant* (pleasure):

Desire for food ⟶ Eating ⟶ Pleasure

If the pleasure traces back to an antecedently existing desire, then the resulting pleasure did not cause the desire (on the assumption that cause must precede effect). However, this unproblematic conclusion does not decide what the relationship is between two *desires*—the *desire for food* and the *desire for pleasure*. In particular, it leaves entirely open what caused the desire for food. If hedonism were true, these desires would be related as follows:

Desire for pleasure ⟶ Desire for food

Hedonism says that people desire food *because* they want pleasure (and believe that food will bring them pleasure). Butler's stone concludes that this causal claim is false, but for no good reason. The crucial mistake in Butler's argument comes from confusing two quite different items—the *pleasure* that results from a desire's being satisfied and the *desire for pleasure*. Even if the occurrence of *pleasure* presupposes that the agent desired something besides pleasure, nothing follows about the relationship between the *desire for pleasure* and the desire for something else (Sober 1992; Stewart 1992).

A hedonist can accept the idea that if an agent wants an external thing and that desire is satisfied, then pleasure is the result. Butler goes further; he claims, not just that this is *one* route to pleasure, but that it is the *only* one. A hedonist could accept this stronger claim as well, but should not, for reasons that Broad identified: "Certain sensations are intrinsically pleasant, e.g., the smell of violets or the taste of sugar. Others are intrinsically unpleasant, e.g., the smell of sulphurated hydrogen or the feel of a burn. We must therefore distinguish between intrinsic pleasures and pains and the pleasures and pains of satisfied or frustrated impulse" (Broad 1965, p. 66).

When Broad says that some experiences are "intrinsically" pleasurable or painful, he seems to mean that this is how they feel, regardless of what the agent thinks or wants. Desiring is a cognitive achievement (Chapter 6) and some sensations apparently can occur without a person's having to construct a propositional representation; you don't have to have any particular belief or desire for a burn to hurt. This point will be important in the next chapter. Broad is right that there are routes to pleasure besides the one that Butler describes.[1]

In any case, Butler's claim about the one road to pleasure is dispensable from his argument. The argument would not be weakened if Butler were to say that satisfying the desire for an external thing is one way, among others, in which people obtain pleasure. The crucial error, woven into the very fabric of the argument, is Butler's idea that hedonism is somehow opposed to the idea that people *do* want external things. This is a mistake. Hedonism attempts to *explain* why people want external things; it does not *deny* that they do so. We conclude that Butler's stone, besides being a fallacy, also has a false premise. It is false that there could be no pleasure unless the agent antecedently desired some external thing. And even if that connection between pleasure and desire were granted, it would not refute the hedonist's contention that people desire external things only because they think those things will satisfy their ultimate desire to gain pleasure and avoid pain.

The Paradox of Hedonism and the Requirements of Reason

Philosophers have sometimes claimed that individuals who obsessively focus on attaining pleasure or happiness inevitably fail to get what they want. Supposedly, pleasure and happiness are attainable only as by-products of becoming absorbed in specific activities. This fact about pleasure and happiness is thought to constitute a paradox for hedonism—the word *paradox* indicating that we are supposed to find here a flaw in hedonism as a psychological theory (Butler 1726; Feinberg 1984).

There is an ounce of truth in this idea. People whose one and only thought is that they want to be happy (or to attain pleasure) will not attend to the task of selecting activities that are means to that end. They will be like stockbrokers who think only that they should buy low and sell high. People who have an end in mind but no means to attain it surely will fail to get what they want. The obvious reply to this criticism is that there is nothing in hedonism that says that people must be monomaniacal in this way. The claim is that people have attaining pleasure and avoiding pain as their only *ultimate* goals; it does not say that attaining pleasure and avoiding pain are the only goals (ultimate *or* proximate) that people ever have. A hedonist will reflect on which activities are most apt to bring pleasure and prevent pain, and will decide what to do on the basis of this reasoning (Sidgwick 1907). Means-end deliberation is not only compatible with hedonism; it is part of the theory's basic logic.

There is a second element in this "paradox" that is equally flimsy. The claim is that hedonists, by virtue of being monomaniacs, must fail to obtain the pleasure they seek. Even if this point entailed that people *should* not be hedonists,[2] it would not follow that people are not hedonists *in fact*. Hedonism as a descriptive thesis needs to be distinguished from hedonism as a normative thesis. This so-called paradox has no bite with respect to the descriptive claim.

A similar distinction needs to be drawn in connection with the principal claims of Thomas Nagel's (1970) *The Possibility of Altruism*. Nagel argues that people who do not take the interests of others into account when they decide what to do are irrational. They attend to their own interests in deliberation, but not to those of others, even though there is no property that they have and others lack that could justify this asymmetry. This is how Nagel's book addresses the subject named in its title: if a person is rational, it will be possible for him or her to act altruistically.

Philosophers disagree about whether "rationality" refers just to the narrow ability to choose efficient means to achieve whatever ends one might have (sometimes called "instrumental" rationality) or should be understood in a fuller sense to mean that the ends themselves are morally defensible ("substantive" rationality).[3] An efficient serial killer might be termed "rational" according to the first approach, but not according to the second. Regardless of how one decides this issue, Nagel's point fails to show that people might have altruistic ultimate motives. If we are talking about *instrumental* rationality, then the narrow self-interest of a thoroughgoing egoist may be morally abhorrent, but it is not necessarily irrational. On the other hand, if rationality is taken to mean *substantive* rationality, then it may be true that egoists are irrational, but this is just another way of saying that their exclusive focus on self-interest is morally wrong.[4] Whichever view one favors, the fact remains that Nagel has not shown that people really are capable of having altruistic motives. This is because Nagel's conclusion is that rationality entails altruism (or its possibility); Nagel offers no argument that people are rational in his sense. Perhaps we *ought* to be rational and maybe we *ought* to be altruistic as well. But neither claim refutes the descriptive thesis of psychological egoism.

The Experience Machine

Even if the experiments carried out by psychologists and the behaviors that each of us observes in everyday life are insufficient to settle the egoism-altruism debate, there is an interesting thought experiment that seems to provide decisive evidence against hedonism. In his book *Anarchy, State, and Utopia,* Robert Nozick (1974, pp. 42–45) describes what he calls an *experience machine.* Nozick's thought experiment does not describe any device that currently exists, but considering the idea of an experience machine does help clarify what our ultimate desires are.

Imagine an elaborate computer that sends complex sequences of electrical stimulation into a person's brain. The machine can be programmed to provide thoroughly convincing simulations of any real-life experience one might choose. If you are plugged into the machine, you can receive experiences that are indistinguishable from the ones you'd have if you were skiing down a mountain, or playing the violin, or giving a speech at the United Nations.

Suppose you were offered the chance to plug into the experience machine for the rest of your life. The machine would be programmed to make you instantly forget that you had chosen to plug in and then would give you whatever sequence of experiences you would find maximally pleasurable and minimally painful. If your wishes involve the experiences associated with being the world leader who finally eradicates world hunger, that can be arranged. If you would like to have the experience of being the greatest athlete who ever lived, that too would be no problem. And if the happiness that comes from deep and sustained relationships of love and friendship is what you crave, the machine can provide that as well. The only catch, of course, is that your beliefs will be illusory. You will *think* you are a great politician, a great athlete, or involved in a loving relationship, but your belief will be *false*. If you choose to plug into the experience machine, you will live your life strapped to a laboratory table with tubes and electrodes sticking into your body. You'll never *do* anything; however, the level of pleasure you'll experience, thanks to the machine, will be extraordinary.

If you were offered the chance to plug into the experience machine, what would you do? One natural reaction is to doubt that the machine will perform as promised; certainly no machine now on the market can deliver what this machine is said to be able to do; virtual reality games don't even come close. However, for the sake of argument, let us set this hesitation aside. Imagine yourself being offered the chance to plug in, and suppose that the machine will deliver as promised—it will provide you with the most pleasurable and least painful sequence of experiences that there can be, ones that are uniquely suited to your temperament. Our guess is that many people would decline the opportunity of plugging in; we'll assume that this is how you'd feel.

This fact about people seems to contradict hedonism. Apparently, many people prefer to have a real life over a simulated one, even if real life brings less pleasure and more pain than the life they'd have if they plugged into the machine. It seems that people care irreducibly about how they are related to the world outside their own minds; it is false that the only things they want as ends in themselves are pleasant states of consciousness.[5]

The hedonist must explain why many people would decline the offer to plug into the machine.[6] One possible explanation is that people aren't really addressing the question posed; although people are asked to assume that the machine will deliver experiences of a certain type, perhaps they find it impossible to adopt this supposition for the sake of argument. Another possible hedonistic reply is that people lose the ability to think rationally when confronted with this problem. Both these suggestions by the hedonist attempt to preserve the hedonistic theory from refutation by abandoning other assumptions concerning how people conceptualize the problem at hand. We don't find either of these suggestions very plausible. After all, people are often pretty good at accepting an assumption for the sake of argument and then reasoning on that basis. When people say they are unwilling to plug into the experience machine, this isn't because they don't understand the problem posed or reason incorrectly in thinking it through.

To describe a more promising hedonistic explanation of why people often decline to plug into the experience machine, we need to map out the sequence of events that will comprise your life if you choose to plug into the experience machine and the sequence of events that will occur if you do not. In both cases, the process begins with deliberation, which terminates in a decision. If you decide to plug in, there is a time lag between your decision and your actually being connected to the machine. Here are the two time lines we need to consider:

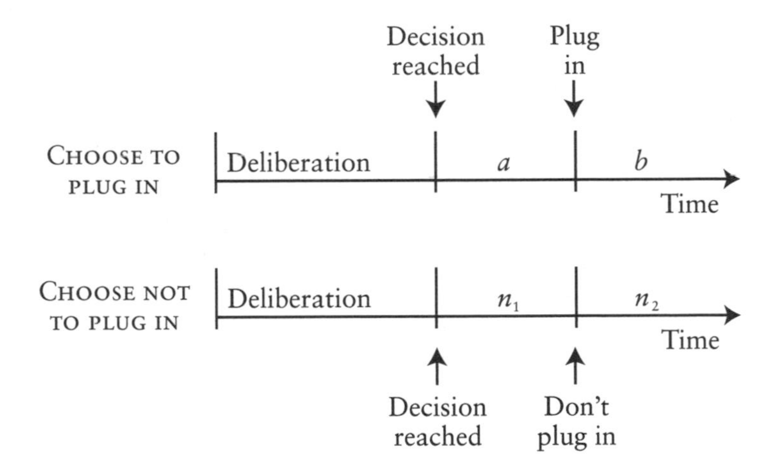

The four letters in these two time lines represent how pleasant your experiences will be during different temporal periods, depending on what you decide. If you choose to plug into the machine, you'll have an immense level of *bliss* (b) after you plug in. This will dwarf the amount of pleasure you'll experience in the same period of time if you decide not to plug in and to lead a normal life instead; $b > n_2$. If this were the only consideration involved, the hedonist would have to predict that people will choose to plug into the machine. How can hedonism explain the fact that many people make the opposite decision?

The hedonist's strategy is to look at earlier events. Let's consider how you'll feel if you decide to plug into the machine, but before you are actually connected. Presumably, you will experience a great deal of *anxiety* (a). You will realize that you will never again see the people you love. You also will realize that you are about to stop leading a real life. It is clear that you will have less pleasure during this period of time than you would if you rejected the option of plugging into the machine and continued with your real life instead; $a < n_1$.

If hedonists are to explain why people choose not to plug into the experience machine, and if they are to do this by considering just the pleasure and pain that subjects expect to come their way *after* they decide what to do, the claim must be that $a + b < n_1 + n_2$. Since b is far greater than n_2, this inequality will be true only if a is far far smaller than n_1. That is, hedonists seem compelled to argue that people reject the option of plugging in because the amount of pain they would experience between deciding to plug in and actually being connected to the machine is *gigantic*—so large that it dwarfs the pleasure they'd experience after they are connected.

This is a very implausible suggestion. The period of time between deciding to plug in and actually doing so can be made very brief, compared with the long stretch of years you'll spend attached to the machine and enjoying a maximally pleasurable ensemble of experiences. We grant that people who decide to connect to the machine will experience sadness and anxiety during the brief interval between deciding to plug in and actually plugging in. But the idea that this negative experience swamps all subsequent pleasures just isn't credible.

To see why, let's consider a second thought experiment.[7] Suppose you were offered $5,000 if you went through 10 seconds of a certain experience. The experience in question is believing that you had just decided to spend the rest of your life plugged into an experience machine. After your 10-second jolt of believing this, you will return to your normal life and receive the money as promised. We expect that most people would choose the 10 seconds just described because it will earn them $5,000. This shows that the hedonist should not claim that the experience of believing you will be plugged into an experience machine for the rest of your life is so horrible that no one would ever choose a life that included it.[8]

The hedonist still has not been able to explain why many people would choose a normal life over a life plugged into the experience machine. The reason is that a hedonistic calculation seems to lead inevitably to the conclusion that $a + b > n_1 + n_2$. Does this mean that the hedonist must concede defeat? We believe that the hedonist has a way out. Quite apart from the amount of pleasure and pain that accrues to subjects *after* they decide what to do, there is the level of pleasure and pain arising in the deliberation process itself. The hedonist can maintain that *deciding* to plug into the machine is so aversive that people almost always make the other choice. When people deliberate about the alternatives, they feel bad when they think of the life they'll lead if they plug into the machine; they feel much better when they consider the life they'll lead in the real world if they decline to plug in. The *idea* of life attached to the machine is painful, even though such a life would be quite pleasurable; the *idea* of real life is pleasurable, even though real life often includes pain.[9]

To see what is involved in this hedonistic explanation of why people refuse to plug in, let's consider in more detail what goes through their minds as they deliberate. They realize that plugging in will mean abandoning the projects and attachments they hold dear; plugging into the machine resembles suicide in terms of the utter separation it effects with the real world. The difference is that suicide means an end to consciousness, whereas the experience machine delivers (literally) escapist pleasures. Hedonism is not betraying its own principles when it claims that many people would feel great contempt for the idea of plugging in and would regard the temptation to do so as loathsome. People who decline the chance to plug in are repelled by the idea of narcissistic escape and find pleasure in the idea of choosing a real life.

One virtue of this hedonistic explanation is that it explains the results obtained in both the thought experiments described. It explains why people often *decline* to plug into the experience machine for the rest of their lives; it also explains why people offered $5,000 often *agree* to have ten seconds of the experience of believing that they have just decided to plug into the machine for the rest of their lives. In both cases, deliberation is guided not so much by beliefs about which actions will bring *future* pleasure, but by the pleasure and pain that accompany certain thoughts *during the deliberation process itself.*

In Chapter 7, we discussed the fact that different formulations of hedonism specify different *time horizons* in describing what people ultimately care about. When Ronald contemplates whether to smoke the cigarette he holds before him, he is choosing between immediate gratification and long-term health. Hedonism is a theory that comes in many forms; it may claim that people want to maximize pleasure and minimize pain over their entire lifetimes or over some shorter period of time. There are many alternatives here; each may be viewed as specifying a "discount rate" that describes the relative import of experiences in different time periods. Hedonism is consistent with the possibility that different people have different discount rates; it also is consistent with a single person's having different discount rates for different types of experience; and it is compatible with people's changing their discount rates on a particular type of experience as they get older. The problem of the experience machine is interesting because it forces hedonism to postulate a time horizon that is extremely short. It is *hedonism of the present moment,* so to speak, that permits hedonism to explain why people choose not to plug into the experience machine; versions of hedonism that posit longer time horizons make false predictions about what people will do in the circumstance described.

The problem of the experience machine resembles a problem that is much more of a fixture in the literature on psychological egoism and altruism. Consider a soldier in a foxhole who throws himself on a live grenade to save the lives of his comrades. How can hedonism explain this act of suicidal self-sacrifice, if the soldier believes that he will not experience anything after he dies? The hedonist suggests that a self-directed benefit accrues *before* the act of self-sacrifice is performed. It is no violation of hedonism to maintain that the soldier decides to sacrifice his life because that decision is less painful than the decision to let his friends die. The problem of suicidal self-sacrifice and the problem posed by the experience machine can be addressed in the same way.

We do not conclude that hedonism provides the right explanation for why people opt for real life instead of the simulated life provided by an experience machine. The nonhedonistic hypothesis—that people care irreducibly about the way the world is, not just about their own states of consciousness—also can explain the choice. Rather, our conclusion is that the experience machine thought experiment does not refute hedonism.

Burden of Proof

Philosophers sometimes maintain that a commonsense idea is innocent until proven guilty. That is, if a question is raised about whether some commonsense proposition is true, and no argument can be produced that justifies or refutes it, then the sensible thing to do is to keep on believing the proposition. Put differently, the burden of proof lies with those who challenge commonsense.

This general attitude sometimes surfaces in discussion of egoism and altruism. The claim is advanced that the egoism hypothesis goes contrary to common sense. The common sense picture of human motivation is said to be pluralistic—people care about themselves, but also care about others, not just as means but as ends in themselves. The conclusion is then drawn that if philosophical and empirical argumentation for and against egoism is indecisive, then we should reject egoism and continue to accept pluralism.[10]

Our first objection to this proposed tie-breaker is that we think it is far from obvious that "common sense" enshrines the pluralistic outlook rather than the egoistic one. What is common sense? Isn't it just what people commonly believe? If so, we suspect that egoism has made large inroads into the realm of common sense; it is now a worldview endorsed by large numbers of people. Philosophers need to be careful not to confuse common sense with what they themselves happen to find obvious. To our knowledge, no empirical survey has determined whether a pluralistic theory of motivation is more popular than psychological egoism.

Regardless of what people commonly believe about psychological egoism and motivational pluralism, we reject the idea that conformity with common sense is a tie-breaker in this debate. It does not have this status in physics or biology, and we see no reason why it should do so when the question happens to be philosophical or psychological in character.[11] In fact, it is arguable that our intuitions in this domain are especially prone to error. People have a picture of their own motives and the motives of others. If certain types of self-deception—either regarding one's own motives or those of others—were advantageous, then evolution might have enshrined these falsehoods in the set of "obvious" propositions we call common sense. A philosophy informed by an evolutionary perspective has no business taking common sense at face value.[12] As it happens, psychology has documented numerous contexts in which people have inaccurate beliefs about their own motives (see, e.g., Nisbett and Wilson 1977; Ross and Nisbett 1991), a point we discussed in the previous chapter. It seems naive to take the introspectively obtained deliverances of common sense at face value when the question is egoism versus altruism.

Before we leave this topic, we want to make a suggestion concerning the status of burden of proof arguments in general. Burden of proof is a legal and procedural concept, designed to protect the rights of this or that party (e.g., the accused). It simply has no place in argumentation whose goal is to figure out what is true. *In law courts, the burden of proof falls on the prosecution; in science and philosophy, it falls on everyone.* When evidence is indecisive, we should admit that it is and not invoke a spurious tie-breaker.

Falsifiability

Egoism is sometimes criticized for being unfalsifiable. Each time a behavior is cited that seems to refute egoism, the egoist can postulate some new self-directed preference that the behavior ultimately subserves. We saw this flexibility in the previous chapter, when we examined Batson's experiments. Each time an experiment refuted one version of egoism, another version cropped up. The egoist's list of self-directed motives seems to be endless.

It might appear that this flexibility in the egoistic hypothesis is one of its strengths. Is it really such a bad thing that the hypothesis can be modified and elaborated so as to accommodate new observations? The standard reply to this suggestion is that scientific hypotheses are supposed to be testable, and testable hypotheses must be falsifiable. A testable hypothesis cannot be logically consistent with all conceivable observations; it must exclude at least one conceivable observational outcome. This widely held view of what makes a hypothesis scientific stems from the enormously influential work of the philosopher Karl Popper (1959). We want to explain briefly why we think it is far from obvious that egoism *is* unfalsifiable. We then will show that the issue of falsifiability cuts no ice in the dispute between egoism and altruism.[13]

We may begin by recalling a point discussed in Chapter 7—a hypothesis about a person's ultimate motives, by itself, predicts nothing about behavior. Additional assumptions are needed—about the person's beliefs and instrumental desires, and about the processes that connect these items to action—before a hypothesis about our ultimate motives can make predictions. This is a feature of many scientific hypotheses; philosophers often call it "Duhem's thesis," after the French historian and philosopher of science who argued that theories in *physics* become testable only when they are supplemented with auxiliary assumptions. Duhem's thesis provides a reason to be circumspect about claiming that observations could never refute the egoism hypothesis. If hypotheses have implications about observations only when additional assumptions are conjoined, how do we know that *new* theories will not be developed and confirmed? These new background theories might furnish auxiliary assumptions that would force the egoism hypothesis to make predictions that could fail to come true.

Even if we withhold assent from the claim that egoism is in principle unfalsifiable, the fact remains that the theory is difficult to test. This is because egoism does not say what ultimate motives people have; it merely says that their ultimate motives are self-directed. In a sense, egoism specifies the principles of a *research program;* it tells you what type of explanation you should try to develop without saying which specific explanations will turn out to be correct. This is the main source of egoism's flexibility; if one explanation of the specified type is disconfirmed, you can always try to construct another.

Propositions that express the central assumptions of a research program often have this feature. For example, consider the hypothesis that natural selection has played a preeminent role in the evolution of most phenotypic traits in most populations. This a reasonable way to formulate adaptationism as a claim about nature; it is distinct from the purely methodological version of adaptationism, discussed in the Introduction, which says merely that constructing and testing adaptation hypotheses is a useful method for investigating evolution (Sober 1993b; Orzack and Sober 1994). Notice that the claim about nature is far more abstract and general than the claim that giraffes have long necks because this helps them reach the canopies of tall trees. Even if this adaptive explanation of the giraffe's neck turns out to be wrong, another can be constructed in its stead. And even if some traits turn out to have nonadaptive explanations, the more general question of how important natural selection has been in evolution overall remains open. This is not the place to explore the ins and outs of testing adaptationism. However, there is one point of analogy between the debate over adaptationism and the egoism-altruism debate that is worth mentioning. When adaptationism is formulated as a claim about nature (rather than as a purely methodological claim about how one should go about investigating nature), adaptationism turns out to be a (relatively) *monistic* doctrine, whereas its chief competitor, evolutionary *pluralism,* views natural selection as one of several important causes of the phenotypic traits found in natural populations (Gould and Lewontin 1979).

In general, pluralism inherits the problems of monism, as far as testability is concerned. If a new adaptationist explanation can always be invented when an old one fails, it also will be true that a new pluralistic explanation can be invented if an old one turns out to be wrong. The reason is that pluralism makes use of the same explanatory variables as monism, and then adds some new ones besides. Adaptationism focuses on natural selection; evolutionary pluralism talks about selection, genetic constraints, drift, and other factors as well. Psychological egoism postulates exclusively self-directed ultimate motives; psychological pluralism postulates self-directed ultimates motives and altruistic ultimate motives as well. However flexible a doctrine monism may be, pluralism is more flexible still.

This is not to deny that monistic doctrines are sometimes poorly defended. Adaptationists sometimes tell just-so stories; egoists do not hesitate to postulate psychological rewards to explain even the most harrowing acts of self-sacrifice. This can be quite exasperating. But the fact that adaptationism and egoism need to be appropriately constrained does not show that they are false. Also, if constraints are needed for monistic doctrines, they are needed *a fortiori* for their pluralistic alternatives. In short, if psychological egoism is untestable, the same will be true of the pluralism within which the altruism hypothesis is embedded. We do not concede that egoism *is* untestable; the present point is just that the issue of testability cannot be used to reject psychological egoism and accept motivational pluralism.[14]

Parsimony

When a defender of the altruism hypothesis cites a behavior as evidence for altruism, advocates of egoism reply by trying to show that the behavior can be explained within their favored framework. If they succeed, the conclusion that egoists usually draw is that egoism is the preferable hypothesis. But why should this be so? If both theories can explain what we observe, why say that egoism is true and motivational pluralism is false? Why not conclude, instead, that the observation fails to discriminate between the two theories?

We earlier examined the argument that pluralism is preferable to egoism because egoism contradicts common sense and therefore has the burden of proof. We now will analyze a different reply, one that leads to the opposite conclusion. This is the claim that egoism should be accepted and pluralism rejected, if both can accommodate the observations, because egoism is more parsimonious. The hypothesis of pure egoism seems to be simpler than a theory that postulates both egoistic and altruistic ultimate motives (Batson 1991).[15] Ockham's razor—the principle of parsimony—seems to furnish a tie-breaking consideration. The point is not that simpler hypotheses are easier to understand or test, or that they are aesthetically more pleasing; the thought is that simpler theories are preferable in the sense that they are more plausibly regarded as true.

We have two objections to this parsimony defense of psychological egoism. First, we don't agree that the simplicity of a motivational theory should be decided just by counting how many ultimate motives it postulates. Second, even if egoism were more parsimonious, the question would remain whether there are other considerations that tip the balance in the other direction. We'll consider these two points in turn.

Consider two motivational theories. The first is monistic; it asserts that the desire for *X* is ultimate and that people desire *Y* only because they believe that attaining *Y* helps one attain *X*. This theory postulates one ultimate and one instrumental desire. The second theory is pluralistic; it says that the desire for *X* and the desire for *Y* are both ultimate. As just noted, if we measure parsimony by counting ultimate motives, then the monistic theory is more parsimonious. However, why focus on this feature of the theories to the exclusion of others? For example, if we count *all* desires, not just *ultimate* desires, the two theories tie.

There is a third way to measure how parsimonious the theories are, and this procedure leads to the surprising conclusion that the pluralistic theory is simpler than its monistic rival. When the monistic theory says that people desire *Y because they think it will help them obtain X,* it thereby attributes to people a certain *causal belief.* In contrast, the pluralistic theory, which says that people view both *X* and *Y* as ends in themselves, does not have to postulate any such belief. If we measured parsimony just by counting beliefs of this type, we would reach the conclusion that motivational pluralism is more parsimonious than egoism.[16] In the absence of a principled decision about how to do the counting, we conclude that there is no reason to think that psychological egoism is more parsimonious than motivational pluralism.

The drawing by Rube Goldberg that appears on page 196 of this book illustrates the point we are making here. Goldberg depicts a device that automatically wipes your face with a napkin after you eat a spoonful of soup. When you lift the spoon to your lips, you pull a string that flips a cracker to a parrot; when the parrot grabs the cracker, a dish of sand tips into a little bucket, which ignites a lighter, thus firing a rocket and cutting the string that holds back the pendulum of a clock, to which a napkin is attached. When the pendulum is released, it swings across your face and wipes your lips with the napkin. The device is funny because of its gratuitous complexity. An automatic mustache wiper might seem to be a "labor-saving device," but think of the labor that goes into building and maintaining the machine. There obviously are much more straightforward ways for people to keep their faces clean when they eat. Let's explore this thought in more detail.

Suppose you are an engineer whose job is to build an organism that not only eats soup but keeps its mustache clean as well. One strategy you might use is pluralistic—give the organism two desires, one for eating soup and one for staying clean. An alternative is to get the organism to perform both behaviors by giving it just one desire. Let the organism have the desire to eat soup. Then equip it with a device that ensures that its mustache gets wiped each time it eats soup. Which of these setups is simpler? If you just count desires, you'll conclude that the monistic arrangement is simpler; the fact that this design involves a Rube Goldberg device has not affected your assessment. But surely this is a blinkered evaluation; the fact that the monistic arrangement involves a Rube Goldberg device should be taken into account. The simplicity of a psychological mechanism isn't determined just by how many ultimate desires it involves. We will return to this point in the next chapter.

This point about simplicity may have come to mind as you read about the psychological experiments described in the previous chapter. Egoists try to explain the results of the different experiments that Batson performed in different ways: they offer the aversion-arousal hypothesis, the empathy-specific punishment hypothesis, the empathic joy hypothesis, and others. These different hypotheses postulate a number of specific beliefs. People think they will feel guilty if they exit from a needy other, they think they will feel bad if they refuse information about a needy other, and so on. The altruism hypothesis, on the other hand, does not make such claims; it provides the same, fairly simple explanation of why high-empathy subjects behave differently from low-empathy subjects across the different experiments. In a sense, the altruism hypothesis provides a more *unified* treatment of this finding than the egoism hypothesis. Should we accept psychological egoism, on the grounds that it postulates fewer ultimate motives, or should we accept the altruism hypothesis instead, on the grounds that it is more unified and postulates fewer beliefs? Parsimony may be desirable in theory, but in this instance at least, the principle of parsimony is hard to put into practice.

We reject the parsimony argument that favors egoism over motivational pluralism for a second reason. If it could be shown that egoism is more parsimonious than motivational pluralism, this would mean that we should accept egoism, all else being equal. What remains to be determined is whether all other considerations *are* equal. Are there other considerations that might favor pluralism? To illustrate the type of reservation we have in mind, consider a somewhat parallel evolutionary problem. Suppose some paleobiologists are trying to figure out what internal organs a newly discovered species of dinosaur possessed. The dinosaur's skeleton can be inferred from fossils, but how are these scientists supposed to ascertain what the organ systems were like if these soft parts left no fossil traces?

One of the scientists, whom we'll call Bill (because of his Ockhamist leanings), suggests that it makes sense to assume that the organism had one heart, one lung, and one kidney. Bill grants that the dinosaur needed each of these organs to perform some function; however, he insists that it is pointless to postulate more than the minimum number that is strictly necessary. Bill's point is not that we should remain agnostic about whether there was *just* one of each organ or *more* than one. Rather, his claim is that parsimony favors postulating *exactly* one rather than any *greater* number.

Bill's colleagues are not convinced; they suspect that his use of Ockham's razor is ham-fisted and produce two reasons to back up their skepticism. First, we know by observation that many contemporary organisms have pairs of organs, rather than singletons. If the new species of dinosaur is closely related genealogically to these contemporary organisms, this may support the conclusion that it resembled them in those respects.[17] The second reason to reject Bill's appeal to parsimony derives from the biological advantages of duplication. Perhaps organs often come in pairs because this arrangement is a useful form of insurance; if an individual loses one of the organs in a pair, it can sometimes make do with the other. If selection favors duplication, perhaps we should be less willing to assume that the dinosaur had one lung rather than two.

The details of why Bill's parsimony argument is unconvincing don't much matter here. The important point is that Bill's colleagues are not satisfied by his appeal to a principle according to which abstract numerology settles the issue. They adduce specifically evolutionary considerations, which, in the case at hand, suggest that a hypothesis that postulates fewer entities is not always more reasonable than a hypothesis that postulates more.[18] Parsimony and plausibility don't always go hand-in-hand.

This is the second hesitation we have about the parsimony argument for egoism. We are not suggesting that motivational pluralism is justified by exactly the same line of argument that makes it reasonable for the paleobiologists to claim that the dinosaur had two lungs instead of one. We concede that there are disanalogies between the two problems. For example, paleobiologists can *observe* that extant species closely related to the dinosaur have two lungs; this evidence licenses the inference that the dinosaur species had two lungs as well. In contrast, we can't directly observe that motivational pluralism is true of species closely related to humans; what is hidden from observation of human behavior also is hidden when we observe the behavior of nonhuman primates. Nonetheless, there is a more general point that unites these two problems. We need to test theories of motivation by seeing whether they are plausible in the light of evolutionary considerations. Although egoism postulates fewer ultimate motives than pluralism, we need to ask whether it makes sense to expect a purely egoistic motivational mechanism to have evolved. This will be the focus of our next chapter.

Leveling the Playing Field

In the previous chapter, we argued that observed behavior does not favor psychological egoism over motivational pluralism. In the present chapter, we reviewed a range of philosophical arguments and concluded that they don't support psychological egoism either. An objective judge would have to conclude that egoism does not deserve its position of strength as the dominant theory in the intellectual pecking order, and never did. If so, it is a fascinating problem in the history of ideas to determine how egoism achieved its position of prominence.[19] Apparently, the egoism hypothesis was able to rise without any genuine means of support. If these two chapters have the effect of leveling the playing field on which the two theories compete, that must count as significant progress. Yet, verdicts of "not proven" are unsatisfying, however much they may be warranted by the evidence at hand. For this reason, there is an incentive to try for more. In addition to experimental social psychology and philosophy, the theory of evolution must enter the egoism-altruism debate. Perhaps it can break the deadlock.[20]

➤ 10 ➤

The Evolution of Psychological Altruism

In this chapter, we will bring together some of the issues that we so far have endeavored to keep apart. We have urged the importance of not confusing psychological egoism and altruism with the evolutionary concepts that go by the same names. We now will try to show how evolutionary considerations bear on the question of psychological motivation.

In Chapters 6 and 7, we described three motivational theories. Psychological *hedonism* says that attaining pleasure and avoiding pain are the only ultimate concerns that people have. Hedonism views people as motivational solipsists; it says that people care ultimately about states of their own consciousness and about nothing else. Psychological *egoism,* the second theory we presented, is more liberal than hedonism. Hedonists are egoists, but not all egoists are hedonists. Egoists may care ultimately about attaining pleasure and avoiding pain, but they also may have ultimate desires that embrace the world outside their own minds. Egoists, for example, may have their own survival as an end in itself; they also may have the irreducible desire to accumulate wealth or to scale Mount Everest. According to egoism, all ultimate desires are *self-directed;* when people care about the situations of others, they do so for purely instrumental reasons. The third theory we considered—motivational *pluralism*—says that the ultimate desires that people have include both egoistic and altruistic motives. People may want to avoid pain as an end in itself, and they also may have their own survival as an ultimate goal, but, in addition, they sometimes care irreducibly about the welfare of others.

In our discussion of egoism and pluralism, we pointed out that these theories can be fleshed out in different ways. Each describes the *types* of motives that people have without saying much about what those specific motives are. Egoism says that our ultimate desires are self-directed, but it does not describe what it is we want for ourselves; pluralism says that some of our ultimate desires are other-directed, but it leaves open who we care about and what we want them to have. The fact that these theories are open-textured in this way makes them difficult to test, since evidence against one version of a theory may leave others unscathed.

We will focus in what follows on hedonism as the main competitor that the altruism hypothesis must confront. The reason is that defenders of egoism inevitably invoke the ultimate desire to attain pleasure and avoid pain to save egoism from refutation. For example, if they maintain that people are motivated exclusively by *external* rewards (such as money), it is easy enough to describe behaviors that cannot be explained within that framework. In order to avoid defeat, egoists then appeal to *internal* rewards to do the explanatory work. Indeed, the egoist could have cited those internal rewards from the start; they can explain the behaviors that external rewards explain, and then some. By pitting altruism against hedonism, we are asking the altruism hypothesis to reply to the version of egoism that is most difficult to refute.[1]

Although hedonism is a special variety of egoism, we believe that our argument against hedonism has more general implications. We will maintain that *no* version of egoism is plausible for organisms such as ourselves. Hedonism exemplifies the kinds of evolutionary implausibility into which egoism inevitably must fall.

What the Evolutionary Framework Assumes

Our discussion in Chapter 8 of psychological evidence concerning the egoism-altruism debate led to the verdict of *not proven.* The philosophical arguments we explored in Chapter 9 did nothing to displace this conclusion. The behaviors we considered could have been produced by altruistic motives or by purely egoistic motives. Even though these behavioral data were unable to discriminate between the motivational hypotheses, we will try to show that evolutionary ideas do better. Our strategy is to shift the focus from behavioral effects to evolutionary causes:

Evolution ⟶ Motives ⟶ Behavior

Even if two motivational mechanisms both are capable of generating a certain type of behavior, it remains possible that one of them was more likely than the other to have evolved.

The causal chain from evolution to motives to behavior just depicted may suggest that we think that *all* behaviors are *completely* explainable in evolutionary terms. We think nothing of the kind. Some questions about behavior can be answered by appeal to evolutionary considerations; others cannot be. All that we require is that evolution is *part* of the explanation of *some* facts about human behavior.

Discussion of the pros and cons of an evolutionary approach to human behavior has often been fixated on questions that are too general and too vague. The question of whether evolution explains why we think and act as we do should not receive a yes or a no answer; rather, the question should be rejected as ill-formed. Which features of mind and behavior are we interested in discussing? Evolution may be relevant to some but not to others. Consider, for example, the very broad category of diet. It is a mistake to ask whether evolution can explain why we eat what we do. Which proposition do we wish to explain? There are many facts about what people eat; these facts differ in the types of explanation they demand.

If you ask why pasta is eaten more in Italy than in France, it is hard to see that biological evolution contributes much to the answer. The explanation of this dietary pattern derives from facts about cultural history. Saying that it is culture, not biological evolution, that provides the explanation is quite consistent with the fact that having a culture requires a big brain, and biological evolution is responsible for our having big brains. Having a big brain may help explain why human beings have highly developed cultural forms; it does not explain why the Italians consume more pasta than the French.

If we shift to another fact about diet, biological considerations become more pertinent. Why are human beings so fond of foods that are high in fat and sugar? These proclivities give rise to a host of medical problems in affluent societies. Why are we so inclined to like things that so often do us no good? A plausible answer is that the taste for fat and sugar evolved in conditions of scarcity. When there was no dependable and regular supply of nutrition, it made sense to maximize caloric intake when food was available. Our ancestors did not possess the concept of calorie, but they possessed taste buds that evolved to ensure that organisms get calories when the getting is good.[2]

In saying that biological evolution explains the inclination to eat foods that are high in sugar and fat, we are not saying that all people have this proclivity to the same degree. Nor are we saying that it cannot be modulated by the culture in which we live. The present diet of human beings, which varies from person to person and from culture to culture, is a consequence of the complex interaction of our biological inheritance, the physical environments in which we live, and the cultural forms that we experience.

The pair of simple examples just described—sugar and fat on the one hand and pasta on the other—illustrates how biological evolution explains some facts about human diet while human culture explains others. There is no need to opt for a purely biological or a purely cultural explanation of the totality of facts that comprise human behavior; different facts demand different types of explanation. Just as we favored pluralism over monism in our discussion of the units of selection in the first part of this book, we here find pluralism more plausible than monism as a view about the explanatory power of biological and cultural evolution.

One other general *caveat* is needed. It would be a mistake to assume that biological evolution is able to explain only those behaviors that are universal among human beings while culture explains all individual differences. Traits that vary within our species may vary in part because biological evolution has made human behavior plastic. A fairly trivial but illustrative example is provided by the fact that people in warm climates tend to eat spicier foods than people in cold climates. It may seem that a perfectly adequate explanation of this pattern of variation is that food spoils faster in warm climates and spices function as preservatives. This account may be correct as far as it goes, but it fails to address the question of why people want their food to be unspoiled. Surely this desire has an evolutionary explanation.

An evolutionary biologist might propose to explain the fact that the spiciness of food varies across societies by hypothesizing that human beings are inclined to adjust the amount of spice they use so as to maximize their fitness. The conjecture need not be formulated by saying that people consciously consider which behavior will maximize fitness; rather, the idea would be that the human mind evolved to select behaviors that conform to this pattern. Our point here is not to endorse this evolutionary hypothesis but to point out a fact about what it says. It is important not to confuse *evolutionary* explanation with *genetic* explanation. If biological evolution helps explain why the behavior varies, this line of reasoning does not require that societies in which people eat bland food differ genetically from societies in which people eat spicy food. In fact, it is quite consistent with the evolutionary hypothesis that behavior varies for purely environmental reasons. Evolutionary explanation is quite consistent with the position of radical environmentalism in the nature-nurture controversy (Dunbar 1982; Tooby and Cosmides 1992; Sober 1993b; Buss 1994, 1995).[3]

This simple example allows a further point to be made. Doubtless there are factors that influence how much spice people use in their food besides the desire to avoid getting sick. People who live in the same climate differ in this feature of their diets, and the evolutionary hypothesis just mentioned says nothing about why this is so. A critic of the evolutionary approach might conclude that biological evolution is the wrong explanation of the phenomenon; however, this criticism would be misplaced. The evolutionary hypothesis does not say that the goal of staying healthy is the *only* factor that influences how much spicing people want in their diet, but that it is *one* factor. Here again, the error to avoid is thinking that evolutionary hypotheses must be embedded in a monistic framework in which biological evolution is the *only* explanatory principle.

These points are easy enough to bear in mind when the examples concern pasta and curry, but people often lose sight of them when the characteristics considered are more emotionally charged. The main example we will discuss in this chapter *is* emotionally charged, so a special effort is required to understand the problem in the right way. Returning now to psychological egoism and altruism, let us ask how evolution could be relevant to this debate. We are not suggesting that it is possible to determine what the motivation is for *every* instance of helping just by thinking about evolutionary considerations. However, we do think that it is fruitful to pursue an evolutionary line of inquiry for *some* types of helping. The main example we will consider in this chapter is parental care. Although organisms take care of their young in many species, human parents provide a great deal of help, for a very long time, to their children. We expect that when parental care evolves in a lineage, natural selection is relevant to explaining why this transition occurs. Assuming that human parents take care of their children because of the desires they have, we also expect that evolutionary considerations will help illuminate what the desires are that play this motivational role.

This is not to say that human beings are all alike in this respect. If we look just within our own culture, we can see plenty of variation. Some parents neglect and abuse their children. And among parents who take care of their children, some provide more care than others. Variation across cultures also is striking. Infanticide is and has been a widespread practice and it has come in many forms. Daly and Wilson (1988) used the Human Relations Area Files, the anthropological data base that we discussed in Chapter 5, to construct a random sample of 60 societies belonging to different language families and geographical regions. Infanticide was documented in 39 of the societies described; for 35 of these societies, the circumstances of infanticide are noted, yielding a total of 112 descriptions. Of these, 21 entries said that the child was deformed or ill, 56 said the economic prospects for rearing the baby were dim (because the mother had died, for example, or there was no male support), and 20 said the baby was the result of adultery, or from the mother's first husband, or that the father was from a different tribe.

These cases make up 97 out of 112 entries, and it would be easy to construct evolutionary hypotheses to consider as explanations of these findings. Other patterns of infanticide, however, provide a more difficult challenge to the evolutionary point of view. Widespread infanticide has sometimes been an adjunct of religious ritual, as in ancient Carthage (Soren, Khaden, and Slim 1990) and Athens (Golden 1981). And the practice of widespread female infanticide, especially within social elites, has existed in many times and places, and often when resources were abundant (Dickemann 1979; Hrdy 1981). Perhaps these patterns of variation *among* human beings and *across* cultures also can be explained by natural selection; we take no stand on that matter here. Our claim is that the behavioral difference *between* human beings and many other species has an evolutionary explanation, one that may tell us something about the kinds of desires that people have. We conjecture that human parents typically *want* their children to do well—to live rather than die, to be healthy rather than sick, and so on.[4] The question we will address is whether this desire is merely an instrumental desire in the service of some egoistic ultimate goal, or part of a pluralistic motivational system in which there is an ultimate altruistic concern for the child's welfare. We will argue that there are evolutionary reasons to expect motivational pluralism to be the proximate mechanism for producing parental care in our species.

In arguing that parents have the well-being of their children as an ultimate goal, we will not deny that they have *other* desires that may influence how they treat their children. For example, parents sometimes kill one child to save another. This behavior is not evidence that the parents did not care about the child they killed. At most, it shows that they cared about the child they saved *more* than they did about the one they killed.[5] Similarly, the occurrence of infanticide in conditions of affluence does not prove that parents do not care about their children; the most one can conclude is that they cared about something else *more*. Daly and Wilson (1988) report instances of infanticide in which mothers express *regret;* regret implies the existence of a stronger desire that trumps a weaker one.

To see how our thesis about parental care is not committed to the idea of "genetic determinism," let's connect this thesis with what we said earlier about the taste for fat and sugar. The fact that biological evolution has given us a taste for fat and sugar does not mean that it is impossible for us to change what we eat. It is entirely consistent with the existence of this taste that people the world over will someday severely curtail their consumption of fat and sugar. The fact that evolution has given us a particular food *preference* leaves entirely open the degree to which eating *behavior* may be modified environmentally. A taste for fat and sugar can come into conflict with a culturally fostered desire for health, and there is no *a priori* reason to assume that the taste must always be more compelling than the desire.

The same point applies to parental care. Our claim is that evolution has influenced the set of desires that people have in such a way that parents typically *want* their children to do well. However, this thesis about human motivation leaves open the possibility that people may acquire *other* motives with the net result that they change their *behavior*. There obviously are circumstances that lead people to abuse, neglect, and kill their children. A desire produced by natural selection can come into conflict with another desire that is culturally induced, and there is no way to determine *a priori* which desire will be stronger.

Thus, our claim that evolution has led parents to *desire* that their children do well does not entail that parents all *behave* the same way. In addition, our evolutionary hypothesis is consistent with the fact, if it is a fact, that there are parents who simply do not have the desire that their children do well. If such parents exist, it might be necessary to appeal to the concept of "developmental noise" to explain why. When selection favors a given phenotype, it favors the evolution of developmental processes that cause organisms to exhibit the favored phenotype. Such developmental processes are not foolproof, however. Even if they cause organisms to exhibit the selected phenotype in the range of environments that were present ancestrally, there is no guarantee that they will ensure that *all* organisms exhibit the phenotype across *all* conceivable environments. Fetuses who develop in normal environments end up with opposable thumbs; however, if their mothers take a drug that causes birth defects, they do not. The fact that not all people have opposable thumbs does not falsify the hypothesis that opposable thumbs evolved by natural selection.

Even if parents, barring developmental accidents, want their children to do well, it still is possible that parents should differ in how strong this desire is. There is evidence that cooperativeness and empathy vary among individuals and that at least some of this variation is due to genetic differences (Rushton et al. 1986; Zahn-Waxler et al.

1992; Segal 1993, 1997; Davis, Luce, and Kraus 1994). If this is correct, it would not be surprising if concern for one's children conformed to the same pattern. We will have nothing to say about whether such patterns of variation exist or about how they should be explained. Our focus will be on a trait that we think most parents have in common; we are not trying to explain individual differences.

We have chosen parental care as our main example of a helping behavior because it is very hard to deny that natural selection has played a role in shaping its character. However, the argument we will construct about its motivational basis generalizes to helping directed to individuals other than one's offspring. The psychological motives that underlie these two forms of helping can be investigated in the same way.[6] Indeed, helping one's children and helping individuals other than one's offspring are probably not totally separate traits—like body weight and eye color—that evolved independently of each other. The quality of the mother-child bond appears to be a crucial predictor of the empathy and prosocial behavior that the child exhibits later in life (Main, Kaplan, and Cassidy 1985; Grusec 1991).[7] This suggests that when selection favored parents who took care of their children, it thereby favored children who provided help to others. If parental care is motivated, at least in part, by altruistic motives, the same may be true of helping that is directed to nonrelatives.

Predicting Proximate Mechanisms

In Chapter 6, we described a marine bacterium that needs to avoid oxygen. In principle, the organism might solve this problem either directly or indirectly. The direct strategy would be to have an oxygen detector; an indirect strategy would be to detect some environmental variable that is correlated with oxygen level. Many marine bacteria have adopted the indirect strategy of using magnetosomes to orient their swimming to the earth's magnetic field; these organisms swim downward, toward deeper water, where there is usually less oxygen (Blakemore and Frankel 1981). In addition to monistic direct and indirect strategies, a third possible type of solution to the problem is the pluralistic strategy, wherein the organism has two (or more) detectors that control where it swims.

Let's consider the parallel problem posed by human parental care. How might the desires that parents have be arranged so as to produce caring behavior? A relatively direct solution to the design problem would be for parents to be psychological altruists—let them care about the well-being of their children as an end in itself. A more indirect solution would be for parents to be psychological hedonists—let them care only about attaining pleasure and avoiding pain, but let them be so constituted that they feel good when their children do well and feel bad when their children do ill. And of course, there is a pluralistic solution to consider as well—let parents have altruistic *and* hedonistic ultimate motives, both of which motivate them to take care of their children.

In the case of the marine bacterium, we can observe its behavior and examine its internal anatomy to determine what proximate mechanism it uses to avoid oxygen. In the case of human psychological motivation, neither observed behavior nor currently available anatomical information seems to be of much help. Nonetheless, we can use the marine bacterium to identify principles that are relevant to predicting what type of internal mechanism will evolve to allow an organism to produce a given behavior. Consider a species of marine bacterium that is in the process of evolving the ability to avoid oxygen. What factors will influence whether this organism will evolve an oxygen detector or a magnetosome as a solution to the design problem?

Broadly speaking, there are three considerations that bear on this question (Sober 1994b). First, we need to consider the range of variant traits that are present in the population. Oxygen detectors might be advantageous in principle, but if they are not present in fact, then natural selection will be unable to cause the trait to increase in frequency. For oxygen detectors to evolve, they must be *available*.

If both oxygen detectors and magnetosomes are available, the next question to consider is whether one of them is more *reliable* than the other in causing the organism to avoid oxygen. It might seem that an oxygen detector must do a better job than a magnetosome if the task is to get the organism to avoid areas in which there is lots of oxygen. However, this isn't necessarily so. An accurate magnetosome might be a more successful guide to where oxygen is than an inaccurate oxygen detector, regardless of how precisely oxygen concentration is correlated with depth. Nonetheless, there is a special circumstance in which the direct strategy will do a more reliable job than the indirect strategy. Consider the following figure, which depicts the fact that a bacterium's fitness is correlated with the oxygen level of the water it is in, and also the fact that oxygen level is correlated with the organism's elevation:

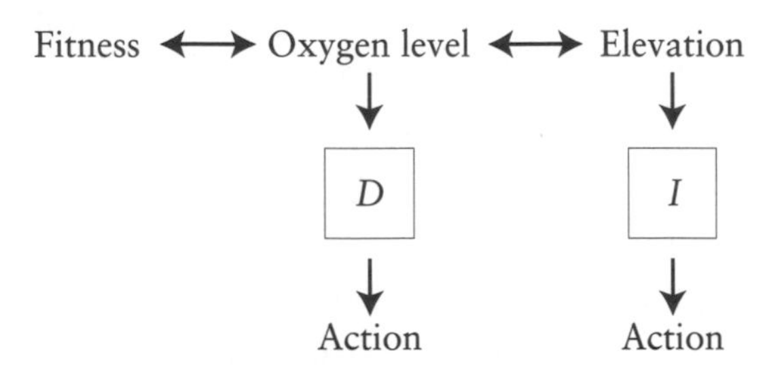

Let's suppose that the organism's elevation—whether it is at the top of the body of water or at the bottom—affects its fitness only because elevation is associated with oxygen level.[8] *D* is a detector of oxygen concentration, and *I* is a device that detects elevation. *D* is more direct and *I* is more indirect, in virtue of the way that each is connected with fitness. In this figure, the double arrows represent correlations and the single arrows represent causation. Oxygen level and elevation are inputs to the detectors; these devices have behaviors as their outputs. Here is an important fact about the reliability of direct and indirect strategies:

> (D/I) *D* will be a more reliable guide than *I* concerning which behaviors are fitness-enhancing, if *D* detects oxygen at least as well as *I* detects elevation and oxygen and elevation are less than perfectly correlated.

For future reference, we will call this principle the *D/I asymmetry*. Although an accurate magnetosome can be more reliable than an inaccurate oxygen detector, an oxygen detector will serve the organism better than an equally accurate magnetosome.[9]

The D/I asymmetry provides a principle for comparing two control devices in terms of how reliable they are. An additional principle about proximate mechanisms also needs to be stated. After all, a marine bacterium may have an oxygen detector, it may have a magnetosome, *or it may have both*. There is a quite general circumstance in which two devices will be more reliable than one:

> (TBO) The two devices *D* and *I* acting together will be a more reliable guide concerning which behaviors are fitness-enhancing than either *D* or *I* acting alone, if each device is positively, though imperfectly, correlated with oxygen level, and if the two devices operate with a reasonable degree of independence of each other.

Why is this principle—that Two Are Better Than One (TBO)—stated with the qualification that the two detection devices operate somewhat independently of each other? Suppose that an organism equipped just with an oxygen detector does a pretty good job of seeking out places that have low oxygen concentration, and that an organism equipped just with a magnetosome also does fairly well in this task. It is nonetheless possible that these two devices would perform very poorly if they were placed together in a single organism. This might happen if the two devices interfere with each other. This is why the TBO principle requires that the two devices operate with a reasonable degree of mutual independence. With this proviso understood, the fact of the matter is that two sources of evidence are better than one, as far as reliability is concerned.[10]

The D/I asymmetry and the TBO principle pertain to the issue of reliability. Reliability, recall, is the second consideration that is relevant to predicting what proximate mechanism will evolve to control a given behavior. The first was availability. We now need to consider a third. Suppose that magnetosomes and oxygen detectors are both represented in the population and that they are equally reliable indicators of the presence of oxygen. Even so, it is possible that the two mechanisms have other consequences for the fitnesses of organisms besides telling the organism where to swim. For example, maybe magnetosomes take fewer calories to build and maintain than oxygen detectors. Or perhaps the bacterium already uses magnetosomes for other tasks. If so, it might be more efficient for the organism to use this device for the additional purpose of navigation, rather than add a second piece of machinery to achieve this end. The adaptations of organisms, no less than the machines that people build for their own use, require energy to construct and maintain. A prediction concerning what proximate mechanism will evolve must take account of *energetic efficiency*.

The mind is a proximate mechanism that causes human beings to produce different behaviors in different circumstances, just as the magnetosome is a proximate mechanism that causes bacteria to swim away from oxygen. To be sure, minds and magnetosomes differ in numerous respects. However, from the point of view of predicting what motives evolved as devices for producing behavior, the same list of considerations pertains. If human parental care is a product of evolution, then so are the motives that lead parents to take care of their children. These motives cannot be ascertained by direct observation. How do considerations of availability, reliability, and efficiency help us predict which motivational mechanism is most likely to have evolved?

Two Forms of Pluralism

Motivational pluralism is the view that we have both egoistic and altruistic ultimate desires. There are two ways that an organism can conform to this theory. One possibility is that some of the actions the organism performs are caused solely by altruistic ultimate motives while others are caused just by egoistic ultimate motives. Alternatively, an organism may be pluralistic because some of its actions are caused by both altruistic and egoistic ultimate motives. If organisms take care of their children solely because they have altruistic ultimate motives, but avoid poisonous snakes just because they care about their own survival, then, as so far described, they are pluralists of the first type. On the other hand, if parents take care of their children because they care irreducibly about their children's welfare *and* because they take pleasure in helping their children, then they are pluralists of the second sort.

Although both types of pluralism are incompatible with egoism, they raise separate evolutionary questions. Hedonism claims that the desire to experience pleasure and avoid pain is the one and only ultimate motive that regulates all behavior. With respect to the first type of pluralism, we need to ask if an organism might evolve that has some of its behaviors regulated, not by pleasure and pain, but just by an altruistic ultimate motive. With respect to the second type of pluralism, we need to ask if an organism might evolve that regulates some types of behavior by having two ultimate motives, rather than just one. The first type of pluralism restricts the *scope* of hedonistic ultimate desires and substitutes an altruistic ultimate desire as the sole control device for a certain kind of behavior; the second *supplements* the hedonistic desires that are universally present with an altruistic ultimate motive.

A Continuum of Cognitive Capacities

In order to apply considerations of availability, reliability, and efficiency to the problem of psychological egoism and altruism, let's begin with a science fiction example, one that describes no real organism but helps bring the present question about behavioral control mechanisms into better focus. Consider a creature who has unlimited perceptual and computational abilities. It can accurately detect any state of the present environment that is worth knowing about, and it can use that information to predict what the future will be like. This creature is not a human being; it is the hypothetical "demon" that Laplace once described to clarify what he meant by the doctrine of determinism. Let us now pose a Darwinian question about Laplace's demon. What set of ultimate desires would be fitness-maximizing for this creature?

The answer is that Laplace's demon cannot go wrong by having its only ultimate desire be the maximization of fitness; there can be no better strategy than focusing on the evolutionary bottom line. This is a fictional example, but its relevance to real organisms becomes apparent when we ask why human beings and other living things *failed* to evolve the design solution that would be so serviceable for Laplace's demon. The reason is that human beings, and other organisms as well, are cognitively limited. We can't always detect features of our present environment that are relevant to future survival and reproductive success. And even when we possess relevant information about the present, we often are unable to use that data to predict which actions will have the best consequences for us in the future. Expert chess players are able to calculate only a few moves ahead, and the game of life is more complicated than the game of chess. The proximate mechanism that would work best for Laplace's demon would not be a viable design solution for us or for any other creature that now exists.

Laplace's demon lies at one end of a continuum—the continuum of cognitive capacity—depicted below. At the other end, we may imagine a creature that has virtually no cognitive capacities at all. Think of an organism that is conditionable according to the law of effect, which we discussed in Chapter 8. It experiences aversive and pleasurable sensations and has the rudimentary ability to avoid behaviors that were accompanied by the former and repeat behaviors that were accompanied by the latter.

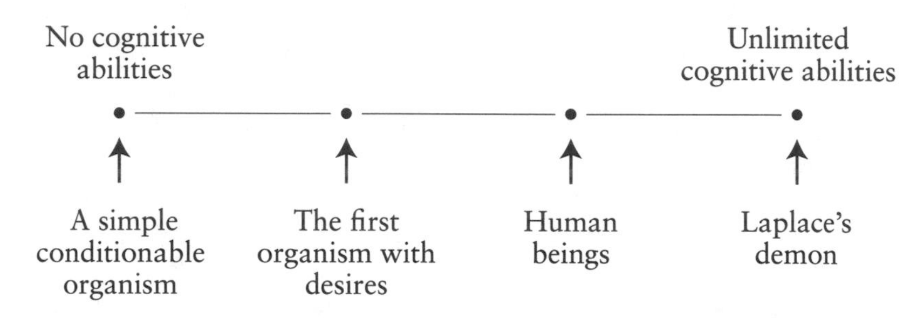

Since this simple conditionable organism has no ability to formulate thoughts, it also has no desires, properly so-called. Desiring is a conceptual achievement, just as much as believing is (Chapter 6). If an organism wants to avoid pain, it must be able to formulate a mental representation with the content *that will be painful.* An organism that totally lacks the ability to conceptualize its surroundings may exhibit tropistic behaviors, but it cannot be said to form desires. For this reason, it would be a mistake to say that this organism is a hedonist.

Nonetheless, this example allows us to pose an important evolutionary question. Let us suppose, for the sake of argument, that cognition began to evolve in organisms that already had the capacity to experience pleasure and pain. If beliefs and desires started to make their appearance in organisms of this sort, it is a good bet that early organisms that formed desires had desires concerning pleasure and pain. And when organisms began to formulate both ultimate and instrumental desires, it makes sense that the desire to attain pleasure and avoid pain should have been on the list of what these organisms had as ends in themselves.[11] What we now need to ask is whether these organisms, or their descendants, ever found an advantage in having a more expansive list of ultimate wants. Even if the first organisms that had desires were hedonists, what circumstances could make it selectively advantageous for organisms to depart from the dictates of hedonism by having some of their behaviors regulated by purely altruistic motives?

To answer this question, let's begin by considering why pain is such a useful device for regulating behavior. Why has evolution contrived to make so many organisms responsive to this sensation? The answer is that pain has two features that make it well-suited for this control task:

(i) *Epistemic access:* Organisms can detect when they are in pain and they can predict what real-world situations will cause pain.

(ii) *Correlation with fitness:* Pain is well correlated with situations that diminish fitness.

From these two considerations, we can see why a cognitively limited organism—one that is unable to detect and predict *other* properties of situations that are well correlated with fitness—might do well by having its behavior regulated by pain and by pain alone. However, if a creature has *greater* cognitive resources, some of its behaviors may come to be keyed to properties of situations in addition to pleasure and pain. If pain evolved its motivational role because it satisfied conditions (i) and (ii), other properties of situations that satisfy these conditions can be expected to do the same.

It is interesting that desiderata (i) and (ii) can conflict. Properties that are easy to detect are sometimes less well correlated with fitness than properties that are harder to detect. If a fruit is *usually* nourishing when it is red and is *usually* unhealthy when it is green, then the ability to distinguish red from green will *usually* serve the organism well. An even higher level of reliability would result if the organism could infallibly detect a chemical constituent of the fruit that is *invariably* associated with its being nourishing or unhealthy. But what if the organism finds it easier to detect color than chemical composition? Superficial characteristics are easier to ascertain than deeper ones; but superficial characteristics are often, well, superficial. Since desiderata (i) and (ii) can come into conflict, the evolution of detection devices can be the result of tradeoffs. If the device D_1 detects environmental parameter E_1 a little less reliably than device D_2 detects parameter E_2, it may still be true that D_1 is a better device for governing behavior than D_2 is; this can happen if $E1$ is more intimately connected with fitness than E_2 is.[12]

Corporeal pain gets extremely high marks on desiderata (i) and (ii). From an evolutionary point of view, it is no surprise that pain matters to us. But from this fact it does not follow that avoiding pain is the *only* thing we care about as an end in itself.

Type-One Pluralism: A Different Ultimate Motive for Each of Two Behaviors

In this section, we will investigate which of two motivational mechanisms does a more reliable job of causing parents to take care of their children:

> (HED) Perform an action if and only if you believe that it will maximize pleasure and minimize pain.

> (ALT) Perform an action if and only if you believe that it will do the best job of improving the welfare of your children.

In representing hedonism and altruism in this way, we are using the idea that desires provide *instructions;* they tell you what to do, given the beliefs you have. This is a useful way to represent the functional role that desires play in the regulation of behavior; an organism's corpus of desires constitutes a device that takes beliefs as inputs and yields behaviors as outputs. Individuals who follow ALT have an altruistic ultimate motive. If they follow this rule when they think their children need help but avoid poisonous snakes for purely egoistic reasons, then they are type-one pluralists; they use purely altruistic motives to regulate some behaviors and purely egoistic motives to regulate others.

Let's imagine that the organism realizes that its children need help and believes that there are several possible actions that it might perform. One of these actions will do the best job of improving the welfare of the organism's children; let's call this best action A^*. The question we need to ask is whether a hedonistic or an altruistic parent is more likely to select this action. A HED organism will choose to perform action A^* precisely when it believes that A^* will maximize its pleasure and minimize its pain. An ALT organism will select action A^* precisely when it believes that A^* will do the best job of improving the welfare of its children. Which of these decision procedures has the higher probability of getting the organism to select action A^*?

That depends on the mental characteristics of the organism. If the ALT organism has outrageously false beliefs about the welfare of its children, then this organism is quite unlikely to perform A*. It is quite possible that a HED organism would do much better. For example, suppose the ALT organism believes that crying is good for children and that cooing is bad for them, whereas the HED organism believes that nursing its children was pleasurable in the past and seeks to repeat that pleasurable experience. In this instance, HED's children will be better off than ALT's. This hypothetical example has the structure depicted in the accompanying figure. There is no *a priori* reason, in this instance, why HED must do worse (or better) than ALT in producing adaptive behavior.

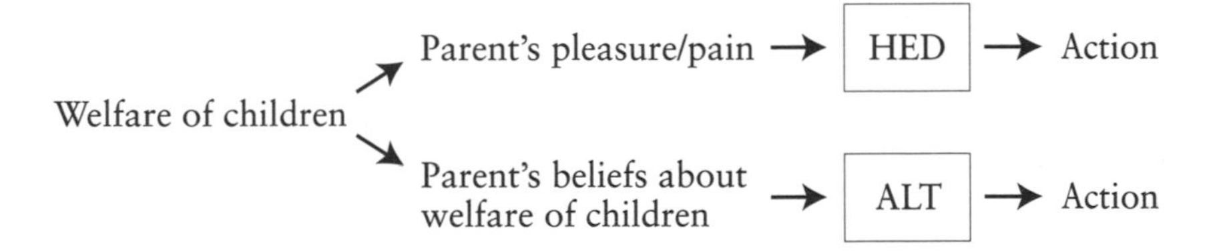

This hypothetical example is unrealistic in several respects. For one thing, we conjecture that human parents are fairly reliable judges of what will help and harm their children. For another, mothers who have experienced mastitis can testify that nursing is not always a source of pleasure. But perhaps the most important defect in this comparison between HED and ALT is the fact that it describes pleasure and pain as occurring independently of the beliefs the mother happens to have about the welfare of her children. It may be true, as a first approximation, that some sensations are pleasurable or painful independent of the beliefs the individual happens to have. When you hit your thumb with a hammer, you feel pain, regardless of what you happen to believe.[13] In contrast, when your neighbor tells you that your child has fallen through the ice on a frozen lake, this causes aversive emotions in you only because you have formed a *belief.* HED as it applies to human beings must postulate cognitive states that connect real-world situations on the one hand and feelings of pleasure and pain on the other:

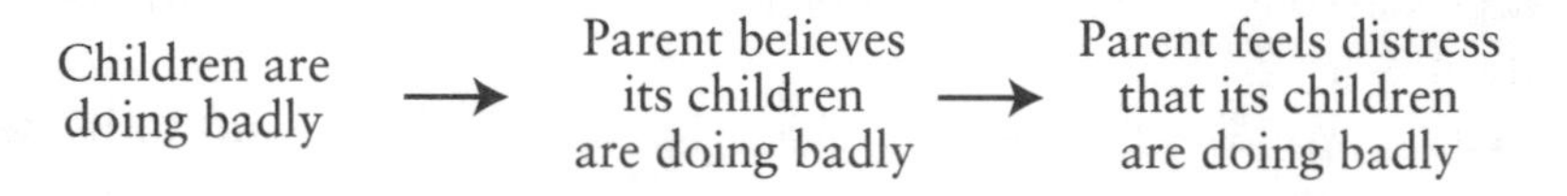

The fact that the pleasures and pains described by hedonism are frequently *mediated by beliefs* has important implications for the adaptive character of a hedonistic device for controlling behavior, as we now will see.

Suppose a hedonistic organism believes on a given occasion that providing parental care is the way for it to attain its ultimate goal of maximizing pleasure and minimizing pain. What would happen if the organism provides parental care, but then discovers that this action fails to deliver maximal pleasure and minimal pain? If the organism is able to learn from experience, it will probably be less inclined to take care of its children on subsequent occasions. Instrumental desires tend to diminish and disappear in the face of negative evidence of this sort. This can make hedonistic motivation a rather poor control device. To avoid this defect, hedonism must ensure not just that the organism performs a *calculation* to decide whether to help its children, and not just that the organism *believes* on a single occasion that taking care of its children is the best act for it to perform. In addition, this belief must actually be *true;* providing parental care must cause the organism more pleasure and less pain than any other available action that would be less consequential in terms of the organism's fitness.

The mischief this can cause for hedonism is illustrated by the Rube Goldberg cartoon we discussed at the end of the previous chapter. This drawing (page 196) illustrates what can happen when an organism regulates two behaviors by having a single ultimate desire rather than by having a different ultimate desire govern each of them. If an organism is going to eat soup and keep its mustache clean, this can easily be accomplished if each action is caused by its own ultimate desire. Alternatively, what would an organism have to be like if it were to perform both actions as a consequence of having a single ultimate desire and a single instrumental desire? One possibility would be to build an organism whose only ultimate desire is to have a napkin wipe its mustache. If the organism believes that the only (or the best) way to achieve this end is to eat soup, the organism will thereby form the instrumental desire to eat soup. This instrumental desire will remain in place only if the organism continues to believe that eating soup causes a napkin to wipe its mustache. Unless the organism is trapped by an unalterable illusion that this is so, it must be *true* that eating soup causes mustache wiping. The device that Rube Goldberg invented effects the requisite connection.

Hedonism assumes that evolution produced organisms—ourselves included—in which psychological pain is strongly correlated with having beliefs of various kinds. In the context of our example of parental care, the hedonist asserts that whenever the organism believes that its children are well off, it tends to experience pleasure; whenever the organism believes that its children are doing badly, it tends to feel pain. What is needed is not just that *some* pleasure and *some* pain accompany these two beliefs. The amount of pleasure that comes from seeing one's children do well must exceed the amount that comes from eating chocolate ice cream and from having one's temples gently massaged to the sound of murmuring voices.[14] This may require some tricky engineering. Rube Goldberg's contraption illustrates the fact that establishing these causal connections is anything but trivial. To achieve simplicity at the level of ultimate desires, complexity is required at the level of instrumental desires. This complexity must be taken into account in assessing the fitness of hedonism.

If the fitness of hedonism depends on how well correlated the organism's pleasure and pain are with its beliefs about the well-being of its children, how strong is this correlation apt to be? It is not easy to study this question experimentally, and introspective reports cannot be taken at face value. However, some indirect guidance on this question is provided by examining a more straightforward case in which pain serves as an indicator of something that affects an organism's fitness. The pain we experience when our bodies are injured is of this type. How reliable an indicator of corporeal injury is pain?

Two phenomena are relevant. The first is the case of bodily injuries that are not painful. Soldiers wounded in battle and people who arrive at hospital emergency rooms with lacerations, fractured bones, and amputated fingers often report that they felt no pain when they received their injury and for some time later. The converse case is pain without bodily injury. There are many documented cases of pain lasting much longer than the bodily injury that initially caused it. And many cases of chronic pain in the lower back occur without there being any detectable signs of injury (Melzack and Wall 1983, pp. 15–23; Fields 1987).

Although pain is an imperfect indication of bodily injury, it plays a crucial role in regulating behavior, as can be seen from the syndrome known as *congenital absence of pain.* People who are born without the capacity to experience pain do a poor job of taking care of their bodies. They damage their joints by bending arms and legs too far; they suffer severe burns because they allow their hands to remain on hot surfaces. Individuals with this problem are usually dead by age thirty (Melzack and Wall 1983). Although a Laplacean demon would have no need of pain, human beings don't seem to be able to get along without it.

When it comes to bodily injury, the verdict seems pretty clear: pain is an extremely useful, but *imperfectly reliable,* indicator of bodily injury. In this light, we think it is quite improbable that the psychological pain that hedonism postulates will be *perfectly* correlated with believing that one's children are doing badly. One virtue of ALT is that its reliability does not depend on the strength of such correlations.

In spite of the imperfect reliability of pain as an indicator of bodily injury, it is not surprising that human beings are constructed so that they sometimes respond directly to this sensation. When a fire burns your fingers, you instantly withdraw them from the flames. You do this without deliberation. The causal chain in this instance has the following structure:

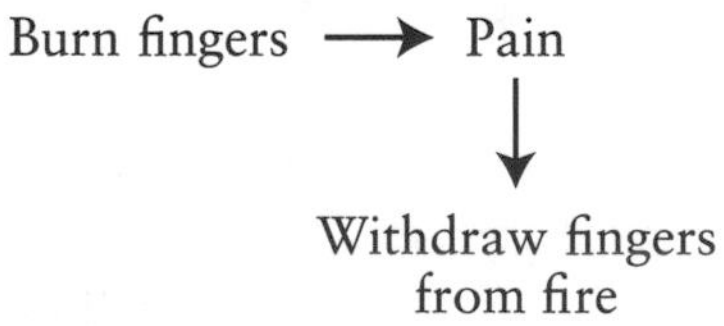

Burn fingers ⟶ Pain

↓

Withdraw fingers
from fire

Since human beings are cognitive agents, the pain produced by the fire causes a belief; however, it would be quite inefficient to have your withdrawal be keyed just to this belief:

Burn fingers ⟶ Pain ⟶ Believe that fingers are injured
↓
Withdraw fingers from fire

We see here an application of the D/I asymmetry. If the choice is between responding to pain and responding to a belief caused by pain, the more reliable strategy is the direct one of responding to pain.[15]

It is important to recognize a difference between the role that pain plays in causing organisms to take care of their bodies and the role that HED says that pain plays in causing parents to take care of their children. When an organism's body is injured, it feels pain without having to form a belief about what has happened. However, when an organism's children are doing badly, the parent typically "feels bad" because it *believes* that its children are doing badly.[16] With respect to bodily injury, an organism will do a better job of taking care of its body if it responds to pain than if it responds only to beliefs that are caused by pain. Similarly, the role that belief formation plays in the hedonistic mechanism for delivering parental care means that ALT can easily do a more reliable job than HED. Consider the accompanying diagram, which mimics the one we used earlier in the chapter to depict the direct and indirect strategies that a marine bacterium might use to avoid oxygen. The ALT parent keys her actions to what she *believes* about her children's welfare. The HED parent keys her behavior to how she *feels*. Notice that this diagram depicts "feeling good" and "feeling bad" as belief-mediated states. Suppose the ALT parent produces parental care in response to her belief that her children need help at least as reliably as the HED parent produces parental care in response to her feeling bad that her children need help. It follows that ALT is more reliable than HED, provided that the parent's feelings of pleasure and pain are less than perfectly correlated with her beliefs about the welfare of her children. This is the D/I asymmetry we discussed before. ALT will do better than HED for the same reason than an oxygen detector will be a more reliable control device than an equally accurate magnetosome.

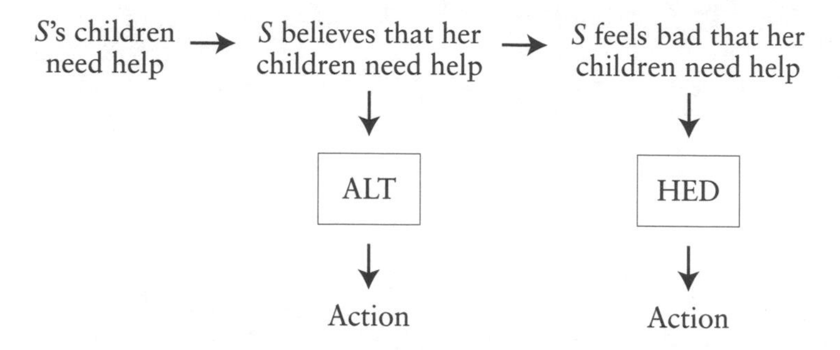

In this analysis, we have represented the hedonist as wanting to alleviate *present* pain, but the argument would remain the same if we thought of the hedonist as acting to avoid *future* pain. The altruist is moved to act by her belief that her children need help. For the future-oriented hedonist to be motivated to help, the belief that her children need help must trigger a further *belief*—that helping will bring pleasure, or that not helping will bring pain. Here again, hedonism deploys a relatively indirect strategy, and altruism is relatively direct.

The hypothesis we have defended in this section is not unprecedented. It is suggested, at least in part, by the following passage from Darwin's *Descent of Man:*

> With respect to the impulse which leads certain animals to associate together, and to aid each other in many ways, we may infer that in most cases they are impelled by the same sense of satisfaction or pleasure which they experience in performing other instinctive actions; or by the same sense of dissatisfaction, as in other cases of prevented instinctive action . . . In many cases, however, it is probable that instincts are persistently followed from the mere force of inheritance, without the stimulus of either pleasure or pain . . . Hence the common assumption that men must be impelled to every action by experiencing some pleasure or pain may be erroneous. (Darwin 1871, pp. 79–80)

Darwin's observation suggests that if an organism gains a fitness advantage by providing parental care, then there may be a disadvantage in having this behavior governed by a hedonic calculation. Unfortunately, Darwin does not describe the proximate mechanisms we should expect instead of hedonism; even if a behavior is produced as a result of "the mere force of inheritance," it still remains to ask what proximate mechanism is doing the work. We have suggested that psychological altruism will be more reliable than psychological hedonism as a device for getting parents to take care of their children.

Type-Two Pluralism: Two Ultimate Desires Influencing a Single Behavior

In the previous section, we did not conclude that human parents provide parental care for purely altruistic reasons. Rather, we argued that parental care would be more reliably produced by purely altruistic motives than by purely hedonistic motives. We now want to add a second argument to our brief against hedonism. This time we'll compare a hedonistic mechanism for regulating parental care with an arrangement in which parents help their children for a mix of hedonistic and altruistic ultimate motives:

> (PLUR) Perform an action if and only if you believe that it will maximize pleasure and minimize pain *or* that it will do the best job of improving the welfare of your children.

This new problem is important—even if HED should somehow be superior to ALT, the question remains whether HED can do better than type-two pluralism.

A very simple argument shows that PLUR will be more reliable than HED at delivering parental care. Suppose, as before, that a parent sees that her offspring need help; suppose further that performing action A^* would do a better job than any of the other available actions of improving their situation. A purely hedonistic parent will perform action A^* if and only if she believes that A^* will maximize her pleasure and minimize her pain. In contrast, a pluralistic organism will perform A^* in a wider range of circumstances. If she thinks that A^* will maximize pleasure and minimize pain, then A^* gets done; however, if she does not believe this, but believes that A^* will do the best job of improving the welfare of her children, then A^* still gets done. Of course, it also is possible that the parent will come to believe that action A^* has *two* properties—that it maximizes pleasure *and* does the best job of improving the children's situation. In this case, the pluralistic parent has two reasons to help.

PLUR postulates two pathways from the belief that one's children need help to the act of providing help. If these operate at least somewhat independently of each other, and each on its own raises the probability of helping, then the two together will raise that probability even more. Unless the two pathways postulated by PLUR hopelessly confound each other, PLUR will be more reliable than HED. PLUR is superior because it is a *multiply connected control device.* Pluralism is a more reliable motivational arrangement than monism for reasons set forth in the TBO principle.

Multiply connected control devices have often evolved. Consider the so-called fight-or-flight response. When an organism believes that it is in danger, this belief triggers a variety of physiological responses. Adrenaline flow increases; so does heart rate. There are psychological consequences as well, wherein the organism explores possible courses of action. All these factors come together to produce the resulting behavior. Another example is provided by the mechanisms that endotherms (warm-blooded organisms) use to regulate body temperature. When the organism is too cold, it starts to shiver, its hairs stand on end, and its blood vessels constrict. These separate pathways conspire to help the organism return to its optimal temperature. Further examples could be supplied from biology, and also from engineering, where intelligent designers supply machines (like the space shuttle) with backup systems. Error is inevitable, but the chance of disastrous error can be minimized by well-crafted redundancy.

The abstract pattern of a multiply connected control device involves a *sensor* (S) that is connected to an *effect* (E) by several pathways (P_1, P_2, . . ., P_n):

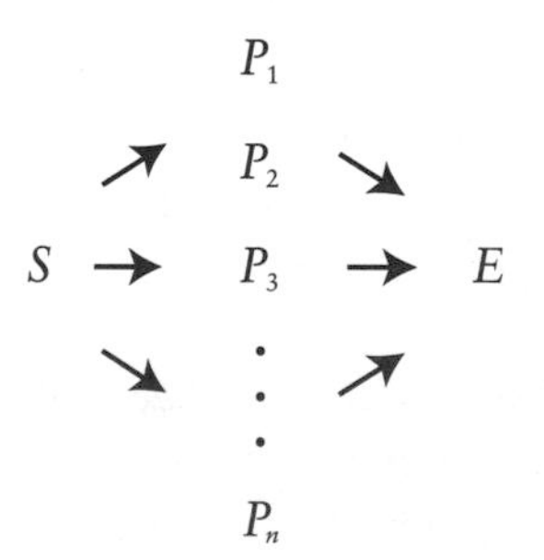

If causal connections are subject to error, then each P_i raises the probability of *E* but never makes the occurrence of *E* an absolute certainty. Multiple connections provide an adaptive advantage, if fitness is enhanced by maximizing the probability that *E* occurs when *S* is triggered.

Availability and Efficiency

In the previous two sections, we argued that hedonism is likely to be less reliable as a device for delivering parental care than either type-one or type-two pluralism. If an organism provides parental care for purely altruistic reasons but avoids poisonous snakes for purely egoistic reasons, this organism will do a more reliable job of delivering parental care than a parent who has a purely hedonistic motivation for doing so. Similarly, an organism who provides parental care because she has both altruistic and hedonistic ultimate motives also will do better than a parent with purely hedonistic motives. Hedonism comes in last in this three-way race. This, by itself, is not enough to prove that human beings are motivational pluralists. One of the lessons we extracted from the example of the oxygen-fleeing marine bacterium was that there are issues besides reliability that need to be considered in predicting which proximate mechanism will evolve to control a given type of behavior. We also need to consider availability and efficiency. How do hedonism and pluralism compare on these dimensions?

Let's begin with the issue of availability. In the problem concerning the marine bacterium, perhaps it is true that many contemporary species lack oxygen detectors because this device was not available in ancestral populations. They have magnetosomes, but not oxygen detectors, because magnetosomes were the only game in town. In contrast, we think it is implausible to reject the hypothesis of motivational pluralism as an explanation of parental care by claiming that pluralism was not available ancestrally. There is every reason to think that pluralism *was* available; it requires the same basic equipment that hedonism demands. A hedonist must formulate beliefs about whether its children are doing well. It must form the ultimate desire to attain pleasure and avoid pain. And the organism must construct the instrumental desire that its children do well. If this is what it takes for a hedonistic organism to provide parental care, what does it take for a pluralistic organism to produce the same behavior? A pluralistic parent also must form beliefs about whether its children are doing well. In addition, the pluralist must formulate the ultimate desire that its children do well. We suggest that constructing this *ultimate desire* involves no great innovation, if the organism is already capable of *believing* that its children are doing well or ill. If the propositions available as objects of belief also are available as objects of desire (Chapter 6), then pluralism will be possible for an organism who is capable of forming beliefs about the welfare of her children. Whereas hedonism requires the organism to form the *instrumental* desire that her children do well, pluralism simply has the organism register this same desire as an end in itself. We conclude that pluralism was probably available ancestrally, if hedonism was.

The issue of efficiency is difficult to assess; little is known about the side effects of the proximate mechanisms here considered or about the energetic costs of building and maintaining the machinery that each hypothesis postulates. Nonetheless, we want to put forward the following conjecture. Although it is energetically more efficient for a bacterium to have a magnetosome than to have both a magnetosome and a separate oxygen detector, it is hard to see why motivational pluralism should be more energetically burdensome than hedonism. The reason for this disanalogy is that pluralism does not require the organism to build and maintain any new *device*. Mechanisms for representing beliefs and desires are required by both hypotheses, both require that the organism be able to experience pleasure and pain, and both require the organism to have both ultimate and instrumental desires. What pluralism requires is that the device for representing the organism's ultimate desires encode an extra *representation*—namely, the desire that its children do well. Since hedonistic organisms already represent this propositional content as a belief and also as an *instrumental* desire, it is hard to see why placing that proposition in an organism's "ultimate desire box" should engender any massive energetic burden. We doubt that people who have more *beliefs* need more calories than people who are less opinionated; the same point applies to *ultimate desires*. Here's an analogy that we find suggestive: Once you buy a computer, the difference in cost between writing one sentence in a file and writing two sentences there is trivial. As a first approximation, it is *devices* that are costly, not the *representations* those devices construct. If so, hedonism is not more energetically efficient than pluralism.

There is another difference between hedonism and pluralism that we need to consider. A pluralistic organism may find itself in situations in which its ultimate desires *conflict*. If the organism is hungry and its child is too, how is the organism to decide what to do if there is not enough food for both? The pluralist needs a mechanism for adjudicating. Apparently, no such requirement attaches to monistic motivational structures; if the hedonist wants only to maximize its pleasure and minimize its pain, the issue never arises as to how this desire should be traded off against some other ultimate end. Does this mean that hedonism will be more efficient than pluralism?

We think not. Hedonism only *seems* to be the more economical mechanism. After all, a hedonistic organism must choose between different types of pleasure and different types of pain. In the situation just described, the organism continues to feel hungry if it gives the food to its child, but it has the agreeable thought that the welfare of its child has just improved. Alternatively, the organism experiences the pleasures of flavor and satiation if it consumes the morsel, but the thought remains that its child is hungry. A hedonistic organism needs a device for rendering commensurable these very different types of pleasure and pain. The fact of the matter is that *both* hedonism and pluralism require the organism to have a mechanism for making comparisons; we see no reason to think that the device required by hedonism will be less costly energetically than the one demanded by pluralism.

The conclusion we draw from these evolutionary considerations is that motivational pluralism is more likely to be the proximate mechanism behind human parental care. Pluralism was just as available as hedonism, it was more reliable, and hedonism provides no advantage in terms of energetic efficiency. We realize that our argument for psychological pluralism contains a number of speculative elements; we don't suggest that this argument is sufficient unto itself to prove conclusively that this motivational hypothesis is true. We hope that other investigators will be able to identify further testable consequences of motivational pluralism. These consequences might be searched out on several fronts; experimental psychologists, evolutionists, and neuroscientists all may be able to throw light on this issue.

In spite of the preliminary character of the argument we have presented, we believe it provides a good *prima facie* case for motivational pluralism, and for the altruistic ultimate motives that form part of that proximate mechanism. Advocates of hedonism must explain why hedonism should have evolved. The fact that hedonism can explain observable behavior is not enough, since pluralism can do the same thing. Defenders of hedonism need to say how their monistic hypothesis is superior with respect to the considerations we have outlined—availability, reliability, and efficiency. Egoism does not deserve to be regarded as a theory that is innocent until proven guilty.

Conclusion

In this chapter, we have presented evolutionary arguments against psychological hedonism. We have suggested that hedonism is a *bizarre* proximate mechanism for creatures such as ourselves; motivational pluralism, we believe, has a higher degree of evolutionary plausibility.

Human beings are capable of forming beliefs about a huge range of propositions. In this respect, we are the most sophisticated organisms on earth. If desires are constructed from the same conceptual resources as beliefs (Chapter 6), then it also is true that human beings are capable of forming desires with respect to a huge range of propositions. Why should creatures with these cognitive capacities have all their desires answerable to the ultimate tribunal of pleasure and pain? It might make sense, from a design standpoint, to have pleasure and pain regulate the trial-and-error learning of an organism whose cognitive abilities are drastically limited. But surely it is possible to think of ways in which a smart organism could find an adaptive advantage in caring about matters that extend beyond feeling good and not feeling bad.

The odd thing about hedonism is that it keys an organism's actions to *sensation,* whereas an organism's evolutionary success depends on its keying its actions to *the real world*—to what happens to its own body, to its offspring, and to its own social group. The obvious evolutionary strategy for an organism that can form reliable beliefs about its own body and about the welfare of relevant others is for it to set its eye on the prize; its ultimate desires should include a concern for something that is far more important in terms of evolutionary success than the states of its own consciousness.

Human beings have succeeded in doing this. Like our ancestors, we find some sensations aversive and others pleasurable. This is not just a feature of how we *feel;* these feelings powerfully influence what we *desire.* People *want* to experience pleasure and to avoid feeling pain. However, human beings have supplemented these hedonistic desires in two ways. What we want for ourselves extends beyond the desire for pleasant states of consciousness. And human beings, we believe, also have ultimate desires concerning the welfare of others.

We suspect that hedonism has been an appealing theory in psychology in part because it seems to have good biological credentials. Pleasure and pain are basic biological reactions that an organism has to the world it encounters; a human psychology built on this basis seems to situate human beings firmly in the midst of nature, establishing our continuity with other living things. What is true in this picture is the idea that human beings possess the capacity to experience corporeal pleasure and pain; these feelings retain a motivational salience for us despite the additional cognitive sophistication that our species has evolved. But it is one thing for pleasure and pain to serve as motivators, something quite different for them to be the *only* things that are able to motivate us.

One of the important insights that emerges from an evolutionary perspective on the problem of psychological egoism and altruism is that the character of an organism's *desires* must be thought of in tandem with a consideration of the organism's capacity to form *beliefs.* What desires it makes sense for an organism to have depends on the types of belief it is able to construct. To our knowledge, the idea that an organism's beliefs are a basis for predicting the structure of its desires has not emerged in psychological investigations that ignore evolutionary issues.

In similar fashion, evolutionary considerations have led us to emphasize the distinction between pleasures and pains that are belief-mediated and pleasures and pains that are not. Hedonism talks about both; there is the pain that come from hammer blows to the thumb and the pain that comes from absorbing bad news about one's children. Hedonism proposes that we think of these pains in the same way. To explain why people don't want to have their thumbs hit with hammers, it suffices to say that they want to avoid pain. The hedonist says the same thing about the explanation of helping behavior—the claim that people want to maximize pleasure and minimize pain is sufficient. An evolutionary perspective makes these two cases look very different, however. Corporeal pain plays a special role in motivation in part because its connection with bodily injury is relatively direct, and bodily injury is manifestly relevant to fitness. But when pleasures and pain are belief-mediated, they are less directly connected with circumstances that are relevant to fitness. The same considerations that make hedonism look plausible as a proximate mechanism for getting organisms to take care of their bodies makes altruism look plausible as a proximate mechanism for getting organisms to take care of their children.[17]

The main example of an altruistic desire discussed in this chapter is the desire that one's children do well. However, the social setting of ancestral human life suggests that concern for others probably embraced a wider circle. Just as selection can promote the evolution of parental care, so it can lead to the evolution of cooperative behaviors in which the beneficiaries are individuals other than one's sons and daughters. If group selection favors the evolution of various forms of cooperative behavior, and the proximate mechanism that produces these behaviors is the organism's beliefs and desires, what types of desire should we expect the organism to possess? This question simply generalizes the problem about parental care. Just as motivational pluralism is a plausible design solution for the problem of getting parents to take care of their children, so pluralism is a plausible design solution for the problem of getting members of a group to take care of each other. Hedonistic design solutions are conceivable, but they have the same drawbacks, whether the helping they control is directed at offspring or unto others.

Parental care is a behavior that most people find pleasing, and the claim that it is driven in part by altruistic motives will also warm the cockles of many hearts. However, if selection promotes an altruistic concern for the welfare of one's near and dear, it also may promote indifference or malevolence toward outsiders. Within-group selection is a competitive process, but so too is between-group selection; it promotes both within-group niceness and between-group nastiness. We mentioned in Chapter 8 that people tend to empathize more with those whom they perceive as similar to themselves (Stotland 1969; Krebs 1975). If empathy elicits altruistic motives with respect to those whom we take to be similar, its absence means that we are less inclined to be altruistically motivated toward those whom we take to be different. It is important not to lose sight of the symmetric logic of selection arguments.

We do not view our discussion in this chapter as describing the full extent of human psychological altruism, but as circumscribing a plausible beginning. Evolution has made us motivational pluralists, not egoists or hedonists. However, this evolutionary legacy has left room for much variation among human beings and enormous flexibility in the desires that a single human being may have in the course of his or her life. Egoistic motives and selfish behavior are no more unalterable elements in the human repertoire than are altruistic motives and helping behavior. Both are elements in the legacy that our past has bequeathed to the present. That legacy is something we can embrace, disown, or try to transform.

Conclusion: Pluralism

We began this book by delineating three broad intellectual traditions that exist inside and outside science—individual-level functionalism, group-level functionalism, and anti-functionalism. These traditions differ so dramatically that they seem to describe different worlds. Indeed, members of the three traditions have to a considerable extent stopped talking to each other. Lack of communication, not to mention lack of consensus, at such a basic level is a major failing.

Of the three traditions, group-level functionalism is the one that has fallen on the hardest times. Once a vigorous perspective in biology and the social sciences, it has come to be regarded as a quaint knick-knack in the antique shop of history. Social scientists who discarded group-level functionalism did so primarily for a combination of methodological and conceptual reasons. Thinking of groups as organic units subject to their own laws of behavior smacked of mysticism, and once holism was purged of mysticism, it was quite unclear how this position could make distinctive predictions that are empirically testable. The demise of holism in the social sciences was driven by general considerations of this type; the reason for individualism's growing attractiveness was not the confirmation of a novel empirical theory that entails that individualism must be true (Wright, Levine, and Sober 1992). The term *methodological individualism* is therefore apt, in that this version of individualism is not grounded in any particular scientific theory. In contrast, evolutionary biologists rejected group-level functionalism on theoretical grounds—not as an inconvenience, but as a near impossibility, according to what seemed to be the testimony of a well-attested theory. Evolutionary biology therefore became the theoretical heart of individualism (Sober 1981). Despite fundamental differences between the evolutionary and psychological concepts of selfishness, the individualistic frameworks constructed in the two fields have seemed to coalesce into an unshakable edifice.

We have shown that the wholesale rejection of group selection—and therefore of group-level functionalism—by evolutionary biologists during the 1960s was misconceived and has not withstood the test of time. We also have argued that evolutionary altruism and psychological altruism must be evaluated separately. Both subjects are evolutionary, but they call on different sets of principles, since one concerns the evolution of behaviors and the other pertains to the proximate mechanisms that evolve to cause those behaviors. In both cases, we concluded that the individualistic edifice is not nearly as strong as it has appeared. The case against evolutionary altruism has already crumbled when judged by normal scientific criteria. Group selection is a significant evolutionary force and its products have been amply documented in nature. The case against psychological altruism has not yet crumbled completely, but the cracks we have identified in the edifice suggest that one may be well advised to stand clear. In retrospect, it seems evident that the case against psychological altruism was never strong; it relied on an intellectual pecking order in which proponents of altruism had to *prove* their case, while proponents of egoism merely had to *imagine* conceivable explanations. Evolutionary theory furnishes an even playing field on which the two theories can compete fairly. When judged in terms of their evolutionary plausibility, psychological egoism turns out to be the inferior theory, or so we have argued.

If we have succeeded in tearing down the edifice of individualism, what stands in its place? The debate about group selection has been repeatedly short-circuited by a confusion—different ways of viewing the same process have often been confused with different processes. There is nothing wrong with computing the average fitness of genes, or of the phenotypic traits exhibited by individuals, if the goal is to predict what will evolve. But the availability of this procedure does nothing to solve the problem of identifying the units of selection. Once this is recognized, it is possible to see that evolution involves at least three different sorts of selection process. There is selection among genes in the same individual, selection among individuals in the same group, and selection among groups in the same metapopulation. The evolution of some traits is driven by one of these processes; the evolution of others is driven by others. And some traits are the joint result of the interaction of several selection processes occurring at once. Still other traits have evolved for reasons having nothing much to do with natural selection.

Thus, there are two forms of pluralism that we endorse concerning evolutionary biology. First is a pluralism of perspectives. There is nothing wrong with representing the same process in different ways, provided that differences in modes of representation are not confused with differences in substantive claims about nature. Second, we have advocated a pluralism concerning the actual causes of evolution. Pluralism in this second sense means that there exists a plurality of causes of evolutionary change, which can and do occur in different combinations.

We think that multilevel selection theory provides the beginning of a unified framework within which the legitimate claims of individual-level functionalism, group-level functionalism, and antifunctionalism can each be given their due. It provides a strategy for assessing functional organization at all levels of the biological hierarchy and for determining which units (if any) have accumulated adaptations in any particular case. It allows ample room for nonfunctional explanations, since there is more to evolution than natural selection. In the case of human beings and perhaps of other species, it emphasizes the importance of culture in addition to genes and shows how behaviors can evolve that make sense only in the context of the cultural system that supports them. The three traditions are thoroughly intertwined in the pages of this book.

Multilevel selection theory does not justify the grandiose claim that higher-level units are always comparable to single organisms. Group-level functionalists must be more cautious and critical in developing their intuitions than they have been in the past. Nevertheless, groups and other higher-level units do sometimes behave in an organismic fashion, for exactly the same reasons that individuals exhibit functional organization. Higher-level adaptations are not everywhere, but they are almost certainly more common than the received wisdom would have us believe. In addition, multilevel selection theory encourages us to look for group-level adaptations where they have been least expected. For all its insights, kin selection theory has played the role of a powerful spotlight that rivets our attention on genetic relatedness. In the center of the spotlight stand identical twins, who are expected to be totally altruistic toward each other. The light fades as genetic relatedness declines, with unrelated individuals standing in the darkness. How can a group of unrelated individuals behave as an adaptive unit when the member individuals have no genetic interest in one another?

Replacing kin selection theory with multilevel selection theory is like shutting off the spotlight and illuminating the entire stage. Genealogical relatedness is suddenly seen as only one of many factors that can influence the fundamental ingredients of natural selection—phenotypic variation, heritability, and fitness consequences. The random assortment of genes into individuals provides all the raw material that is needed to evolve individual-level adaptations; the random assortment of individuals into groups provides similar raw material for group-level adaptations. Mechanisms of nonrandom assortment exist that can allow strong altruism to evolve among nonrelatives. Nothing could be clearer from the standpoint of multilevel selection theory, and nothing could be more obscure from the standpoint of kin selection theory. The implications of seeing the full stage extend to virtually every topic studied in the evolution of social behavior and of multi-species interactions.

For all appearances, our own species seems special when it comes to group-level functional organization. From ancestors who were at best only moderately well-adapted at the group level, our lineage has evolved so that individuals participate in social groups that sometimes invite comparison to bee hives and single organisms. Attempts to explain our groupish nature from an individualistic perspective appear tortured in the light of multilevel selection theory; we think that multilevel selection theory will provide a much more satisfactory explanation of human groups as adaptive units. Human groups are like single genomes, which achieve their unity by being organized to prevent subversion from within as much as possible. It is a great irony that the language of human social control—sheriffs, police, parliaments, rules that enforce fairness, etc.—has been borrowed to describe the social behavior of genes, without the reciprocal conclusion being drawn that human social groups can be like genomes. When we are relieved of the burden of having to explain human behavior without resort to group selection, these connections become obvious and the groupish side of human nature can be taken at face value. Group selection has not been the only force in human evolution—the groupish side is only one side—but in all likelihood it has been a tremendously powerful force.

All adaptations require proximate mechanisms that actually cause those adaptations to be expressed in the lifetimes of individuals. An evolutionary perspective on the motivational question of psychological altruism and egoism changes the nature of this centuries-old debate. Crude versions of egoism are easy to refute, but more subtle versions of the theory, which postulate attaining pleasure and avoiding pain as the ultimate goals of behavior, can be made to conform to the behaviors we observe. This has allowed defenders of egoism to continue to endorse their pet theory in spite of the fact that motivational pluralism—the hypothesis that some of our ultimate desires are altruistic while others are egoistic—also is consistent with the observations. Without an evolutionary perspective, the conflict between these two hypotheses seems unresolvable. Observed behavior does not decide the question, and the conceptual arguments furnished by philosophers have not broken the deadlock either.

One part of our inquiry into this psychological question involved laying to rest a set of fallacious arguments for egoism. These exhibit a striking similarity to the confusing arguments that often made the units of selection problem seem like such a tangle. Just as group selection can be refuted *a priori* by defining as selfish any trait that evolves, so psychological altruism can be refuted *a priori* by defining as egoistic anything that people want. Appeals to the principle of parsimony have played a spurious role in both problems as well; just as group selection has been said to be too complicated to be worth taking seriously as a possibility, so motivational pluralism has seemed too complex, when compared with the egoistic alternative.

Identifying inconclusive arguments and observations gets one only so far. To make further progress, we had to return to the more general setting of evolutionary theory and rethink the problem of predicting proximate mechanisms from first principles. We developed a procedure for organizing this general question, not by addressing the issue of psychological altruism head on, but by considering the example of a marine bacterium that evolves an internal mechanism that steers it away from oxygen. What considerations would permit one to predict the character of this internal mechanism? This may seem like a bizarre question, given that psychological egoism and altruism involve sophisticated beliefs and desires, which there is no reason to think are present in bacteria. However, if the mind is a proximate mechanism for producing adaptive behavior, then the comparison is telling, in spite of the obvious disanalogy. This line of analysis suggests that motivational pluralism is more probable than psychological egoism as an outcome of the evolutionary process.

It is worth noting that the evolutionary pluralism we defended with respect to the units of selection problem precisely parallels the motivational pluralism we defended as a resolution of the debate between psychological egoism and altruism. Just as evolution is driven by processes of different types, so human behavior is caused by different types of ultimate motive.

Our evolutionary argument for the hypothesis that people sometimes have altruistic ultimate motives, we believe, is the beginning of a more detailed story about the impact of evolution on human motivation. In the course of developing this argument, we bracketed a number of important collateral issues, concerning empathy, morality, and the question of how altruistic we are inclined to be, to whom, and in what contexts. We focused on the task of refuting psychological egoism because more detailed questions about psychological altruism require that this preliminary question be addressed. Now that we have climbed to the summit of this modest peak, a vast continent of human thought and feeling lies before us, awaiting evolutionary exploration.

Some of the dimensions of this terrain are suggested by Stephen Crane's short story "The Open Boat," which was based on a real-life event. Crane was a passenger on a ship that sank in a storm off the coast of Florida. He found himself in a tiny dinghy with the captain and two other members of the crew. With the boat riding only a few inches above the water and frequently swamped by heavy seas, Crane took turns rowing toward shore. He endured more pain and fatigue than he thought possible; when he could no longer row, trading places required the most careful movements to avoid upsetting the craft. In the story, Crane provides the following remarkable account of how he felt toward the other members of his group, who were total strangers only a few hours before:

> It would be difficult to describe the subtle brotherhood of men that was here established on the seas. No one said that it was so. No one mentioned it. But it dwelt in the boat, and each man felt it warm him. They were a captain, an oiler, a cook, and a correspondent, and they were friends—friends in a more curiously iron-bound degree than may be common. The hurt captain, lying against the water jar in the bow, spoke always in a low voice and calmly; but he could never command a more ready and swiftly obedient crew than the motley three of the dinghy. It was more than a mere recognition of what was best for the common safety. There was surely in it a quality that was personal and heart-felt. After this devotion to the commander of the boat, there was this comradeship, that the correspondent, for instance, who had been taught to be cynical of men, knew even at the time was the best experience of his life. (Crane 1898, p. 353)

Crane was behaving adaptively and was working to save himself along with the other members of his group. It might seem that he could have given his best effort without caring a shred about the others. An inner voice might have told him—this is a matter of life or death for you and here is what you need to do for yourself. However, Crane takes pains to say that his behavior was not based on recognizing what the situation required merely for himself, but was motivated by such a powerful feeling of devotion toward the members of his group that the comradeship was the best experience of his life. This is an extraordinary statement, given a situation that in every other respect must have been the worst experience of his life. Crane loved the members of his group with an intensity he had never felt before, and this feeling impelled him to work harder than he might have otherwise.

Obviously, a short story proves nothing, but we think it is highly suggestive of the way the human mind really works. Behaving as part of a coordinated group is sometimes a life-or-death matter in which the slightest error—or the slightest reluctance to participate—can result in disaster for all. Situations of this sort—in which the members of a group are bound together by the prospect of a common fate—have been encountered throughout human evolution, with important fitness consequences, so it is reasonable to expect that we are psychologically adapted to cope with them. Crane tells us that the other-oriented psychology triggered by the crisis in the open boat differs from the more individualistic feelings experienced in other contexts.

As the most facultative species on earth, human beings appear willing and able to span the full spectrum from mercilessly exploiting their social partners to sacrificing their lives for others. This stunning plasticity of behavior must somehow be orchestrated by proximate mechanisms that probably include altruistic ultimate motives, but certainly include much more, such as emotions (Frank 1988; Hirshleifer 1987), moral principles, and other mechanisms that we deliberately set to one side to make headway on the subject of psychological altruism (e.g., Eibl-Eibesfeldt 1982; Simon 1990). To the extent that behavior has evolved by group selection, multilevel selection theory will be required to understand the full architecture of human motivation.

Among the three perspectives that readers may have brought to the present work, anti-functionalism is the one that might seem least amenable to evolutionary treatment. It is true that we have not tried to explain the variation in customs that exists across cultures; for example, we have not taken up the task of explaining why the people in one society value highly individualistic clothing while those in another all dress the same. However, our approach does show how important elements in the anti-functionalist position can be developed within an evolutionary context. For example, anti-functionalists often attack evolutionary accounts of human mind and culture on the ground that those accounts assume genetic determinism. We have shown how evolutionary ideas can be developed with more complex conceptions of the relation of gene to behavior, and also without the mediation of genes at all. Even groups that are genetically identical can differ profoundly at the phenotypic level because of cultural mechanisms, and these differences can be heritable in the only sense that matters as far as the process of natural selection is concerned. The fact that culture by itself can provide the ingredients required by the process of natural selection gives culture the status that critics of biological determinism have emphasized. Natural selection based on cultural variation has produced adaptations that have nothing to do with genes. Critics of genetic determinism should study culture as an evolutionary process, rather than abandon the entire evolutionary framework (Boyd and Richerson 1985, 1990a,b, 1992).

We also have accommodated the anti-functionalist's contention that human cultures include numerous details that promote neither the welfare of individuals nor the well-being of society as a whole. Following the lead of Boyd and Richerson, we have shown how powerful social norms can be imposed at low cost. These norms involve rewards and punishments that create selection pressures within societies. The result is that different societies will evolve to different internally stable configurations. If these equilibria differ in their consequences for the fitnesses of groups, then cultural group selection ensues. If they do not, then cultures will exhibit neutral variation and random drift will occur. The human capacity for culture thus sets in motion an evolutionary process in which some of the anti-functionalist's principal claims emerge as results. Cultural transmission opens the way for an elaborate edifice of phenotypes that enhance group-level functional organization. But this mechanism also leads to the evolution of behaviors that make no sense outside the cultural system that promotes them. Nonfunctional and even dysfunctional behaviors can ride along with adaptive behaviors as hitchhikers. These and other sources of nonfunctional behaviors can be understood from an evolutionary perspective. Just as selection does not require genetic determinism, so an evolutionary model of cultural change does not require an exclusive focus on the process of selection.

Our book has been about altruism, but it also has opened the door to a wide range of other subjects. Altruism can be removed from the endangered species list in both biology and the social sciences. Groups can qualify as organismic units. Culture can play a vital role in the evolutionary process. And the study of psychological mechanisms can be as evolutionary as the study of behavior. It is heartening to contemplate the emergence of a legitimate pluralism—for evolutionary theories of social behavior, for theories of psychological motivation, and for the larger intellectual traditions that influence how we think about ourselves and the world around us.

Notes

Introduction

1. Given that Bentham and Mill defended psychological egoism as a descriptive thesis, one may wonder how they could also endorse utilitarianism as a normative claim. If each individual cares ultimately only about his or her own well-being, in what sense are people obliged to advance the greatest good for the greatest number? The ought-implies-can principle, which says that if you ought to perform an action, then it must be possible for you to do so, seems to pose a problem. The principle entails that if people are psychological egoists, then it is false that they ought to care about the welfare of others as an end in itself. Nonetheless, not only did Mill maintain that psychological egoism and utilitarianism are *consistent;* in addition, he heroically attempted to derive the latter from the former in his essay *Utilitarianism.*

2. The resiliency of egoism is illustrated by considering the essays in Jane Mansbridge's (1990) anthology *Beyond Self-Interest.* The authors are social scientists in a variety of disciplines who argue that people have motives that extend beyond the egoistic desire for consumer goods, status, and power; people, they suggest, are influenced by feelings of solidarity and concern for others. The point to notice is that egoists can grant this claim if they assert that these other-directed impulses are mere instruments for attaining pleasure and avoiding pain.

3. In a footnote to *A System of Logic,* Mill (1874, p. 328) does allow that Darwin's "remarkable speculation" is a "legitimate hypothesis." He goes on to claim, however, that Darwin provided no "proof" of his conjecture, by which Mill meant that Darwin adduced no evidence that it is true.

4. In the *Origin of Species,* Darwin (1859, p. 62) points out that he uses the term "Struggle for Existence in a large and metaphorical sense . . . Two canine animals in a time of dearth, may be truly said to struggle with each other which shall get food and live. But a plant on the edge of a desert is said to struggle for life against the drought." The two dogs compete with each other, but so do two plants that differ in their ability to cope with a hostile

environment. Competition can take the form of direct warfare, or it can simply refer to the "contest" among organisms to contribute the greatest number of progeny to the next generation. The same distinction needs to be drawn at the level of group selection. Groups can compete with each other by direct warfare, but they also can compete in the wider sense of reproductive success.

1. Altruism as a Biological Concept

1. The altruism in this example refers to the effect of the brain worm on the other parasites of the same species within a single host and not to the effect of the brain worm on the host, which is clearly detrimental.

2. The equations for average fitnesses in this situation would be $W_A = X + 1 + 2(pN - 1)$ and $W_S = X + 2pN$.

3. If we had constructed our two-group example with less variation between groups, the selfish type would have evolved, but more slowly than in the one-group example. The fact that selfishness evolves does not mean that group selection is absent, but rather that it is a weaker evolutionary force than individual selection. Similarly, if altruism evolves, this does not mean that individual selection favoring selfishness is absent, but merely that it is weaker than beween-group selection for altruism.

4. Assuming that a trait appears fully formed in the population as a mutant is another simplifying assumption that is often used in evolutionary models. Many traits vary in a more continuous fashion and emerge more gradually, which can have important consequences for the evolution of altruism that we explore in Chapter 4.

5. Because the value of q is so close to zero when the A type appears only as a rare mutation, it is common to ignore the first term in the equation (qE_2) when calculating the average fitness of the S type.

6. We are not the first people to recognize that the distinction between component causes and net effect needs to be recognized in evolutionary theory. In a context having nothing to do with discerning the units of selection, Denniston (1978) made fundamentally the same point while criticizing a proposed definition of trait fitness. Suppose the fitness of a trait is defined as its rate of increase (or the mathematical expectation thereof). The problem with this definition is that it fails to distinguish among different possible causes of a trait's increasing in frequency. The increase may be due to natural selection, but it also may be due to mutation and migration; in fact, all three processes may be present simultaneously. An appropriate definition of fitness should isolate the component cause of trait increase that is due to natural selection. It is better to define a trait's fitness as the average number of

offspring an individual with the trait will have. For more philosophical discussion concerning how averaging can obscure causal facts and can lead to mistaken predictions if the averages are not recomputed for each generation, see Wimsatt (1980), Sober and Lewontin (1982), Sober (1984), Lloyd (1988), and Brandon (1990). For a defense of averaging, see Sterelny and Kitcher (1988).

7. Before sex ratio can be considered an adaptation, it must be shown to be an evolvable trait. In the Hymenoptera (ants, bees, and wasps) and many other species of arthropods, sex is determined by a mechanism called haplo-diploidy, in which fertilized eggs become daughters and unfertilized eggs become sons. (The daughters, with chromosomes from both egg and sperm, have a *diploid* set of chromosomes; the sons, with chromosomes only from one source, have a *haploid* set.) Females often store sperm and control its release as they lay their eggs, which enables them to determine the sex of their offspring with precision. Sex ratio is therefore clearly an evolvable trait that can be influenced by natural selection in haplo-diploid species. When sex is determined by a genetic polymorphism, such as the inheritance of *either* two X chromosomes *or* an X and a Y in humans, an even sex ratio might result automatically from the random process of meiosis. It is not obvious how a parent could bias the sex ratio of its offspring, even if it would be adaptive to do so. One possibility is that if multiple eggs are fertilized and their survival is sex-biased, a biased sex ratio at birth would result. Although the mechanisms are poorly understood, adaptive sex-ratio bias has been demonstrated in a number of species in which sex is determined by a genetic polymorphism (Clark, 1978; van Schaik and Hrdy, 1991; and Holekamp and Smale, 1995).

8. Parasites and other disease organisms, such as bacteria and viruses, are distinguished primarily by the size of the organism, which is not relevant to the conceptual issues that we are discussing. We will use the term *disease* to refer to both.

9. Virulence is defined as any negative effect of a disease on its host, from a small decrease in growth to an agonizing death. It has many causes, only some of which can be understood from the standpoint of adaptation and natural selection. See Ewald (1993), Bull (1994), and Frank (1996a) for more general reviews of the subject. Forms of virulence that have no adaptive explanation are especially likely when diseases occur in novel host species. Some disease organisms are able to facultatively adjust their virulence to the state of their host; see our general discussion of facultative behaviors and multilevel selection in Chapter 3.

10. When thinking about the relationship between the virulence and the reproductive capacity of a disease organism, it is important to avoid comparing very different *kinds* of organisms. For example, most strains of *Es-*

cherichia coli are avirulent and propagate more rapidly than *Mycobacterium tuberculosis,* which is virulent. These species reside in different organs, however, and are not in direct competition with each other. The question is whether the population of *M. tuberculosis* (for example) within a single host is genetically variable and whether fast-reproducing strains are more virulent than slow-reproducing strains. Of course, it is also true that the entire multispecies community of organisms within a host are "in the same boat" with respect to their effect on the host. This brings us into the realm of community-level selection, which we discuss in Chapter 3.

11. A detailed history of this fascinating era in evolutionary biology remains to be written. Our discussions of sex ratio and kin selection in this book are the most detailed published accounts. The earlier history of group selection is reviewed by Wilson (1983) and Wade (1978). The work of the Chicago school between the two world wars is examined by Mitman (1992). The sociological aspects of the controversy from the 1960s to the present have never been carefully documented and studied. We encourage historians of science to study the group selection controversy as an example of the problems associated with scientific change.

2. A Unified Evolutionary Theory of Social Behavior

1. Robert Frank is an economist who is interested in the evolution of human behavior. The passage that we quote is the standard portrayal of group selection and kin selection in the social sciences literature. More recently, Frank (1994) has constructively related his views to the multilevel framework.

2. D. C. Williams is G. C. Williams's wife. When asked about her role in the development of the model, she stated that she "helped with the math," since she had more formal mathematical training than her husband (personal communication to DSW).

3. More generally, Hamilton showed that there is a net increase in the number of altruistic alleles when $r > c/b$, where r is the coefficient of relationship ($r = 0.5$ for full siblings). This has become known as Hamilton's rule.

4. Strictly speaking, the fixation of an altruistic gene in a few groups by genetic drift is not an improbable event. Given a sufficient number of groups, it is virtually inevitable, and the expected proportion of groups that become fixed for altruism can be predicted from the laws of probability.

5. Altruism will go extinct within each group if they are isolated for a sufficient number of generations, as we emphasized in Chapter 1. Also, all groups grow exponentially in the example that we present. Because altruistic groups grow at a faster rate, they achieve very large numbers over multiple generations. If each group reaches a carrying capacity, then the differential

productivity of groups (favoring altruism) can be diminished while selection within groups (favoring selfishness) continues unabated. These and other permutations are explored in Wilson (1987). The point of our example is not to show that group selection always prevails over individual selection, but merely that both levels of selection need to be reckoned with. Maynard Smith's (1964) assumption that altruism goes completely extinct within every mixed group between dispersal episodes is completely unwarranted, given his goal of assessing whether altruism can evolve by group selection.

6. Actually, the within- and between-group components of the Price equations do not correspond exactly to within- and between-group selection, as first pointed out by Heisler and Damuth (1987) and elaborated by Goodnight, Schwartz, and Stevens (1992). Traits that increase the fitness of individuals without having any effect on other members of the group do not evolve by group selection, as pointed out by the example of tall and short populations in Sober (1984, pp. 258–262, pp. 314–316). Nevertheless, if individuals that vary for these kinds of traits are placed in groups, groups with the most fit individuals will produce more than groups with the least fit individuals. These differences appear in the group component of the Price equation, even though they should not be regarded as examples of group selection. The statistical technique of contextual analysis (see above-cited authors) partitions gene-frequency change into within- and between-group components that correspond more closely to the evolutionary concepts of within- and between-group selection. Nevertheless, these subtle differences do not alter the general conclusions that Hamilton derived from the Price equations.

7. When a single gene influences more than one trait, it is said to be pleiotropic. Hamilton is suggesting that a single gene might influence both the expression of altruism and some other trait that causes the altruists and nonaltruists to sort nonrandomly into groups.

8. The ease with which Hamilton accepted group selection suggests that his primary objective was to show how altruism evolves, by any mechanism, and not to discover an alternative to group selection.

9. Hamilton includes us in this complaint! We failed to cite his 1975 paper in our recent review of group selection (Wilson and Sober 1994), although it is cited in our previous publications (e.g., Wilson 1977b, 1980).

3. Adaptation and Multilevel Selection

1. Other factors can complicate the interpretation of sex ratios. For example, group selection should promote a maximally female-biased sex ratio, subject to the constraint that all of the females can be fertilized by the minority of males. A sex ratio of 75 percent daughters would reflect pure group selection

if 25 percent of the males were required to fertilize the females. There is no substitute for detailed biological information, so completing steps 1 and 2 of the procedure allows only preliminary tests of group vs. individual selection.

2. Each of the interacting traits may be genetically determined, in the same sense as the trait being selected is assumed to be genetically determined in simpler models. Recognizing that the trait being selected in this example is caused by complex interactions among other traits, that are themselves genetically determined, constitutes a step away from models based on naive genetic determinism.

3. The genetic determination of traits is sometimes referred to as "broad-sense heritability, as opposed to the resemblance between parents and offspring, which is referred to as "narrow-sense heritability" (Falconer 1981). Narrow-sense heritability is required for a response to selection.

4. The plateau that is reached in a selection experiment can be caused by linkage disequilibrium, in addition to the depletion of additive variance. See Falconer (1981) for more on these concepts.

5. The practice of family selection in agricultural genetics is a form of artificial group selection that dates back to eighteenth-century cattle breeding. Darwin (1859) was familiar with the practice of family selection and used it to explain the evolution of altruism in honey bees. Wade was the first to test the modern theory of group selection in the laboratory.

6. The reasons that phenotypic variation can be nonheritable at the individual level and heritable at the group level are too technical to be treated in this book. See Goodnight and Stevens (1997) for a recent review.

7. Trivers (1971) included interactions among species in his discussion of reciprocal altruism.

8. The evolution of facultative behaviors becomes more complicated when we consider the principle of specialization, according to which "a jack of all trades is a master of none." In this case, a fixed behavior can be advantageous even in a variable environment, because the advantages of performing the behavior well in the right situation outweigh the disadvantages of doing poorly in the wrong situation (see Wilson and Yoshimura 1994 for a review).

9. Although *Nasonia*'s ability to change its sex ratio is impressive, it is important to avoid exaggerating the sophistication of this facultative response. In the most comprehensive study to date, Orzack, Parker, and Gladstone (1991) examined facultative adjustment of sex ratios in a number of genetic strains of *Nasonia*. They found that all the strains could distinguish between the presence and absence of other females and modified the sex ratio of their offspring in the direction that theory predicts. However, the strains varied in the degree to which they modified the sex ratio of their offspring. This shows that not all strains have converged on an optimal strategy of sex ratio adjust-

ment. On the other hand, it also provides proof that facultative sex ratio adjustment is a heritable trait in *Nasonia* that can respond to natural selection.

4. Group Selection and Human Behavior

1. In principle, stabilizing selection on a polygenic trait can produce an adaptive mean phenotype without any maladaptive variation around the mean. For example, if a + and − allele exists at 100 loci, and a phenotypic value of 0 is favored by stabilizing selection, zero variance can be achieved if 50 loci become fixed for the + allele and the remaining 50 loci become fixed for the − allele. Selection for this arrangement is weak, however, and will be opposed by mutations at all 100 loci. The balance between mutation and selection results in continuous maladaptive variation around the adaptive mean. See Ridley (1993) for a more detailed discussion.

2. Simon (1990) emphasized docility, or a willingness to follow social norms, as an important mechanism in the evolution of altruism. He interpreted his model as an alternative to group selection but did not focus on relative fitness within and between groups. When Simon's model is translated into the multilevel framework, it is close to the theme that we are developing here. It is important to emphasize, however, that willingness to follow social norms depends on a concensus that the norms are in some sense fair (Boehm 1993). Multilevel selection theory predicts that people should strenuously resist social norms that contribute to exploitation within groups (Wilson and Sober 1996).

3. The internal stability of social norms depends upon the ability to reward and punish. As we discussed in the previous section, these secondary behaviors are not internally stable. For example, the rule "Punish cheaters" is vulnerable to a freeloading strategy in which people don't cheat and also do not punish those who do. The higher-order rule "Punish cheaters and those who do not punish cheaters" is vulnerable to a still higher-order freeloading strategy in which people both abstain from cheating and punish cheaters, but do not punish those who do not punish cheaters. Boyd and Richerson (1992) suggest that a generalized punishment rule can escape this infinite regress, allowing social norms to be truly stable within groups. We are not convinced and suspect that social norms ultimately rest upon weakly altruistic traits. In either case, given the existence of low-cost secondary behaviors, a vast diversity of primary behaviors can be made internally stable.

5. Human Groups as Adaptive Units

1. The HRAF is not a random sample of world cultures. It is obviously biased toward cultures that have been well studied by anthropologists and

also consists mostly of traditional societies rather than modern nations. Despite the fact that it is not a random sample, the traditional societies included in the HRAF are unlikely to be biased with respect to the questions that we are attempting to address. In other words, there is no reason to expect that the societies included in the HRAF are more or less well adapted at the group level than societies that are not included in the HRAF. The comparison between traditional societies and modern nations is another matter. Modern nations are vastly different from traditional societies both in their size and the recency of their cultural history. Biologically evolved mechanisms of group-level adaptations that work well in small face-to-face groups may break down as the size of the society increases. Similarly, group-level cultural adaptations that have accumulated over a relatively long period may break down when modern mass societies come into being. We will attempt to address these issues to the best of our ability, but many will remain unresolved.

2. The HRAF is periodically updated, and some libraries have newer and more extensive versions than others. We used the version available at the the Binghamton University library, which includes 354 cultures.

3. Agriculture and animal husbandry are very recent by evolutionary standards. During most of human evolution, our ancestors foraged for natural food resources. Modern hunter-gatherer societies are therefore thought to approximate the physical and social conditions that existed for most of our evolutionary history. Today, however, hunter-gatherer societies are themselves diverse and are often restricted to marginal habitats by other cultures. Hunter-gatherer societies in more productive habitats may have been quite different.

4. The amount of ethnographic material in the HRAF varies widely among cultures. Thirteen randomly chosen cultures were discarded because of insufficient material (Bedouin, Belorussia, Bihar, Iraq, Kamchadal, Lithuanian, Mohave, Mossi, Pashtun, Talamanca, Telugu, Trucial Oman, and Woleai). Information was sparse for these cultures with respect to most categories, not just social norms, so it is unlikely that their exclusion introduces a sampling bias with respect to the questions that we are asking. Another possible source of sampling bias concerns the historical relationships among the cultures. Cultures that are historically derived from a common ancestral culture cannot be regarded as providing statistically independent observations, in the same way that phylogenetically related species are not statistically independent. Anthropologists have identified a number of "culture clusters" that are similar to higher-level phylogenetic units in biology. To minimize the effects of history, a survey should confine itself to only one culture within each cluster. However, this procedure can introduce other forms of bias, favoring small clusters over large clusters, and we have elected to sample without respect to

clusters. The effect of including historically related cultures is to reduce the effective sample size and does not bias the sample with respect to the questions that we are asking.

5. Ethnographer bias can be reduced by including multiple ethnographies, but it can never be eliminated. To the extent that the ethnographers come from the same culture and read each others' work, their accounts cannot be regarded as statistically independent events.

6. Social norms that allow all men to compete for every eligible woman are likely to be highly dysfunctional at the group level. Rules that determine who can marry whom are therefore likely to benefit the group, even if the rules themselves are arbitrary. To some degree, marriage rules in traditional cultures might resemble driving rules in modern nations; it doesn't matter whether the convention is to drive on the right or left side of the road, as long as everyone stays on the same side.

7. The third case was said to provoke an "energetic reaction from the community" without specifying the details.

8. Group-level benefits can sometimes arise as a coincidental by-product of within-group selection (Williams 1966). Thus, group selection is best invoked when there is a detailed and pervasive pattern of group-level benefit; an isolated group-level benefit is not sufficient.

9. In addition to the need for social systems to become hierarchical as they increase in size, other kinds of nonegalitarian social systems can remain adaptive at the group level. For example, high status can be conferred only on individuals who benefit the group. In this case, within-group selection would remain strong but would be aligned with between-group selection. Group selection would be required to explain why group-advantageous traits, rather than a diversity of other traits, are the criterion for high status.

10. In modern societies, the attention focused on the moral conduct of leaders does not seem a very effective means of promoting moral behavior. The same attention is probably more effective in small-scale societies, illustrating the problems of scale that we discussed earlier. Also, before we conclude that mechanisms of controlling a leader's behavior are totally ineffective in large-scale societies, we should contemplate how leaders would behave if their constituents truly didn't care.

11. The loss of personal autonomy and direct participation in decision making in large-scale societies illustrate further problems of scale. The mechanisms that allow small groups to function as adaptive units simply become unworkable and must be replaced by other institutions for large social groups to remain functional. The mechanisms that operate in large groups may be more difficult to regulate, however, so opportunities for exploitation within groups increase.

12. Varieties of Christianity might differ not only in *how* they promote group-level adaptations but also in the *extent* to which group-level adaptations are promoted. Recall that many group-level adaptations are sometimes bundled together in the same unit (such as an organism or a bee hive) but are sometimes more loosely organized. Religious groups (Christian and otherwise) might also span this continuum.

13. Both the Nuer and the Dinka have a segmented social system that they understand as a nested hierarchy. The minimal segment is the lowest level in the hierarchy, although it is larger than a village, which is the primary "corporate unit" as far as day-to-day economic activities are concerned. See Evans-Pritchard (1940) for a more thorough discussion.

6. Motives as Proximate Mechanisms

1. We are not claiming that *all* human behaviors are products of natural selection; nor are we claiming that *every* belief and *every* desire is an adaptation that evolved to produce adaptive behavior. For now, we are merely explaining what it means to regard the mind as a proximate mechanism for controlling behavior. The evolutionary commitments of our discussion of psychological egoism and altruism will be clarified in Chapter 10.

2. This point bears on Blurton-Jones's (1984, 1987) model of food sharing, which we mentioned briefly in Chapter 4. In this model, the behavior of hunting-and-sharing is said to be evolutionarily selfish if an individual would do better by hunting-and-sharing than he would do if he chose not to hunt. Here evolutionary selfishness is identified with the trait that a psychological egoist would choose. Blurton-Jones's analysis of this problem sometimes fails to accurately predict what trait will evolve. And when it does make the right prediction, it says that individual selection was the process at work when, in fact, the trait is evolutionarily altruistic and requires group selection to evolve (Wilson 1998). Like the averaging fallacy, the heuristic of using psychological egoism to think about evolution has the effect of making altruism and group selection invisible.

3. According to kin selection theory, helping offspring and helping other relatives are similar because the donor shares genes with the recipient in both cases. According to multilevel selection theory, altruism among relatives evolves by group selection, thus raising the possibility that parental care also includes an element of group selection. This suggestion may sound strange, but it is conceptually well motivated and may lead to some interesting insights. If both parents can provide care, but one provides more than the other, these behaviors may be instances of altruism and exploitation. The altruism in parental care derives not from the mere fact that offspring receive benefits,

but from the benefits that one parent confers on the other. If parents are not genetic relatives, this is a benefit conferred upon *non*kin. Parental pairs are groups of size two, in which there may be zero, one, or two altruists. Thinking through the implications of this idea is an interesting task for the future; for now, we will regard parental care conventionally, as an individual-level adaptation. To see our point in the present context—that evolutionary selfishness can be motivated by proximate mechanisms that are psychologically altruistic—imagine that the species in question is asexual.

4. We owe this example to Dretske (1981).

5. Earlier in this century, certain forms of behaviorism denied that inner mental states should be regarded as causes of behavior. More recently, Stich (1983) and P. M. Churchland (1984) have argued that the psychology of beliefs and desires should be rejected and replaced by a quite different framework (e.g., connectionism). We agree that there is no *a priori* guarantee that beliefs and desires exist and cause behavior. However, we see no reason *now* to reject the idea that a science of the causes of behavior should postulate inner representational states.

6. In what follows, we do not attempt to give a full theory of what beliefs and desires are or of how they differ; we merely want to point out some of their salient features.

7. Although a full development of the view that beliefs and desires are propositional attitudes would require a theory about how propositions are individuated, we will not attempt to provide such an account here. The claims we will defend can be formulated so as not to depend on this issue.

8. Even if "John made the apples fall" means that John is related in a certain way to the proposition *the apples fall,* the sentence does not imply that John forms a mental representation. This is why the idea that believing and desiring are propositional *attitudes* goes beyond the point that ascriptions of belief and desire posit relations between individuals and propositions.

9. We do not exclude the possibility that the lifeguard formulated a plan in the past for dealing with drowning people, so that when the moment arises she merely sets the plan into motion without having to engage in detailed deliberation.

10. This is a conceptual point—desire should not be *defined* to demand a concomitant affective state. This leaves open the empirical possibility that many or even all episodes of desiring in fact have an affective valence—whether positive or negative.

11. It may be more accurate to say that the proposition that Jane wants true she represents to herself as *I have a glass of water.* We will discuss this point shortly.

12. McCawley (1988, p. 146) provides an additional argument for the idea

that desires are propositional attitudes. The following sentence is well-formed with *it,* but not with *her:* "John wants a mistress, but his wife won't allow *it/her.*" This suggests that the grammatical object of *wants* is not a feminine singular, such as *a mistress; he has a mistress* is the obvious candidate.

13. The existence of *de re* belief and desire locutions is perfectly compatible with the claim that belief and desire are always propositional attitudes. Suppose we say of Fido, "he wants *that,*" as we point to a bone. This remark fails to indicate how Fido conceptualizes what he wants, but the truth of the remark does not show that there are nonconceptual desires.

14. This speculation entails a small irony. People use the concept of "I" to formulate the thought that they are unique. Yet, part of the reason that people have this concept is that they are *not* unique, at least not with respect to simple schemes of concepts that may have been present ancestrally.

15. More exactly, *S* has *U* as an ultimate desire precisely when *S* desires *U,* and, for all the other desires that *S* has, it is false that *S* desires *U* solely because *U* is a means to satisfying one or more of these other desires. *S* may recognize that *U* contributes to the fulfillment of some of these other desires, but this can't be the sole reason that *S* desires *U,* if *U* is ultimate.

16. Although it is an open question whether people have one ultimate desire or more than one, we think it is clear that people must have at least one. We can mimic Aquinas's "Five Ways" to see why. Consider how means-end desires can be strung together in a chain: *S* wants *A* solely as a means to attaining *B, S* wants *B* solely as a means to attaining *C,* and so on. If these chains cannot circle back on themselves, and if people don't have infinitely many desires, then these chains must be finite and each must trace back to a first member.

17. We must distinguish ongoing desires from desires for particular things, which may be fully satisfied on a single occasion. Proposition (3) applies to the latter, not to the former. Consider Arthur, who wants praise from others; receiving it on a single occasion is not enough to (completely) satisfy the desire, since the desire is ongoing, not specific. If Arthur wants to help others solely because doing so will elicit praise from third parties, he may continue to want to help even if, on a single occasion, he receives praise without his having helped. The reason is that a single experience of praise without having helped may not be enough to erase the belief that helping raises the probability of praise. For human and nonhuman subjects alike, a series of "extinction trials" may be needed (Slote 1964).

18. A further assumption is needed for either prediction to be valid. To see why, imagine that Sally wants *M* solely as a means to obtaining *E,* but that giving her *E* while withholding *M* makes Sally *change.* Suppose obtaining *E* without *M* somehow causes her to start believing that *E* is worthless and that

M has a value she had not recognized before. The fact that *E* turns to dust and ashes once it is attained does not mean that *E* was not sincerely desired earlier. And the fact that *M* comes to be desired independent of its contribution to *E* does not mean that this is how the two goals were related before.

It is a quite general fact about experiments that they involve *intervention* and *manipulation*—they inevitably change the object on which the experiment is carried out. However, this property of experiments is not enough to prevent us from using experiments to find out what objects were like before the experimental intervention occurred. For example, suppose a thermometer raises the temperature of the thing being measured by 2 degrees; if we know that the thermometer has this effect, we can use the thermometer to determine what the object's temperature was before the measurement was made. The problematic case, which is relevant to our discussion of propositions (2) and (3), is that a measurement procedure may change the system being measured *in ways we don't anticipate.* Since this *caveat* pertains to all experiments, it plays no special role in our discussion of what it means for one desire to be purely instrumental with respect to another.

19. This is the kind of transformation that Hoffman (1976, 1981b) postulates in his developmental theory. Eisenberg and her associates have attempted to document this change; see, for example, Eisenberg, Lennon, and Roth (1983), Eisenberg (1986), and Eisenberg and Strayer (1987).

20. If altruism can emerge from hedonism, presumably the reverse process is also possible. People who help purely because they care about the welfare of others may find that they get pleasure from this activity, and this may give rise to the hedonistic desire to secure this type of pleasure in the future; here the initial altruistic desire does not disappear, but is supplemented. Perhaps this idea applies to the motivational changes that occur when people repeatedly donate blood, on which see Piliavin and Callero (1991).

7. Three Theories of Motivation

1. Our characterizations of egoism and altruism will share this feature.

2. Kavka (1986, p. 41) notes that it is unclear how "the desire for posthumous fame" can be compatible with egoism.

3. Kavka (1986, p. 41) takes the problem posed by relational desires to indicate that egoism is an inherently vague theory; one can list examples of desires that the egoist is prepared to regard as ultimate, but there is no way to delimit these options in a principled manner. Kavka thinks this constitutes a defect in the egoistic theory; we see it as a problem that egoism and altruism both must address.

4. Skeptical views about the concept of personal identity (Hume 1739;

Parfit 1984) may suggest that hedonism does not make sense when it goes beyond the time frame of the present moment. But here we must be careful. If the objective existence of a temporally enduring self is an illusion, this may mean that people *ought* not to be hedonists; however, this philosophical result would not show that people fail to be hedonists *in fact*. Note also that altruistic desires often concern the welfare of temporally enduring persons, so the skeptical claim impinges equally on egoism and altruism.

5. The term *welfare* is sometimes used in a way that is narrower than we intend. When altruists have ultimate desires concerning the "welfare" of others, this may or may not mean that they want others to have their desires satisfied. For us, "welfare" just means faring well.

6. In describing egoism as monistic, we are counting *types* of ultimate desire—all are self-directed. This, of course, is consistent with the idea that people may have a number of different self-directed ultimate goals. Egoism, at this level of resolution, may be quite pluralistic.

7. It should not be thought that concern for the welfare of a group, a religion, or a cultural tradition always "reduces" to concern for the individuals in the group, the religion, etc. For example, a reduction in salary may be good for a corporation but bad for its employees. Just as is true in the multilevel selection theory discussed in the first part of this book, it is important to recognize the conflicts of interest that exist between wholes and their parts.

8. This definition construes "empathize" as a success verb; *S* can't empathize with *O*'s grief unless *O* really is experiencing grief. Sympathy is different, or so we shall suggest. The definitions can be modified by readers who find this asymmetry implausible.

9. Indeed, psychologists discuss what they call the "empathic joy hypothesis," which we will examine in Chapter 8.

10. Daniel Batson (personal communication) has drawn our attention to another example that illustrates this point. The Good Samaritan sympathizes with someone who is unconscious.

11. The distinction among empathy, sympathy, and personal distress does not entail that an individual experiences only one of these emotions at a time; nor do we assume that they are causally unrelated to each other. For example, our framework is consistent with the conjecture of Eisenberg et al. (1994) that personal distress may sometimes be caused by excessive empathy.

12. Whereas Piaget thought that children younger than 5 or 6 rarely are able to take the perspectives of others (Piaget and Inhelder 1971), Radke-Yarrow and Zahn-Waxler (1984) and Zahn-Waxler, Radke-Yarrow, Wagner, and Chapman (1992) provide evidence that empathy and sympathy start emerging in the second year of life; these emotions and the incidence of helping behavior gradually increase from age 2 to 6. Perhaps children younger

than 2 fail to be altruistically motivated because they lack the cognitive abilities that are needed to understand the situations and experiences of others (Eisenberg and Miller 1987; Eisenberg and Fabes 1991; Schroeder, Penner, Dovidio, and Piliavin 1995).

13. In their discussion of monkeys and apes, Cheney and Seyfarth (1990, p. 236) point out that "an animal who aids a wounded companion . . . might recognize that his companion cannot walk properly without also knowing that his companion is experiencing pain." If the helper feels bad because his companion is injured, then the term "sympathy" applies.

14. A number of works in recent moral philosophy argue that there is more to morality than the general principles that we are considering here. See, for example, Williams (1981), Nagel (1986), Stocker (1989), and Wolf (1990). Our discussion of moral principles is consistent with this point of view.

15. In saying that a general principle covers a range of objects, we are not claiming that it is true of them; rather, we are describing its intended scope.

16. Sidgwick (1907) thought that this idea is self-evident and that it is part of commonsense morality. We take no stand on whether the principle is correct nor on whether it is part of common sense; our point is just that it follows from what moral principles assert.

The present view of what a moral principle is leaves open how useful general principles are in thinking through real-world moral problems; maybe the principles we manage to state often fail to fully capture the morality we actually embrace. However, the descriptive inadequacy of slogans and maxims (not to mention current philosophical theories) does not decisively establish that general moral principles do not exist.

Also, we hope it is clear that the notion of universalizability we are describing is quite different from Kant's universalizability criterion, which was intended to provide a criterion for determining the moral rightness of an action. We are advancing no such normative proposal.

17. Universalizability is a necessary feature of moral principles, but it is not unique to them. Rules of etiquette and pointers for playing a sport also have this feature.

18. Although the ultimate desire that one's children do well is *psychologically* altruistic, it is an example of *evolutionary* selfishness, if having the desire enhances the parent's reproductive success. We discussed this in Chapter 6.

19. For examples of nonutilitarian moral theories that entail that people are not always obliged to sacrifice self-interest to benefit others, see Rawls's (1971) theory of justice and Nozick's (1974) libertarianism.

20. A probabilistic formulation would be better here—that individuals *maximize expected utility*. This fine point will not affect the questions to be raised, however.

21. William James (1890, p. 558) points out that the attainment of pleasure may accompany an action without being the action's goal. To confuse the two, James says, is like arguing that the goal of ocean voyages is for ships to burn coal because that is something that ships always do when they cross the ocean. James's idea is represented by the preference structure of the Pure Altruist.

22. Our notion of pluralism is intended to represent what economists sometimes call "dual preference structures." See Margolis (1982) for elaboration of this idea. Margolis points out that Arrow (1963), Buchanan (1954), and Harsanyi (1955) argued for the importance of representing individuals as having desires other than ones that are purely selfish. Sen (1978) provides a more recent and influential articulation of this point of view.

23. Precisely intermediate between E-over-A and A-over-E pluralism is a brand of pluralism in which self and other are viewed as equally important. And, of course, there are further preference structures that this 2-by-2 format can describe, and there are choice situations additional to the main diagonal and anti-diagonal problems upon which we have focused.

24. Individual desires are related to behavioral dispositions in much the same way that component forces are related to net forces, as discussed in Chapter 1. Just as an exclusive attention to net causes provides an impoverished picture of the causal facts, so an exclusive focus on behavioral dispositions yields a limited representation of psychological motives.

25. McNeilly (1968, p. 99) and Kavka (1986, p. 36) suggest that a forced-choice experiment identifies the "real object of desire." In our formulation, however, there need not be just one thing that the agent "really" (i.e., ultimately) wants. By placing self-interest and the welfare of others in conflict, it is possible to discover which desire is stronger; the procedure does not guarantee that the stronger desire is the only desire.

26. These four outcomes exhaust the possibilities for *additive* relations between the two factors; in each data set, the effect of shifting from one row (column) to another does not depend on what column (row) one considers. We ignore cases of *non*additive relations, though they often occur in nature, because they are not relevant to understanding the four preference structures described in the previous section.

27. The inferences here described are not infallible. For example, it may be that G1 and G2 in fact make no causal difference in plant height but are themselves correlated with other genetic factors that do make a difference. In addition, such inferences must take account of the amount of variation that exists within each treatment cell. These complications, whose ramifications are worked out in the statistical methodology called "the analysis of variance" (ANOVA), do not affect the points at issue here. See Sober (1988a) for further

discussion of philosophical issues concerning the problem of apportioning causal responsibility.

8. Psychological Evidence

1. The psychological literature on egoism and altruism spans a much broader range of subjects than we will consider here. For useful surveys, see Piliavin and Charng (1990) and Schroeder, Penner, Dovidio, and Piliavin (1995).

2. The philosophy classroom provides a natural laboratory for studies of this type. If one professor teaches introduction to philosophy by arguing against psychological egoism, while a second professor teaches the course by arguing for that position, why not have the students introspect at the end of the course? We suspect that the two classes will produce different introspective reports, if the teachers are sufficiently charismatic.

3. In using these terms, we are not gainsaying the truism that every behavior is a joint consequence of the organism's genes and environment. The behaviors we are talking about here are ones that are fairly invariant over environmental variation.

4. In other experiments, subjects in the low-empathy treatment were asked to consider certain factual and objective pieces of information about the needy other, while subjects in the high-empathy manipulation were asked to consider how it felt to be the person described. In each case, it was argued that there is independent evidence that the manipulation influences empathy.

5. Batson says that the two hypotheses make additional predictions about the frequencies of offers to help by subjects in the four experimental treatments. These are summarized in the following table, reproduced from Batson (1991, p. 111):

		Aversive-Arousal Reduction Hypothesis		Empathy-Altruism Hypothesis	
		ESCAPE IS		ESCAPE IS	
		Easy	Difficult	Easy	Difficult
EMPATHY IS	Low	Low	High	Low	High
	High	Low	High/Very High	High	High

We think this understanding of what the empathy-altruism hypothesis predicts is at variance with Batson's description of what the hypothesis asserts.

What is it in the empathy-altruism hypothesis that predicts that empathy level will make no difference when escape is difficult? And what feature of the hypothesis entails that low-empathy subjects will help more when escape is difficult than when it is easy? Neither of these predictions follows from the claim that empathy causes altruistic motivation. Also, if the aversive-arousal reduction hypothesis predicts that empathy promotes helping when escape is difficult (pushing its probability from high to very high), why does the empathy-altruism hypothesis deny this? Perhaps these questions indicate that we should understand the empathy-altruism hypothesis as asserting some form of motivational pluralism, according to which altruistic and egoistic motives both play a role; however, the details of this pluralistic hypothesis are not provided.

We have similar interpretive disagreements with Batson concerning another experiment that we will review in this chapter. They are not fundamental to our argument, however, since we agree with Batson's overall conclusion that the experiments disconfirm the versions of egoism under test, but not the empathy-altruism hypothesis.

6. That is, the difference in observed frequencies was statistically significant, given the sample size.

7. As was true for the first experiment we discussed, questions can be raised concerning Batson's interpretation of the different hypotheses. Why does the empathy-specific punishment hypothesis predict not only that the justification treatment influences helping but that the level of empathy does too? In the previous experiment, the aversive-arousal reduction hypothesis was said to *deny* that empathy makes a difference when escape is easy. Aspects of Batson's interpretation of the empathy-altruism hypothesis also are puzzling. What is it in this hypothesis that entails that justification for not helping affects low-empathy subjects but does not affect high-empathy subjects? Where does the hypothesis say that a high degree of empathy is so strong a motivator that it makes us totally unresponsive to what other people think?

8. The empathy-altruism hypothesis has this implication if we interpret it as saying that high empathy triggers an altruistic motive and no egoistic motives; the hypothesis does not entail this prediction if it says merely that empathy triggers an altruistic motive. The stronger reading of the hypothesis is the one that Batson (1991, pp. 87–88) prefers.

9. Philosophical Arguments

1. Just as there can be pleasure without an antecedent desire, so the antecedent desire can fail to be a desire for an external thing. For example, people want to be rid of headaches and to experience various sorts of physical

pleasure (Penelhum 1985, p. 51; Henson 1988). This is another respect in which Butler's second premise is overstated.

2. Whether a goal is worth attaining and whether adopting that goal as a conscious end will undermine your chances of attaining it are quite different questions (Parfit 1984). Suppose archers inevitably fail to hit the bull's eye when they aim at it; you have to aim high to hit the center. It does not follow that hitting the bull's eye is an inappropriate goal; all that follows is that the way to achieve this goal is to think about something else. Hedonism as a normative proposal—that people should do what they can to maximize their pleasure and minimize their pain—could live with a "paradox" of this sort.

3. See Gibbard (1990) for discussion of this issue and a defense of the latter point of view.

4. Nagel (1970, pp. 90–91) claims that if *S* has a reason to perform an action *A,* then everyone else has a reason to promote *S*'s doing *A* or to want *S* to do *A*. We find this claim implausible; for example, when two individuals have conflicting interests, each may have a reason to act as he does, but it hardly follows that each has a reason to help the other or to want the other to act as he does (Sturgeon 1974). We suspect that Nagel's claim, at best, describes what is true when reasons reflect *moral obligations;* perhaps it is true that when *S* is morally obligated to perform an action, then others have a moral obligation not to prevent *S* from doing so. Of course, this normative remark does nothing to show that altruism is psychologically possible.

5. The experience machine thought experiment describes a situation in which one's preference for a real life and one's preference for a pleasurable life *conflict,* in the sense described at the end of Chapter 7. Notice that if people chose to plug into the experience machine, this would not prove that they are hedonists; rather, what would follow is that their desire for certain types of experience is *stronger* than any conflicting nonsolipsistic motives they might have.

6. The argument we are considering asserts that hedonism as a general thesis is false if even *one* person would decline to plug into the experience machine; this is consistent with the possibility that many people lead lives of such suffering and deprivation that they would elect to plug into the machine if the offer were made.

7. This second thought experiment was suggested to us by William Talbott.

8. The movie "Total Recall" depicts a future society in which people "take vacations" by plugging into experience machines, after which they return to their real lives. Although many of us would decline to spend the rest of our lives connected to such a machine, it is entirely plausible that many of us would elect to plug in for brief periods of time. Even if people value having

real lives, they apparently are prepared (for hedonic reasons) to reduce the percentage of their lives that are real.

9. We are here exploiting a distinction that Schlick (1939) drew between the idea of a pleasant state and the pleasant idea of a state. He denied that we always choose actions because we expect them to produce the most pleasant experiences, though he affirmed that deliberation always terminates with the agent choosing the most pleasant plan of action. Schlick denied that this latter doctrine is a form of hedonism. He would be right, if "pleasant" just meant *preferred;* the idea that people choose the action they prefer isn't distinctive of any particular motivational theory. However, if the suggestion is that people choose an action because that act of choosing makes them feel good (where feeling good is what they ultimately want), then the claim entails hedonism.

10. It isn't just philosophers who advance burden of proof arguments against egoism. For example, Hoffman (1981b) chooses to interpret his data as favoring the existence of altruism, even though he admits that they also could be interpreted from the point of view of egoism.

11. If the mind were transparent to introspection, then there would be an asymmetry between the authority of intuition in psychology and the authority of intuition in biology and physics. This is the idea behind Hume's (1751) argument in his essay "Of Self-Love," that the seeming obviousness of motivational pluralism is a sign of its truth.

12. This general point is discussed in terms of a quantitative model in Sober (1994a).

13. We will restrain ourselves from enumerating standard criticisms of Popper's falsifiability criterion. For a round-up of the usual suspects, see Sober (1993b, pp. 47–54).

14. This helps clarify the claim by Krebs (1975, pp. 1134–1135) that if altruism is defined in terms of "helping behavior that is not motivated by expectations of reward . . . the existence of altruism can never be proved . . . because it requires proving the null hypothesis (i.e., establishing the absence of expectations of reward)." The first point is that the altruism hypothesis, as we understand it, does not say that some actions are produced by purely altruistic motives (Batson 1991, p. 51); as argued in Chapter 7, the hypothesis is quite consistent with a form of pluralism in which an altruistic ultimate motive is always accompanied by egoistic ultimate motives in the production of behavior. In reality, it is egoism that should be regarded as a "null hypothesis," since this hypothesis says that altruistic ultimate motives don't exist. Defending pluralism requires that one undermine this null hypothesis.

15. In his essay "Of Self-Love," Hume (1751) suggests that the attraction of the egoism hypothesis traces to "that love of *simplicity* which has been the

source of much false reasoning in philosophy." Hume goes on to claim that pluralism is in fact the simpler theory; he defends this contention by presenting an argument that resembles Butler's argument against hedonism, which we discussed earlier in the present chapter.

16. We argued in Chapter 2 that a model that postulates individual selection alone is more parsimonious than a model that postulates both individual and group selection. We are here denying that psychological egoism is more parsimonious than motivational pluralism. Parsimony has been invoked as a solution to both problems, but the underlying logic of these arguments is different.

17. This inference might appeal to the principle of cladistic parsimony, on which see Sober (1988b).

18. A similar evolutionary perspective helps clarify the status of Lloyd Morgan's "canon"—that "higher" psychological mechanisms should be postulated to explain a behavior only if "lower" mechanisms are insufficient. See Sober (1998a) for discussion.

19. Macpherson (1962) argues that "possessive individualism" became the dominant view of human nature in Western Europe during the rise of capitalism, and because of it.

20. Readers who extract a different lesson from the present chapter and the previous one may wish to interpret the chapter that follows differently. We see it as an effort to break a deadlock, but those who think they already possess a successful argument for the existence of psychological altruism may still want an explanation of why this motivational arrangement evolved. And for those who think that psychological egoism is true, Chapter 10 can be viewed as posing a puzzle: Why did egoism evolve, if psychological altruism has the evolutionary plausibility that we claim?

10. The Evolution of Psychological Altruism

1. Since hedonism entails egoism, but not conversely, any proposition that refutes egoism also refutes hedonism. From this logical point of view, it therefore appears that hedonism is easier to refute than egoism, not harder. We agree. However, our point is that hedonistic egoism is harder to refute than nonhedonistic egoism.

2. Forty percent of the average American's calories comes from fat; the figure for contemporary hunter-gatherers is 20 percent. The difference between the fat content of wild game and of meat from animals raised on farms is equally striking. Also, skeletons from preagricultural societies show no tooth decay. This evidence strongly suggests that we eat more fat and sugar than our ancestors did. A plausible explanation is that this change in diet is

due to a change in food availability interacting with taste preferences that people have long had (Nesse and Williams 1994, pp. 147–151).

3. The process of evolution by natural selection requires heritable variation in the trait under selection. However, this requirement does not entail that the phenotypic variation that remains after the selection process has run its course has a genetic basis. In fact, selection often *destroys* genetic variation; see the discussion of heritability in Chapter 3. Here, as elsewhere, we must not confuse the *process* of natural selection with its *product*.

4. One of the most obvious patterns of within-species variation is that between the sexes. Human mothers, on average, provide more child care than fathers do. In addition, maternal child care in humans is continuous with a long evolutionary history, whereas paternal child care is probably a more recent innovation, given the traits that we may infer were present in the ancestors that human beings share with closely related species. Nonetheless, the argument we will develop asserts that males and females typically want their children to do well as an end in itself. This is consistent with the possibility that males and females differ with respect to their other motives, and also in terms of how strong their altruistic regard for offspring is, compared with other desires.

5. Of the eleven societies in Daly and Wilson's survey in which it is noted that infanticide occurs when the interval between births is too short, the practice is always to kill the newborn, never the older sibling. Daly and Wilson (1988, p. 75) suggest that the motivational basis for this choice is that maternal attachment typically deepens as the infant develops.

6. In Part I, we argued that helping behaviors directed to individuals other than one's offspring require group selection to evolve. In Chapter 6 (note 3), we pointed out that parental care also may require group selection to evolve when organisms have two parents. In this instance, the behavior involves an altruistic donation, not to one's offspring, but to the other parent. Neither of these points about the evolutionary process are assumed in our inquiry into the proximate psychological mechanisms that underlie helping behavior.

7. Oliner and Oliner (1988, pp. 184, 297–298) compared Christians who rescued Jews from the Nazis during World War II with nonrescuers who were otherwise similar on a number of dimensions; they found that rescuers were more likely than nonrescuers to have been close to their parents. Rosenhan (1970) found a similar pattern when he compared individuals who participated for a long time in the U.S. civil rights movement of the 1960s with individuals whose participation was of shorter duration.

8. This is not inevitable. For example, if predators are more abundant at the bottom of ponds than at the top, the organism's elevation will affect its

fitness for reasons additional to the way that elevation is associated with oxygen concentration.

9. In a causal chain from *X* to *Y* to *Z*, in which *Y* screens off *Z* from *X* (i.e., renders them conditionally independent of each other), the information that *Y* provides about *X* must be at least as great as the information that *Z* provides about *X*, if information is understood in terms of R. A. Fisher's concept of *mutual information*. See Sober (1993a) for further discussion.

10. The TBO Principle does not require that oxygen concentration screen off elevation from fitness, which was required in our argument for the D/I Asymmetry. Both these principles can be given straightforward representations and justifications by using the resources of path analysis.

11. There are substantive problems concerning how one might determine whether an organism has beliefs and desires, and whether it is able to experience pleasure and pain. Our argument does not assume that either of these questions is easy to answer; nor does it require that sensation must have preceded cognition in evolution. If our task is to assess hedonism, however, it is useful to imagine that early cognizers were hedonists and then ask what would lead their descendants to evolve away from this ancestral condition.

12. Thus, the fact that human beings are unable to calculate infallibly which action will be fitness-maximizing does not decisively show that the desire to maximize fitness would not have been advantageous. Rather, the argument must be that better tradeoffs were available ancestrally.

13. Of course, the pain may be greater or less depending on your cognitive state; we do not deny that this may be true.

14. If parents feel pain when they believe their children are doing badly, what is to prevent the HED organism from pursuing the "ostrich strategy" of avoiding discomfiting information? This is yet another design problem that hedonism must solve if it is to do a reasonable job of delivering parental care.

15. To understand why human beings respond to pain when they burn their fingers, we must take account of the fact that human beings inherited many behavioral control devices from ancestors who were less cognitively sophisticated. The point made in the text is that this ancestral arrangement is more reliable than the hypothetical mechanism of having the organism respond only to beliefs that are caused by pain, not to pain itself. In fact, human beings are pluralistic in this context—they respond both to sensations of pain *and* to beliefs about the states of their bodies.

16. Typically, but not always. Perhaps hearing a baby cry makes one feel bad without one's having to form a belief; the sound may be intrinsically disconcerting, in something like the way that the sound of chalk squeaking on a blackboard can induce shivers. Our point is not that *all* feelings are

mediated by belief but that *vast numbers* are; hedonism relies heavily on belief-mediated feelings to explain behavior.

17. In fact, we don't think that pure hedonism *is* the behavioral control mechanism that people use to take care of their bodies. We suggest that people not only want to avoid pain; they also have the ultimate desire to avoid bodily injury. The imperfect reliability of pain as an indicator of bodily injury, which we noted earlier, made it advantageous for hedonism to be supplemented by a nonhedonistic egoism.

References

Alexander, R. D. (1974). "The Evolution of Social Behavior." Annual Review of Ecology and Systematics 5: 325–383.

——— (1979). Darwinism and Human Affairs. Seattle: University of Washington Press.

——— (1987). The Biology of Moral Systems. Hawthorne, N.Y.: Aldine De Gruyter.

——— (1992). "Biological Considerations in the Analysis of Morality." In M. H. Nitecki and D. V. Nitecki, Evolutionary Ethics, pp. 163–196. Albany: State University of New York Press.

Alexander, R. D., and G. Borgia. (1978). "Group Selection, Altruism and the Levels of Organization of Life." Annual Review of Ecology and Systematics 9: 449–474.

Allee, W. (1951). Cooperation among Animals. New York: Henry Shuman.

Ammar, H. (1954). Growing Up in an Egyptian Village: Silaw, Province of Aswan. London: Routledge and Kegan Paul.

Arrow, K. J. (1963). Social Choice and Individual Values, 2nd edition. New York: John Wiley and Sons.

Axelrod, R. (1980a). "Effective Choice in the Prisoner's Dilemma." Journal of Conflict Resolution 24: 3–25.

——— (1980b). "More Effective Choice in the Prisoner's Dilemma." Journal of Conflict Resolution 24: 379–403.

Axelrod, R., and W. D. Hamilton. (1981). "The Evolution of Cooperation." Science 211: 1390–1396.

Barkow, J. H. (1992). "Beneath New Culture Is Old Psychology." In J. H. Barkow, L. Cosmides, and J. Tooby (eds.), The Adapted Mind, pp. 627–637. Oxford: Oxford University Press.

Barkow, J. H., L. Cosmides, and J. Tooby, (eds.). (1992). The Adapted Mind: Evolutionary Psychology and the Generation of Culture. New York: Oxford University Press.

Barth, F. (1989). Cosmologies in the Making. Cambridge: Cambridge University Press.

Basehart, H. W. (1974). Mescalero Apache Subsistence Patterns and Sociopolitical Organization. New York: Garland Publishing.

Batra, S. W. T. (1979). Insect-Fungus Symbiosis: Nutrition, Mutualism, and Commensalism. New York: John Wiley.

Batson, C. D. (1991). The Altruism Question: Toward a Social-Psychological Answer. Hillsdale, N.J.: Lawrence Erlbaum Associates.

Batson, C. D., J. G. Batson, C. A. Griffitt, S. Barrientos, J. R. Brandt, P. Sprengelmeyer, and M. J. Bayly. (1989). "Negative-State Relief and the Empathy-Altruism Hypothesis." Journal of Personality and Social Psychology 56: 922–33.

Batson, C. D., J. G. Batson, J. K. Slingsby, K. L. Harrell, H. M. Peekna, and R. M. Todd. (1991). "Empathic Joy and the Empathy-Altruism Hypothesis." Journal of Personality and Social Psychology 61: 413–426.

Batson, C. D., J. L. Dyck, J. R. Brandt, J. G. Batson, A. L. Powell, M. R. McMaster, and C. A. Griffitt. (1988). "Five Studies Testing Two New Egoistic Alternatives to the Empathy-Altruism Hypothesis." Journal of Personality and Social Psychology 55: 52–77.

Batson, C. D., J. Fultz, P. A. Schoenrade, and A. Paduano. (1987). "Critical Self-Reflection and Self-Perceived Altruism: When Self-Reward Fails." Journal of Personality and Social Psychology 53: 594–602.

Batson, C. D., and L. Shaw. (1991). "Evidence for Altruism: Toward a Pluralism of Prosocial Motives." Psychological Inquiry 2: 107–122.

Bentham, J. (1789). An Introduction to the Principles of Morals and Legislation. University of London: The Athlone Press, 1970.

Bernal Villa, S. (1953). "Aspects of Paez Culture: The Fiesta of San Juan in Calderas, Tierradentro." Revista Colombiana de Antropología 1: 177–221.

Betzig, L. (1997). Human Nature: A Critical Reader. Oxford: Oxford University Press.

Betzig, L., M. Borgerhoff Mulder, and P. Turke. (1988). Human Reproductive Behavior: A Darwinian Perspective. Cambridge: Cambridge University Press.

Bhagwat, A. L., and J. V. Craig. (1977). "Selection for Age at First Effect on Social Dominance." Poultry Science 56: 362–363.

Blakemore, R. P., and R. B. Frankel. (1981). "Magnetic Navigation in Bacteria." Scientific American 245: 58–65.

Block, J. (1978). The Q-sort Method. Palo Alto, Calif.: Consulting Psychologists Press.

Blurton-Jones, N. G. (1984). "A Selfish Origin for Human Food Sharing: Tolerated Theft." Ethology and Sociobiology 5: 1–3.

——— (1987). "Tolerated Theft: Suggestions About the Ecology and Evolution of Sharing, Hoarding and Scrounging." Social Science Information 26: 31–54.

Boehm, C. (1978). "Rational Preselection from Hamadryas to *Homo sapiens:* The Place of Decisions in Adaptive Process." American Anthropologist 80: 265–296.

——— (1981). "Parasitic Selection and Group Selection: A Study of Conflict Interference in Rhesus and Japanese Macaque Monkeys." In A. B. Chiarelli and R. S. Corruccini (eds.), Primate Behavior and Sociobiology. Berlin: Springer-Verlag.

——— (1983). Montenegrin Social Organization and Values: Political Ethnography of a Refuge Area Tribal Adaptation. New York: AMS Press.

——— (1984). Blood Revenge. Philadelphia: University of Pennsylvania Press.

——— (1992). "Segmentary 'Warfare' and the Management of 'Conflict' Comparison of East African Chimpanzees and Patrilineal-patrilocal Humans." In A. H. Harcourt and F. B. M. DeWaal (eds.), Coalitions and Alliances in Humans and Other Animals, pp. 139–173. Oxford: Oxford University Press.

——— (1993). "Egalitarian Society and Reverse Dominance Hierarchy." Current Anthropology 34: 227–254.

——— (1996). "Emergency Decisions, Cultural Selection Mechanics and Group Selection." Current Anthropology 37: 763–793.

——— (1997a). "Egalitarian Behavior and the Evolution of Political Intelligence." In R. W. Byrne and A. Whiten (eds.), Machiavellian Intelligence II (in press). Cambridge: Cambridge University Press.

——— (1997b). "Impact of the Human Egalitarian Syndrome on Darwinian Selection Mechanics." American Naturalist 150: S100–S121.

Bourke, A., and N. Franks. (1995). Social Evolution in Ants. Princeton: Princeton University Press.

Boyd, R., and P. J. Richerson. (1980). "Effect of Phenotypic Variation on Kin Selection." Proceedings of the National Academy of Sciences 77: 7506–7509.

——— (1985). Culture and the Evolutionary Process. Chicago: University of Chicago Press.

——— (1990a). "Group Selection among Alternative Evolutionarily Stable Strategies." Journal of Theoretical Biology 145: 331–342.

——— (1990b). "Culture and Cooperation." In J. J. Mansbridge (ed.), Beyond Self-Interest, pp. 111–132. Chicago: University of Chicago Press.

——— (1992). "Punishment Allows the Evolution of Cooperation (or Anything Else) in Sizable Groups." Ethology and Sociobiology 13: 171–195.

Brandon, R. (1990). Organism and Environment. Princeton: Princeton University Press.

Breden, F. J., and M. J. Wade. (1989). "Selection Within and Between Kin Groups of the Imported Willow Leaf Beetle." American Naturalist 134: 35–50.

Breland, K., and M. Breland. (1961). "The Misbehavior of Organisms." American Psychologist 16: 681–684.

Bremermann, H. J., and J. Pickering. (1983). "A Game-Theoretical Model of Parasite Virulence." Journal of Theoretical Biology 100: 411–426.

Broad, C. D. (1952). Ethics and the History of Philosophy. London: Routledge and Kegan Paul.

——— (1965). Five Types of Ethical Theory. Totowa, N.J.: Littlefield, Adams.

Buchanan, J. M. (1954). "Individual Choice in Voting and the Market." Journal of Political Economy 62: 334–343.

Bull, J. J. (1994). "Virulence." Evolution 48: 1423–1437.

Buss, D. M. (1994). The Evolution of Desire: Strategies of Human Mating. New York: Basic Books.

——— (1995). "Evolutionary Psychology: A New Paradigm for Psychological Science." Psychological Inquiry 6: 1–30.

Buss, L. W. (1987). The Evolution of Individuality. Princeton: Princeton University Press.

Butler, J. (1726). Fifteen Sermons upon Human Nature. Reprinted in L. A. Selby-Bigge (ed.), British Moralists: Being Selections from Writers Principally of the Eighteenth Century, vol. 1, pp. 180–241. New York: Dover Books, 1965.

Camacho, J. P. M., M. W. Shaw, M. D. Lopez-Leon, M. C. Pardo, and J. Cabrero. (1997). "Population Dynamics of a Selfish B Chromosome Neutralized by the Standard Genome in the Grasshopper *Eyprepocnemis plorans.*" American Naturalist 149: 1030–1050.

Campbell, D. T. (1974). "Evolutionary Epistemology." In P. A. Schilpp (ed.), The Philosophy of Karl Popper. LaSalle, Ill.: Open Court Publishing.

——— (1994). "How Individual and Face-to-Face-Group Selection Undermine Firm Selection in Organizational Evolution." In J. A. C. Baum and J. V. Singh (eds.), Evolutionary Dynamics of Organizations, pp. 23–38. New York: Oxford University Press.

Cartwright, N. (1978). "Causal Laws and Effective Strategies." Nous 13: 419–437.

Cavalli-Sforza, L. L., and M. W. Feldman. (1981). Cultural Transmission and Evolution: A Quantitative Approach. Princeton: Princeton University Press.

Charnov, E. L. (1982). The Theory of Sex Allocation. Princeton: Princeton University Press.

Cheney, D. L., and R. M. Seyfarth. (1990). How Monkeys See the World: Inside the Mind of Another Species. Chicago: University of Chicago Press.

Churchland, P. M. (1984). Matter and Consciousness: A Contemporary Introduction to the Philosophy of Mind. Cambridge, Mass.: MIT Press/A Bradford Book.

Cialdini, R. B., M. Schaller, D. Houlihan, K. Arps, J. Fultz, and A. L. Beaman. (1987). "Empathy-Based Helping: Is It Selflessly or Selfishly Motivated?" Journal of Personality and Social Psychology 52: 749–758.

Clark, A. B. (1978). "Sex Ratio and Local Resource Competition in a Prosimian Primate." Science 201: 163–165.

Cohan, F. M. (1984). "Can Uniform Selection Retard Random Genetic Divergence between Isolated Conspecific Populations?" Evolution 38: 495–504.

Colwell, R. K. (1981). "Group Selection Is Implicated in the Evolution of Female-Biased Sex Ratios." Nature 190: 401–404.

Cosmides, L., and J. Tooby. (1992). Cognitive Adaptations for Social Exchange. In J. Barkow, L. Cosmides, and J. Tooby (eds.), The Adapted Mind, pp. 163–225. New York: Academic Press.

Coyne, J. A., N. H. Barton, and M. Turelli. (1997). "Perspective: A Critique of Sewall Wright's Shifting Balance Theory of Evolution." Evolution 51: 643–671.

Craig, J. V., M. L. Jan, C. R. Polley, A. L. Bhagwat, and A. D. Dayton. (1975). "Changes in Relative Aggressiveness and Social Dominance Associated with Selection for Early Egg Production in Chickens." Poultry Science 54: 1647–1658.

Craig, J. V., and W. M. Muir. (1995). "Group Selection for Adaptation to Multiple-Hen Cages: Beak-related Mortality, Feathering, and Body Weight Responses." Poultry Science 75: 294–302.

Crane, S. (1898). "The Open Boat" In T. A. Gullason (ed.), Complete Short Stories and Sketches of Stephen Crane. Garden City, N.Y.: Doubleday, 1963.

Cronin, H. (1991). The Ant and the Peacock: Altruism and Sexual Selection from Darwin to Today. Cambridge: Cambridge University Press.

Crow, J. F. (1979). "Genes That Violate Mendel's Rules." Scientific American 240 (2): 134–146.

Daly, M., and M. Wilson. (1988). Homicide. Hawthorne, N.Y.: Aldine De Gruyter.

Darwin, C. (1859). On the Origin of Species. Cambridge, Mass.: Harvard University Press, 1964.

——— (1871). The Descent of Man and Selection in Relation to Sex. London: Murray.

Davis, M. H., C. Luce, and S. J. Kraus. (1994). "The Heritability of Characteristics Associated with Dispositional Empathy." Journal of Personality 62: 369–391.

Dawkins, R. (1976). The Selfish Gene. New York: Oxford University Press.

——— (1980). "Good Strategy or Evolutionary Stable Strategy?" In G. W. Barlow and J. Silverberg (eds.), Sociobiology: Beyond Nature/Nurture?, pp. 331–367. Boulder, Colo.: Westview Press.

——— (1982). The Extended Phenotype: The Long Reach of the Gene. New York: Oxford University Press.

——— (1989). The Selfish Gene, 2nd edition. New York: Oxford University Press.

Dennett, D. C. (1975). "Why the Law of Effect Will Not Go Away." Journal of the Theory of Social Behavior, 5 (2): 169–187. Reprinted in Brainstorms: Philosophical Essays on Mind and Psychology, pp. 71–89. Cambridge, Mass.: MIT Press/A Bradford Book, 1978.

Denniston, C. (1978). "An Incorrect Definition of Fitness Revisited." Annals of Human Genetics 42: 77–85.

Dickemann, M. (1979). "Female Infanticide, Reproductive Social Stratification: A Preliminary Model." In N. Chagnon and N. Irons (eds.), Evolutionary Biology and Human Social Behavior: An Anthropological Perspective, pp. 321–367. North Scituate, Mass.: Duxbury Press.

Dobzhansky, T. (1937). Genetics and the Origin of Species. New York: Columbia University Press.

Dretske, F. (1981). Knowledge and the Flow of Information. Cambridge, Mass.: The MIT Press.

Drucker, P. (1951). The Northern and Central Nootkan Tribes. Washington, D.C.: Government Printing Office.

Dugatkin, L. A., and M. Alfieri. (1991a). "Tit-for-Tat in Guppies *(Poecilia reticulata):* The Relative Nature of Cooperation and Defection during Predator Inspection." Evolutionary Ecology 5: 300–309.

——— (1991b). "Guppies and the Tit for Tat Strategy: Preference Based on Past Interaction." Behavioral Ecology and Sociobiology 28: 243–246.

Dugatkin, L. A., and H. K. Reeve. (1994). "Behavioral Ecology and Levels of Selection: Dissolving the Group Selection Controversy." Advances in the Study of Behavior 23: 101–133.

Dunbar, R. (1982). "Adaptation, Fitness, and the Evolutionary Tautology."

In King's College Sociobiology Group (eds.), Current Problems in Sociobiology. Cambridge: Cambridge University Press.

Durham, W. H. (1991). Coevolution: Genes, Culture and Human Diversity. Palo Alto, Calif.: Stanford University Press.

Ehrenpreis, A. (1650). "An Epistle on Brotherly Community as the Highest Command of Love." In R. Friedmann (ed.), Brotherly Community: The Highest Command of Love, pp. 9–77. Rifton, N.Y.: Plough Publishing Co., 1978.

Eibl-Eibesfeldt, I. (1982). "Warfare, Man's Indocrinability and Group Selection." Zeitschrift für Tierpsychologie 60: 177–198.

Eigen, M., and P. Schuster. (1977). "The Hypercycle: A Principle of Natural Self-Organization. A: Emergence of the Hypercycle." Naturwissenschaften 64: 541–565.

——— (1978a). "The Hypercycle: A Principle of Natural Self-Organization. B: The Abstract Hypercycle." Naturwissenschaften 65: 7–41.

——— (1978b). "The Hypercycle: A Principle of Natural Self-Organization. C: The Realistic Hypercycle." Naturwissenschaften 65: 341–369.

Eisenberg, N. (1986). Altruistic Emotion, Cognition, and Behavior. Hillsdale, N.J.: Lawrence Erlbaum Associates.

Eisenberg, N., and R. A. Fabes. (1991). "Prosocial Behavior and Empathy: A Multimethod Developmental Perspective." In M. S. Clark (ed.), Prosocial Behavior, pp. 34–61. Newbury Park, Calif.: Sage Publications.

Eisenberg, N., R. A. Fabes, B. Murphy, M. Karbon, P. Maszk, M. Smith, C. O'Bouyle, and K. Suh. (1994). "The Relations of Emotionality and Regulation to Dispositional and Situational Empathy-related Responding." Journal of Personality and Social Psychology 66: 776–797.

Eisenberg, N., R. Lennon and K. Roth. (1983). "Prosocial Development: A Longitudinal Study." Developmental Psychology 19: 846–855.

Eisenberg, N., and P. A. Miller. (1987). "The Relation of Empathy to Prosocial and Related Behaviors." Psychological Bulletin 101: 91–119.

Eisenberg, N., and J. Strayer. (1987). "Critical Issues in the Study of Empathy." In N. Eisenberg and J. Strayer (eds.), Empathy and Its Development. New York: Cambridge University Press.

Ellickson, R. C. (1991). Order without Law: How Neighbors Settle Disputes. Cambridge, Mass.: Harvard University Press.

Enç, B. (1996). "Hume's Unreasonable Passions." History of Philosophy Quarterly 13: 239–254.

Endler, J. A. (1986). Natural Selection in the Wild. Princeton: Princeton University Press.

Evans-Pritchard, E. E. (1940). The Nuer: A Description of the Modes of

Livelihood and Political Institutions of a Nilotic People. Oxford: Oxford University Press.

Ewald, P. W. (1993). Evolution of Infectious Disease. Oxford: Oxford University Press.

Falconer, D. S. (1981). Introduction to Quantitative Genetics, 2nd edition. London: Longman.

Feinberg, J. (1984). "Psychological Egoism." In S. Cahn, P. Kitcher, and G. Sher (eds.), Reason at Work, pp. 25–35. San Diego: Harcourt Brace and Jovanovich.

Fields, H. (1987). Pain. New York: McGraw-Hill.

Findlay, C. S. (1992). "Phenotypic Evolution under Gene-Culture Transmission in Structured Populations." Journal of Theoretical Biology 156: 387–400.

Fodor, J. (1983). Modularity of Mind: An Essay on Faculty Psychology. Cambridge, Mass.: MIT Press/A Bradford Book.

Forster, M., and E. Sober. (1994). "How to Tell When Simpler, More Unified, and Less Ad Hoc Theories Will Provide More Accurate Predictions." British Journal for the Philosophy of Science 45: 1–35.

Fortes, M. (1945). The Dynamics of Clanship among the Tallensi; Being the First Part of an Analysis of the Social Structure of a Trans-Volta Tribe. London: Oxford University Press for the International African Institute.

Fox, R. (1975). Biosocial Anthropology. London: Malaby Press.

Frank, R. H. (1988). Passions within Reason: The Strategic Role of the Emotions. New York: W. W. Norton.

——— (1994). "Group Selection and 'Genuine' Altruism." Behavioral and Brain Sciences 17: 620–621.

Frank, R. H., T. Gilovich, and D. Regan. (1993). "Does Studying Economics Inhibit Cooperation?" Journal of Economic Perspectives 7: 159–171.

Frank, S. A. (1986). "Hierarchical Selection and Sex Ratios. I. General Solutions for Structured Populations." Theoretical Population Biology 29: 312–342.

——— (1991). "Divergence of Meiotic Drive-Suppression Systems as an Explanation for Sex-Biased Hybrid Sterility and Inviability." Evolution 45: 262–267.

——— (1994). "Genetics of Mutualism: The Evolution of Altruism between Species." Journal of Theoretical Biology 170: 393–400.

——— (1995a). "George Price's Contributions to Evolutionary Genetics." Journal of Theoretical Biology 175: 373–388.

——— (1995b). "Mutual Policing and Repression of Competition in the Evolution of Cooperative Groups." Nature 377: 520–522.

——— (1995c). "The Origin of Synergistic Symbiosis." Journal of Theoretical Biology 176: 403–410.
——— (1996a). "Models of Parasite Virulence." Quarterly Review of Biology 71: 339–344.
——— (1996b). "Host Control of Symbiont Transmission: The Separation of Symbionts into Germ and Soma." American Naturalist 148: 1113–1124.
——— (1996c). "Host-Symbiont Conflict over the Mixing of Symbiotic Lineages." Proceedings of the Royal Academy of London B 263: 339–334.
——— (1997). "Models of Symbiosis." American Naturalist 150: S80–S89.
Fulton, R. M. (1969). The Kpelle of Liberia: A Study of Political Change in the Liberian Interior. Ph.D. dissertation, University of Connecticut.
Gallistel, C. R. (1980). The Organization of Action: A New Synthesis. Hillsdale, N.J.: Lawrence Erlbaum Associates.
Garcia, J., and R. Koelling. (1966). "Relation of Cue to Consequence in Avoidance Learning." Psychonomic Science 4: 123–124.
Ghiselin, M. (1974). The Economy of Nature and the Evolution of Sex. Berkeley: University of California Press.
Gibbard, A. (1990). Wise Choices, Apt Feelings. Cambridge, Mass.: Harvard University Press.
Gleick, J. (1987). Chaos: The Making of a New Science. New York: Penguin Books.
Godelier, M. (1986). The Making of Great Men: Male Dominance and Power among the New Guinea Baruya. Cambridge: Cambridge University Press.
Goldberg, R. (1940). The Best of Rube Goldberg. Englewood Cliffs, N.J.: Prentice Hall.
Golden, M. (1981). "Demography and the Exposure of Girls at Athens." Phoenix 35: 316–331.
Goodenough, W. H. (1951). Property, Kin and Community on Truk. New Haven, Conn.: Yale University Press.
Goodnight, C. J. (1989). "Population Differentiation and the Correlation among Traits at the Population Level." American Naturalist 133: 888–900.
——— (1990a). "Experimental Studies of Community Evolution, I: The Response to Selection at the Community Level." Evolution 44: 1614–1624.
——— (1990b). "Experimental Studies of Community Evolution, II: The Ecological Basis of the Response to Community Selection." Evolution 44: 1625–1636.

Goodnight, C. J., J. M. Schwartz, and L. Stevens. (1992). "Contextual Analysis of Models of Group Selection, Soft Selection, Hard Selection, and the Evolution of Altruism." American Naturalist 140: 743–761.

Goodnight, C. J., and L. Stevens. (1997). "Experimental Studies of Group Selection: What They Tell Us about Group Selection in Nature." American Naturalist 150: S59–S79.

Goodnight, K. F. (1992). "The Effect of Stochastic Variation on Kin Selection in a Budding-Viscous Population." American Naturalist 140: 1028–1040.

Gould, S. J. (1982). "The Uses of Heresy: An Introduction to Richard Goldschmidt's 'The Material Basis of Evolution.'" In R. Goldschmidt, The Material Basis of Evolution, pp. xiii–xlii. New Haven, Conn.: Yale University Press.

Gould, S. J., and R. C. Lewontin. (1979). "The Spandrels of San Marco and the Panglossian Paradigm: A Critique of the Adaptationist Programme." Proceedings of the Royal Society of London B205: 581–598. Reprinted in E. Sober (ed.), Conceptual Issues in Evolutionary Biology. Cambridge, Mass.: MIT Press, 1994.

Grafen, A. (1984). "Natural Selection, Kin Selection and Group Selection." In J. Krebs and N. Davies (eds.), Behavioral Ecology: An Evolutionary Approach, pp. 62–84. Oxford: Blackwell Scientific Publications.

Grant, P. R. (1986). Ecology and Evolution of Darwin's Finches. Princeton: Princeton University Press.

Grattan, F. J. H. (1948). An Introduction to Samoan Custom. Apia, Western Samoa: Samoa Printing and Publishing Co.

Griffing, B. (1977). "Selection for Populations of Interacting Genotypes." In E. Pollack, O. Kempthorne, and T. B. Bailey (eds.), Proceedings of the International Congress on Quantitative Genetics, August 16–21, 1976, pp. 413–434. Ames: Iowa State University Press.

Grusec, J. E. (1991). "The Socialization of Empathy." In M. Clark (ed.), Prosocial Behavior, pp. 9–33. Newbury Park, Calif.: Sage Publications.

Hamilton, W. D. (1963). "The Evolution of Altruistic Behavior." American Naturalist 97: 354–356.

——— (1964a). "The Genetical Evolution of Social Behavior I." Journal of Theoretical Biology 7: 1–16.

——— (1964b). "The Genetical Evolution of Social Behavior II." Journal of Theoretical Biology 7: 17–52.

——— (1967). "Extraordinary Sex Ratios." Science 156: 477–488.

——— (1971a). "Selection of Selfish and Altruistic Behavior in Some Extreme Models." In J. F. Eisenberg and W. S. Dillon (eds.), Man and

Beast: Comparative Social Behavior, pp. 57–91. Washington, D.C.: Smithsonian Institution Press.
——— (1971b). "Geometry for the Selfish Herd." Journal of Theoretical Biology 31: 295–311.
——— (1975). "Innate Social Aptitudes of Man: An Approach from Evolutionary Genetics." In R. Fox (ed.), Biosocial Anthropology, pp. 133–155. New York: John Wiley and Sons.
——— (1979). "Wingless and Fighting Males in Fig Wasps and Other Insects." In M. S. Blum and N. A. Blum (eds.), Sexual Selection and Reproductive Competition in Insects, pp. 167–220. New York: Academic Press.
——— (1987). "Discriminating Nepotism: Expectable, Common, Overlooked." In D. J. C. Fletcher and C. D. Michener (eds.), Kin Recognition in Animals, pp. 417–437. New York: John Wiley and Sons.
——— (1996). The Narrow Roads of Gene Land. Oxford: W. H. Freeman/Spektrum.
Harsanyi, J. (1955). "Cardinal Welfare, Individualistic Ethics, and Interpersonal Comparisons of Utility." Journal of Political Economy 63: 315–321.
Hawkes, K. (1993). "Why Hunter-Gatherers Work: An Ancient Version of the Problem of Public Goods." Current Anthropology 34: 341–361.
Heckathorn, D. D. (1990). "Collective Sanctions and Compliance Norms: A Formal Theory of Group-Mediated Social Control." American Sociological Review 55: 366–384.
——— (1993). "Collective Action and Group Heterogeneity: Voluntary Provision vs. Selective Incentives." American Sociological Review 58: 329–350.
Hegel, G. W. F. (1807). The Phenomenology of Mind. Translated by J. B. Baillie. New York: Harper and Row, 1967.
Heisler, I. L., and J. Damuth. (1987). "A Method of Analyzing Selection in Hierarchically Structured Populations." American Naturalist 130: 582–602.
Henry, J. (1951). "The Economics of Pilaga Food Distribution." American Anthropologist 53: 187–219.
Henson, R. (1988). "Butler on Selfishness and Self-Love." Philosophy and Phenomenological Research 49: 31–57.
Herre, E. A. (1993). "Population Structure and the Evolution of Virulence in Nematode Parasites of Fig Wasps." Science 259: 1442–1446.
Hineline, P., and H. Rachlin. (1969). "Escape and Avoidance of Shock by Pigeons Pecking a Key." Journal of the Experimental Analysis of Behavior 12: 533–538.

Hirshleifer, J. 1987. "On the Emotions as Guarantors of Threats and Promises." In J. Dupré (ed.), The Latest on the Best: Essays on Evolution and Optimality, pp. 307–326. Cambridge, Mass: MIT Press.

Hoffman, M. L. (1976). "Empathy, Role-Taking, Guilt, and Development of Altruistic Motives." In T. Lickona (ed.), Moral Development and Behavior: Theory, Research, and Social Issues, pp. 124–143. New York: Holt, Rinehart and Winston.

——— (1981a). "Is Altruism Part of Human Nature?" Journal of Personality and Social Psychology 40: 121–137.

——— (1981b). "The Development of Empathy." In J. P. Rushton and R. M. Sorrentino (eds.), Altruism and Helping Behavior: Social Personality and Developmental Perspectives, pp. 41–63. Hillsdale, N.J.: Lawrence Erlbaum Associates.

——— (1991). "Is Empathy Altruistic?" Psychological Inquiry 2: 131–133.

Hoijer, H., and M. E. Opler. (1938). Chiricahua and Mescalero Apache Texts, by Harry Hoijer, with Ethnological Notes by Morris Edward Opler. Chicago: University of Chicago Press.

Holekamp, K. E., and L. Smale. (1995). "Rapid Change in Offspring Sex Ratios after Clan Fission in the Spotted Hyena." American Naturalist 145: 261–278.

Hornstein, H. (1991). "Empathic Distress and Altruism: Still Inseparable." Psychological Inquiry 2: 133–135.

Hostetler, J. A. (1980). Amish Society, 3rd edition. Baltimore: Johns Hopkins University Press.

Howell, S. (1984). Society and Cosmos: Chewong of Peninsular Malaya. Singapore: Oxford University Press.

Hrdy, S. (1981). The Woman That Never Evolved. Cambridge, Mass.: Harvard University Press.

Hudson, R. A. (1980). Sociolinguistics. Cambridge: Cambridge University Press.

Hull, D. (1980). "Individuality and Selection." Annual Review of Ecology and Systematics 11: 311–332.

Hume, D. (1739). A Treatise of Human Nature. New York: Oxford University Press, 1978.

——— (1751). An Enquiry Concerning the Principles of Morals. Indianapolis: Hackett, 1970.

Hurst, L. D., Atlan, A., and Bengtsson, B. O. (1996). "Genetic Conflicts." Quarterly Review of Biology 71: 317–364.

James, W. (1890). The Principles of Psychology. Cambridge, Mass.: Harvard University Press, 1981.

Jorgenson, J. (1967). Salish Language and Culture. Bloomington: Indiana University Press.

Joseph, A., R. B. Spicer, and J. Chesky. (1949). The Desert People: A Study of the Papago Indians. Chicago: Chicago University Press.

Kahnemann, D., P. Slovic and A. Tversky (eds.). (1982). Judgment under Uncertainty: Heuristics and Biases. New York: Cambridge University Press.

Kaplan, H., and K. Hill. (1985a). "Hunting Ability and Reproductive Success among Male Ache Foragers." Current Anthropology 26: 131–133.

——— (1985b). "Food Sharing among Ache Foragers: Tests of Explanatory Hypotheses." Current Anthropology 26: 223–245.

Kaplan, H., K. Hill, and A. Hurtado. (1984). "Food Sharing among the Ache Hunter-Gatherers of Eastern Paraguay." Current Anthropology 25: 113–115.

Karniol, R. (1982). "Settings, Scripts, and Self-Schemata: A Cognitive Analysis of the Development of Prosocial Behavior." In N. Eisenberg (ed.), The Development of Prosocial Behavior, pp. 251–278. New York: Academic Press.

Karsten, S. R. (1923). The Toba Indians of the Bolivian Gran Chaco. Abo, Finland: Abo Akademi.

Kavka, G. (1986). Hobbesian Moral and Political Theory. Princeton: Princeton University Press.

Kelly, J. K. (1992). "The Evolution of Altruism in Density Regulated Populations." Journal of Theoretical Biology 157: 447–461.

Kelly, R. (1985). The Nuer Conquest: The Structure and Development of an Expansionist System. Ann Arbor: University of Michigan Press.

Kettlewell, H. B. D. (1973). The Evolution of Melanism. Oxford: Clarendon.

Kitcher, P. (1993). "The Evolution of Human Altruism." Journal of Philosophy 110: 497–516.

Kluckhohn, C. (1952). "Values and Value-Orientations in the Theory of Action: An Exploration in Definition and Classification." In T. Parsons and E. Shils (eds.), Toward a General Theory of Action, pp. 395–418. Cambridge, Mass.: Harvard University Press.

Knauft, B. M. (1991). "Violence and Sociality in Human Evolution." Current Anthropology 32: 391–428.

Krebs, D. L. (1975). "Empathy and Altruism." Journal of Personality and Social Psychology 32: 1134–1146.

Kroeber, A. L. (1948). Anthropology. New York: Harcourt-Brace.

Kuhn, T. (1970). The Structure of Scientific Revolutions, 2nd edition. Chicago: University of Chicago Press.

Lacey, J. I. (1967). "Somatic Response Patterning and Stress: Some Revisions of Activation Theory." In M. Appley and R. Trumbell (eds.), Psychological Stress, Issues in Research, pp. 14–37. New York: Appleton-Century-Crofts.

LaFollette, H. (1988). "The Truth in Psychological Egoism." In J. Feinberg (ed.), Reason and Responsibility, 7th edition, pp. 500–507. Belmont, Calif.: Wadsworth.

Lamphere, L. (1977). To Run after Them: Cultural and Social Bases of Cooperation in a Navajo Community. Tucson: University of Arizona Press.

Lancy, D. F. (1975). Work, Play and Learning in a Kpelle Town. Ph.D. dissertation, University of Pittsburgh.

Latané, B., and J. M. Darley. (1970). The Unresponsive Bystander: Why Doesn't He Help? New York: Appleton-Century-Crofts.

Latané, B., S. A. Nida and D. W. Wilson. (1981). "The Effects of Group Size on Helping Behavior." In J. P. Rushton and R. M. Sorrentino (eds.), Altruism and Helping Behavior: Social Personality and Developmental Perspectives, pp. 287–313. Hillsdale, N.J.: Lawrence Erlbaum Associates.

Leigh, E. G. J. (1977). "How Does Selection Reconcile Individual Advantage with the Good of the Group?" Proceedings of the National Academy of Sciences 74: 4542–4546.

Levine, D. N. (1965). Wax and Gold: Tradition and Innovation in Ethiopian Culture. Chicago: University of Chicago Press.

Lewontin, R. C. (1970). "The Units of Selection." Annual Review of Ecology and Systematics 1: 1–18.

Lifton, R. J. (1986). The Nazi Doctors: Medical Killing and the Psychology of Genocide. New York: Basic Books.

Lloyd, E. (1988). The Structure and Confirmation of Evolutionary Theory. New York: Greenwood Press.

Lorenz, K. (1965). Evolution and the Modification of Behavior. Chicago: Chicago University Press.

Lyttle, T. W. (1991). "Segregation Distorters." Annual Review of Genetics 25: 511–557.

Macaulay, J., and L. Berkowitz. (1970). Altruism and Helping Behavior. New York: Academic Press.

MacDonald, K. B. (1994). A People That Shall Dwell Alone: Judaism as a Group Evolutionary Strategy . Westport, Conn.: Praeger.

Macpherson, C. (1962). The Political Theory of Possessive Individualism. Oxford: Clarendon Press.

Main, M., N. Kaplan, and J. Cassidy. (1985). "Security in Infancy, Childhood and Adulthood: A Move to the Level of Representation." Monographs of the Society for Research in Child Development 50: 66–104.

Mansbridge, J. (ed.). (1990). Beyond Self-Interest. Chicago: University of Chicago Press.

Margolis, H. (1982). Selfishness, Altruism, and Rationality: A Theory of Social Choice. New York: Cambridge University Press.

Margulis, L. (1970). Origin of the Eukaryotic Cells. New Haven, Conn.: Yale University Press.

Maynard Smith, J. (1964). "Group Selection and Kin Selection." Nature 201: 1145–1146.

——— (1976). "Group Selection." Quarterly Review of Biology 51: 277–283.

——— (1982). Evolution and the Theory of Games. New York: Cambridge University Press.

——— (1987a). "How To Model Evolution." In J. Dupré (ed.), The Latest on the Best: Essays on Evolution and Optimality, pp. 119–131. Cambridge, Mass.: MIT Press/A Bradford Book.

——— (1987b). "Reply to Sober." In J. Dupré (ed.), The Latest on the Best: Essays on Evolution and Optimality, pp. 147–150. Cambridge, Mass.: MIT Press/A Bradford Book.

Maynard Smith, J., and E. Szathmary. (1995). The Major Transitions of Life. New York: W. H. Freeman.

Mayr, E. (1961). "Cause and Effect in Biology." Science 134: 1501–1506.

McCauley, D. E., and M. J. Wade. (1980). "Group Selection: The Genotypic and Demographic Basis for the Phenotypic Differentiation of Small Populations of *Tribolium castaneum*." Evolution 34: 813–821.

McCawley, J. P. (1988). The Syntactic Phenomena of English, volume 1. Chicago: University of Chicago Press.

McKim, F. (1947). "San Blas: An Account of the Cuna Indians of Panama." In H. Wassen (ed.), The Forbidden Land: Reconnaissance of Upper Bayano River, R.P., in 1936, p. 186. Goteborg: Etnografiska Museet.

McNeilly, F. S. (1968). The Anatomy of Leviathan. New York: St. Martin's Press.

Mead, M. (1930). Social Organization of Manua. Honolulu: Pernice P. Bishop Museum.

Melzack, R., and P. D. Wall. (1983). The Challenge of Pain. New York: Basic Books.

Michod, R. E. (1982). "The Theory of Kin Selection." Annual Review of Ecology and Systematics 13: 23–55.

——— (1983). "Population Biology of the First Replicators: On the Origin of the Genotype, Phenotype and Organism." American Zoologist 23: 5–14.

——— (1996). "Cooperation and Conflict in the Evolution of Individuality. II. Conflict Mediation." Proceedings of the Royal Society of London B, 263: 813–822.

——— (1997a). "Cooperation and Conflict in the Evolution of Individuality.

I. Multilevel Selection of the Organism." American Naturalist 149: 607–645.

——— (1997b). "Evolution of the Individual." American Naturalist 150: S5–S21.

Mill, J. S. (1874). A System of Logic. New York: Harper, 1900.

Mitman, G. (1992). The State of Nature: Ecology, Community and American Social Thought, 1900–1950. Chicago: University of Chicago Press.

Moss, M. K., and R. A. Page. (1972). "Reinforcement and Helping Behavior." Journal of Applied Psychology 2: 360–371.

Muir, W. M. (1995). "Group Selection for Adaptation to Multiple-Hen Cages: Selection Program and Direct Responses." Poultry Science 75: 447–458.

Murdock, G. P. (1967). Ethnographic Atlas. Pittsburgh: University of Pittsburgh Press.

Nagel, T. (1970). The Possibility of Altruism. Oxford: Oxford University Press.

——— (1986). The View from Nowhere. Oxford: Oxford University Press.

Naik, T. B. (1956). The Bhils: A Study. Delhi: Bharatiya Adimjati Sevak Sangh.

Nesse, R. M., and G. C. Williams. (1994). Why We Get Sick: The New Science of Darwinian Medicine. New York: Times Books.

Nisbett, R., and T. D. Wilson. (1977). "Telling More Than We Can Know: Verbal Reports on Mental Processes." Psychological Review 84: 231–259.

Nozick, R. (1974). Anarchy, State, and Utopia. New York: Basic Books.

Obrist, P. A., R. A. Webb, J. R. Sutterer, and J. L. Howard. (1970). "The Cardiac-Somatic Relationship: Some Reformulations." Psychophysiology 6: 569–587.

Ohmman, A., and U. Dimberg. (1978). "Facial Expressions as Conditional Stimuli for Electrodermal Responses: A Case of 'Preparedness.'" Journal of Personality and Social Psychology 36: 1251–1258.

Oldenquist, A. (1980). "The Possibility of Selfishness." American Philosophical Quarterly 17: 25–33.

Oliner, S. P., and P. M. Oliner. (1988). The Altruistic Personality: Rescuers of Jews in Nazi Europe. New York: Free Press.

Orzack, S., E. Parker, and J. Gladstone. (1991). "The Comparative Biology of Genetic Variation for Conditional Sex Ratio Adjustment in a Parasitic Wasp, *Nasonia vitripennis.*" Genetics 127: 583–599.

Orzack, S., and E. Sober. (1994). "Optimality Models and the Test of Adaptationism." American Naturalist 143: 361–380.

Parfit, D. (1984). Reasons and Persons. Oxford: Oxford University Press.

Parker, I. (1996). "Richard Dawkins's Evolution." The New Yorker, September 9, 1996: 41–45.
Peck, J. R. (1992). "Group Selection, Individual Selection, and the Evolution of Genetic Drift." Journal of Theoretical Biology 159: 163–187.
Penelhum, T. (1985). Butler. London: Routledge and Kegan Paul.
Perry, J. (1979). "The Problem of the Essential Identical." Noûs 13: 3–21.
Piaget, J., and B. Inhelder. (1971). Mental Imagery in the Child. New York: Basic Books.
Pianka, E. (1983). Evolutionary Ecology, 3rd edition. New York: Harper and Row.
Piliavin, J., and P. Callero. (1991). Giving Blood—The Development of an Altruistic Identity. Baltimore: Johns Hopkins University Press.
Piliavin, J., and H. Charng. (1990). "Altruism—A Review of Recent Theory and Research." Annual Review of Sociology 16: 27–65.
Pollock, G. B. (1983). "Population Viscosity and Kin Selection." American Naturalist 122: 817–829.
Popper, K. (1959). The Logic of Scientific Discovery. London: Hutchinson.
Powdermaker, H. (1933). Life in Lesu: The Study of a Melanesian Society in New Ireland. New York: Norton.
Price, G. R. (1970). "Selection and Covariance." Nature 277: 520–521.
——— (1972). "Extension of Covariance Selection Mathematics." Annals of Human Genetics 35: 485–490.
Provine, W. B. (1986). Sewall Wright and Evolutionary Biology. Chicago: University of Chicago Press.
Queller, D. C. (1991). "Group Selection and Kin Selection." Trends in Ecology and Evolution 6: 64.
——— (1992a). "Quantitative Genetics, Inclusive Fitness and Group Selection." American Naturalist 139: 540–558.
——— (1992b). "Does Population Viscosity Promote Kin Selection?" Trends in Ecology and Evolution 7: 322–324.
Rachman, S. J. (1990). Fear and Courage, 2nd edition. New York: W. H. Freeman.
Radke-Yarrow, M., and C. Zahn-Waxler. (1984). "Roots, Motives, and Patterns in Children's Prosocial Behavior." In E. Staub, D. Bar-Tal, J. Karylowski, and J. Reykowski (eds.), Origins and Maintenance of Prosocial Behaviors, pp. 81–100. New York: Plenum.
Rapoport, A. (1991). "Ideological Commitments and Evolutionary Theory." Journal of Social Issues 47: 83–99.
Ratnieks, F. L. (1988). "Reproductive Harmony via Mutual Policing by Workers in Eusocial Hymenoptera." American Naturalist 132: 217–236.

Ratnieks, F. L., and P. K. Visscher. (1989). "Worker Policing in the Honeybee." Nature 342: 796–797.

Rawls, J. (1971). A Theory of Justice. Cambridge, Mass.: Belknap Press of Harvard University Press.

Reddy, R. (1980). "Individual Philanthropy and Giving Behavior." In D. Smith and J. Macaulay (eds.), Participation in Social and Political Activities, pp. 370–399. San Francisco: Jossey-Bass.

Ridley, M. (1993). Evolution. Oxford: Blackwell Scientific.

Rissing, S. W., G. B. Pollock, M. R. Higgins, R. H. Hagen, and D. R. Smith. (1989). "Foraging Specialization without Relatedness or Dominance among Co-founding Ant Queens." Nature 338: 420–422.

Rollo, D. C. (1994). Phenotypes: Their Epigenetics, Ecology and Evolution. London: Chapman and Hall.

Rosenhan, D. L. (1970). "The Natural Socialization of Altruistic Autonomy." In J. Macaulay and L. Berkowitz (eds.), Altruism and Helping Behavior: Social Psychological Studies of Some Antecedents and Consequences, pp. 251–268. New York: Academic Press.

Ross, L., and R. Nisbett. (1991). The Person and the Situation—Perspectives on Social Psychology. New York: McGraw-Hill.

Rousseau, J. J. (1767). A Treatise on the Social Contract. London: Beckett and DeHondt.

Runes, D. (1959). A Pictorial History of Philosophy. New York: Bramhall Press.

Rushton, J., D. Fulker, M. Neale, D. Nias, and H. Eysenek. (1986). "Altruism and Aggression: The Heritability of Individual Differences." Journal of Personality and Social Psychology 50: 1192–1198.

Schaller, M., and R. B. Cialdini. (1988). "The Economics of Empathic Helping: Support for a Mood Management Motive." Journal of Experimental Social Psychology 24: 163–181.

Schlick, M. (1939). Problems of Ethics. New York: Prentice Hall.

Schroeder, D. A., J. F. Dovidio, M. E. Sibiky, L. L. Matthews, and J. L. Allen. (1988). "Empathic Concern and Helping Behavior: Egoism or Altruism?" Journal of Experimental Social Psychology 24: 333–353.

Schroeder, D. A., L. Penner, J. F. Dovidio, and J. Piliavin. (1995). The Psychology of Helping and Altruism. New York: McGraw-Hill.

Seeley, T. D. (1985). Honeybee Ecology: A Study of Adaptation in Social Life. Princeton: Princeton University Press.

——— (1996). The Wisdom of the Hive. Cambridge, Mass.: Harvard University Press.

Segal, N. (1993). Twin, Sibling, Adoption Methods: Tests of Evolutionary Hypotheses. American Psychologist 1993: 943–956.

——— (1997). "Twin Research Perspective on Human Development." In N.

Segal, G. Weisfeld, and C. Weisfeld (eds.), Uniting Psychology and Biology: Integrative Perspective on Human Development. Washington, D.C.: American Psychological Association.
Sen, A. (1978). "Rational Fools—A Critique of the Behavioral Foundations of Economic Theory." In H. Harris (ed.), Scientific Models and Men. London: Oxford University Press.
Sevenster, P. (1973). "Incompatibility of Response and Reward." In R. A. Hinde and J. Stevenson-Hinde (eds.), Constraints on Learning, pp. 265–284. New York: Academic Press.
Shepardson, M. T. (1963). Navajo Ways in Government: A Study in Political Process. Menasha, Wis.: American Anthropological Association.
Shepardson, M. T., and B. Hammond. (1970). The Navaho Mountain Community: Social Organization and Kinship Terminology. Berkeley: University of California Press.
Sherman, P. W., J. U. M. Jarvis, and R. D. Alexander. (1991). The Biology of the Naked Mole Rat. Princeton: Princeton University Press.
Shternberg, L. I. (1933). The Gilyak, Orochi, Goldi, Negidal, Ainu: Articles and Materials. Khabarovsk: Dal'giz.
Sidgwick, H. (1907). The Methods of Ethics, 7th edition. London: Macmillan, 1922.
Simmons, R. G., S. D. Klein, and R. L. Simmons. (1977). Gift of Life: The Social and Psychological Impact of Organ Transplantation. New York: John Wiley and Sons.
Simner, M. L. (1971). "Newborn's Response to the Cry of Another Infant." Developmental Psychology 5: 136–150.
Simon, H. A. (1962). "The Architecture of Complexity." Proceedings of the American Philosophical Society 106: 467–482.
——— (1981). The Sciences of the Artificial, 2nd edition. Cambridge, Mass.: MIT Press.
——— (1983). Reason in Human Affairs. Palo Alto, Calif.: Stanford University Press.
——— (1990). "A Mechanism for Social Selection and Successful Altruism." Science 250: 1665–1668.
Simpson, E. H. (1951). "The Interpretation of Interaction in Contingency Tables." Journal of the Royal Statistical Society B 13: 238–241.
Slote, M. (1964). "An Empirical Basis for Psychological Egoism." Journal of Philosophy 61: 530–537. Reprinted, with revisions, in R. Milo (ed.), Egoism and Altruism, pp. 100–107. Belmont, Calif.: Wadsworth, 1970.
Smith, K., J. Keating, and E. Stotland. (1989). "Altruism Revisited: The Effect of Denying Feedback on a Victim's Status to Empathic Witnesses." Journal of Personality and Social Psychology 57: 641–650.
Sober, E. (1981). "Holism, Individualism, and the Units of Selection." In R.

Giere and P. Asquith (eds.), PSA 1980, volume 2, pp. 93–121. East Lansing, Mich.: Philosophy of Science Association.
——— (1984). The Nature of Selection—Evolutionary Theory in Philosophical Focus. Cambridge, Mass.: MIT Press. 2nd edition, Chicago: University of Chicago Press, 1994.
——— (1985). "Methodological Behaviorism, Evolution, and Game Theory." In J. Fetzer (ed.), Sociobiology and Epistemology, pp. 181–200. Dordecht: Reidel.
——— (1988a). "Apportioning Causal Responsibility." Journal of Philosophy 85: 303–318. Reprinted in From a Biological Point of View: Essays in Evolutionary Philosophy, pp. 184–200. New York: Cambridge University Press, 1994.
——— (1988b). Reconstructing the Past: Parsimony, Evolution, and Inference. Cambridge, Mass.: MIT Press.
——— (1990). "Let's Razor Ockham's Razor." In D. Knowles (ed.), The Limits of Explanation, pp. 73–94. New York: Cambridge University Press. Reprinted in From a Biological Point of View: Essays in Evolutionary Philosophy, pp. 136–157. New York: Cambridge University Press, 1994.
——— (1992). "Hedonism and Butler's Stone." Ethics 103: 97–103.
——— (1993a). "Epistemology for Empiricists." In P. French, T. Uehling, and H. Wettstein (eds.), Midwest Studies in Philosophy 18: 39–61. Notre Dame, Ind.: University of Notre Dame Press.
——— (1993b). Philosophy of Biology. Boulder, Colo.: Westview Press.
——— (1994a). "The Adaptive Advantage of Learning and A Priori Prejudice." In From a Biological Point of View: Essays in Evolutionary Philosophy, pp. 50–70. New York: Cambridge University Press.
——— (1994b). "Did Evolution Make Us Psychological Egoists?" In From a Biological Point of View: Essays in Evolutionary Philosophy, pp. 8–27. New York: Cambridge University Press.
——— (1998a). "Morgan's Canon." In C. Allen and D. Cummins (eds.), The Evolution of Mind. Oxford: Oxford University Press.
——— (1998b). "Three Differences between Deliberation and Evolution." In P. Danielson (ed.), Modeling Rational and Moral Agents. Oxford: Oxford University Press.
Sober, E., and R. C. Lewontin. (1982). "Artifact, Cause and Genic Selection." Philosophy of Science 49: 157–180.
Soren, D., A. Khaden, and H. Slim. (1990). Carthage. New York: Simon and Shuster.
Stampe, D. (1994). "Desire." In S. Guttenplan (ed.), A Companion to the Philosophy of Mind, pp. 244–250. Cambridge, Mass.: Basil Blackwell.

Steele, E. J. (1979). Somatic Selection and Adaptive Evolution, 2nd ed. Chicago: University of Chicago Press.

Sterelny, K. (1996). Return of the Group. Philosophy of Science 63: 562–584.

Sterelny, K., and P. Kitcher. (1988). "The Return of the Gene." Journal of Philosophy 85: 339–361.

Stewart, R. M. (1992). "Butler's Argument against Psychological Hedonism." Canadian Journal of Philosophy 22: 211–221.

Stich, S. (1983). From Folk Psychology to Cognitive Science: The Case against Belief. Cambridge, Mass.: MIT Press.

Stocker, M. (1989). Plural and Conflicting Values. New York: Oxford University Press.

Stotland, E. (1969). "Exploratory Investigations of Empathy." In L. Berkowitz (ed.), Advances in Experimental Social Psychology 4, pp. 271–314. New York: Academic Press.

Sturgeon, N. L. (1974). "Altruism, Solipsism, and the Objectivity of Reasons." Philosophical Review 83: 374–402.

Taylor, P. D. (1988). "An Inclusive Fitness Model of Dispersal of Offspring." Journal of Theoretical Biology 130: 363–378.

——— (1992). "Altruism in Viscous Populations—An Inclusive Fitness Model." Evolutionary Ecology 6: 352–356.

Tetlock, P. E., R. S. Peterson, C. McGuire, S. Chang, and P. Feld. (1992). "Assessing Political Group Dynamics: A Test of the Groupthink Model." Journal of Personality and Social Psychology 63: 403–425.

Thomas, L. (1972). "Notes of a Biology-Watcher: Germs." The New England Journal of Medicine 287: 553–555.

Thompson, R. A. (1987). "Empathy and Emotional Understanding: The Early Development of Empathy." In N. Eisenberg and J. Strayer (eds.), Empathy and Its Development, pp. 119–145. New York: Cambridge University Press.

Titherington, G. W. (1927). The Raik Dinka of Bahr el Ghazal Province. Sudan Notes and Records 10: 159–209.

Tooby, J., and L. Cosmides. (1992). "The Psychological Foundations of Culture." In J. H. Barkow, L. Cosmides, and J. Tooby (eds.), The Adapted Mind: Evolutionary Psychology and the Generation of Culture, pp. 19–136. New York: Oxford University Press.

Trivers, R. L. (1971). "The Evolution of Reciprocal Altruism." Quarterly Review of Biology 46: 35–57.

Turnbull, C. M. (1965). The Mbuti Pygmies: An Ethnographic Survey. New York: American Museum of Natural History.

Uyenoyama, M. K., and M. W. Feldman. (1980). "Theories of Kin and

Group Selection: A Population Genetics Perspective." Theoretical Population Biology 17: 380–414.
van Schaik, C. P., and S. B. Hrdy. (1991). "Intensity of Local Resource Competition Shapes the Relationship between Maternal Rank and Sex Ratios at Birth in Cercophithecine Primates." American Naturalist 138: 1555–1562.
Von Neumann, J., and O. Morgenstern. (1947). Theory of Games and Economic Behavior. Princeton: Princeton University Press.
Wade, M. J. (1976). "Group Selection among Laboratory Populations of *Tribolium*." Proceedings of the National Academy of Sciences 73: 4604–4607.
——— (1977). "An Experimental Study of Group Selection." Evolution 31: 134–153.
——— (1978). "A Critical Review of Models of Group Selection." Quarterly Review of Biology 53: 101–114.
——— (1979). "The Primary Characteristics of *Tribolium* Populations Group Selected for Increased and Decreased Population Size." Evolution 33: 749–764.
——— (1985). "Soft Selection, Hard Selection, Kin Selection and Group Selection." American Naturalist 125: 61–73.
Wallach, L., and M. Wallach. (1991). "Why Altruism, Even Though It Exists, Cannot Be Demonstrated by Social Psychological Experiments." Psychological Inquiry 2: 153–155.
Wegner, D. M. (1986). "Transactive Memory: A Contemporary Analysis of the Group Mind." In B. Mullen and G. R. Goethals (eds.), Theories of Group Behavior. New York: Springer-Verlag.
Wickler, W. (1976). "Evolution-Oriented Ethology, Kin Selection, and Altruistic Parasites." Zeitschrift für Tierpsychologie 42: 206–214.
Williams, B. (1981). Moral Luck. Cambridge: Cambridge University Press.
Williams, G. C. (1966). Adaptation and Natural Selection: A Critique of Some Current Evolutionary Thought. Princeton: Princeton University Press.
——— (1985). "A Defence of Reductionism in Evolutionary Biology." Oxford Surveys in Evolutionary Biology 2: 1–27.
——— (1992). Natural Selection: Domains, Levels and Challenges. New York: Oxford University Press.
Williams, G. C., and R. M. Nesse. (1991). "The Dawn of Darwinian Medicine." Quarterly Review of Biology 66: 1–22.
Williams, G. C., and D. C. Williams. (1957). "Natural Selection of Individually Harmful Social Adaptations among Sibs with Special Reference to Social Insects." Evolution 11: 32–39.
Wilson, D. S. (1975). "A Theory of Group Selection." Proceedings of the National Academy of Sciences 72: 143–146.

——— (1977a). "How Nepotistic Is the Brain Worm?" Behavioral Ecology and Sociobiology 2: 421–425.
——— (1977b). "Structured Demes and the Evolution of Group Advantageous Traits." American Naturalist 111: 157–185.
——— (1980). The Natural Selection of Populations and Communities. Menlo Park, Calif.: Benjamin/Cummings.
——— (1983). "The Group Selection Controversy: History and Current Status." Annual Review of Ecology and Systematics 14: 159–187.
——— (1987). "Altruism in Mendelian Populations Derived from Sibling Groups: The Haystack Model Revisited." Evolution 41: 1059–1070.
——— (1989). "Levels of Selection: An Alternative to Individualism in Biology and the Human Sciences." Social Networks 11: 257–272. Reprinted in E. Sober (ed.), Conceptual Issues in Evolutionary Biology, pp. 143–154. Cambridge, Mass.: MIT Press, 1994.
——— (1991). "On the Relationship between Evolutionary and Psychological Definitions of Altruism and Egoism." Biology and Philosophy 7: 61–68.
——— (1992). "Complex Interactions in Metacommunities, with Implications for Biodiversity and Higher Levels of Selection." Ecology 73: 1984–2000.
——— (1997). "Incorporating Group Selection into the Adaptationist Program: A Case Study Involving Human Decision Making." In J. Simpson and D. Kendrick (eds.), Evolutionary Approaches to Personality and Social Psychology. Hillsdale, N.J.: Lawrence Erlbaum Associates, 345–386.
——— (1998). "Hunting, Sharing and Multi-level Selection: The Tolerated Theft Model Revisited." Current Anthropology, in press.
Wilson, D. S., and R. K. Colwell. (1981). "Evolution of Sex Ratio in Structured Demes." Evolution 35: 882–897.
Wilson, D. S., and L. A. Dugatkin. (1991). "Tit-for-tat vs. Nepotism, Or, Why Should You Be Nice to Your Rotten Brother?" Evolutionary Ecology 5: 291–299.
——— (1992). Altruism: Contemporary Debates. In E. F. Keller and E. A. Lloyd (eds.), Keywords in Evolutionary Biology, pp. 29–33. Cambridge, Mass.: Harvard University Press.
——— (1997). "Group Selection and Assortative Interactions." American Naturalist 149: 336–351.
Wilson, D. S., and W. G. Knollenberg. (1987). "Adaptive Indirect Effects: The Fitness of Burying Beetles with and without Their Phoretic Mites." Evolutionary Ecology 1: 139–159.
Wilson, D. S., D. Near, and R. R. Miller. (1996). "Machiavellianism: A Synthesis of the Evolutionary and Psychological Literatures." Psychological Bulletin 119: 285–299.

Wilson, D. S., G. B. Pollock, and L. A. Dugatkin. (1992). "Can Altruism Evolve in Purely Viscous Populations?" Evolutionary Ecology 6: 331–341.

Wilson, D. S., and E. Sober. (1994). "Reintroducing Group Selection to the Human Behavioral Sciences." Behavioral and Brain Sciences 17: 585–654.

——— (1996). "Continuing Commentary on Reintroducing Group Selection to the Human Behavioral Sciences." Behavioral and Brain Sciences 19: 777–787.

Wilson, D. S., and J. Yoshimura. (1994). "On the Coexistence of Specialists and Generalists." American Naturalist 144: 692–707.

Wilson, E. O. (1975). Sociobiology: The New Synthesis. Cambridge, Mass.: Harvard University Press.

Wimsatt, W. (1980). "Reductionistic Research Strategies and Biases in the Units of Selection Controversy." In T. Nickles (ed.), Scientific Discovery. Dordrecht: Reidel.

Wispé, L. G. (1987). "History of the Concept of Empathy." In N. Eisenberg and J. Strayer (eds.), Empathy and Its Development, pp. 17–37. New York: Cambridge University Press.

Wolf, S. (1990). Freedom within Reason. New York: Oxford University Press.

Wright, E., A. Levine and E. Sober. (1992): "Marxism and Methodological Individualism." In Reconstructing Marxism—Essays on Explanation and the Theory of History, pp. 107–128. London: Verso Books.

Wright, S. (1945). "Tempo and Mode in Evolution: A Critical Review." Ecology 26: 415–419.

Wynne-Edwards, V. C. (1962). Animal Dispersion in Relation to Social Behavior. Edinburgh: Oliver and Boyd.

Zahn-Waxler, C., M. Radke-Yarrow, E. Wagner, and M. Chapman. (1992). "Development of Concern for Others." Developmental Psychology 28: 126–136.

Zahn-Waxler, C., J. Robinson, and R. Emde. (1992). "The Development of Empathy in Twins." Developmental Psychology 28: 1038–1047.

Index